OPERATIONS RESEARCH

OPERATIONS RESEARCH

A Decision-Making Tool for Engineers and Managers

Dr. B.J. Ranganath

Dean—Administration
and
Professor and Head (Retd)
Industrial Production Engineering
The National Institute of Engineering, Mysore
Karnataka, India

Dr. Srikantappa A.S.

Principal
Cauvery Institute of Technology
Sundahally Mandya
Karnataka, India

yesdee

Published by ***Yes Dee Publishing Pvt Ltd***
D-8, Fourth Street, Sector One, Ambattur Industrial Estate
Chennai – 600 058
Tamil Nadu, INDIA

Tel : +91 44 4508 2085
E-Mail : suresh@yesdee.com
Web : www.yesdee.com

ISBN: 978-93-80381-99-2

CONTENTS

PREFACE

Decision-making is an important function of the management of any organisation. Operations research is an analytical tool that helps managers in decision-making. This book deals with the application of quantitative techniques in operations research through a large number of examples. An attempt has been made to discuss the commonly used techniques/topics, viz., linear programming problems with graphical solutions, simplex method, transportation problems, assignment problems, game theory, network analysis, queuing theory, sequencing, integer programming problems, decision analysis, metaheuristics, simulation and project management techniques. These techniques help managers to tackle different problems associated with decision-making such as determination of optimum quantity of products to be manufactured, quantity of products to be transported, assignment of work, selection of strategy, determination of optimum time, and so on.

This book is intended for a one semester course on "Operations Research" for different disciplines like mechanical engineering, industrial production engineering, electrical and electronics engineering, electronics and communication engineering, and computer science engineering. This book will be useful to students, practising engineers and managers in decision-making in various situations.

The authors are thankful to Yes Dee Publishing, for publishing this book on time.

B.J. Ranganath
Srikantappa A.S.

ABBREVIATIONS

CC	Crashing Cost
CP	Critical Path
CPM	Critical Path Method
CS	Cost Slope
CT	Crashing Time
Dn	Decision Variable
DPP	Dynamic Programming Problems
EF	Early Finish
ES	Early Start
FCFS	First Come First Served
FS	Free Slack
GT	Game Theory
IPP	Integer Programming Problems
LCFS	Last Come First Served
LCL	Lower Control Limit
LCM	Least Cost Method
LF	Late Finish
LPP	Linear Programming Problems
LS	Late Start
MODI	Modified Distribution
NC	Normal Cost
NPE	New Pivot Equation
NT	Normal Time
NWCR	North West Corner Rule
OR	Operations Research
P	Probability
PC	Pivot Column
PERT	Performance Evaluation and Review Technique
PM	Project Management
PR	Pivot Row
QT	Queuing Theory
R-Chart	Range Chart
Rn	Return Function
RN	Random Number
SIRO	Selection of Service in Random Order
Sn	State Variable
ST	Standard Deviation
T	Time
T_e	Estimated Time
T_m	Most Likely Time
T_o	Optimistic Time
T_p	Pesimistic Time
TP	Transportation Problem
TS	Total Slack
UCL	Upper Control Limit
VAM	Vogle's Approximation Method
VT	Variance

Chapter 1

INTRODUCTION

1.1 ORIGIN AND DEVELOPMENT OF OPERATIONS RESEARCH

It is a fact that in times of natural crisis which may be political, economic or cultural, the talents from all walks of life join together to overcome the situation and solve the problems. These combined efforts always result in new discoveries and techniques. Operations Research (OR) is also the outcome of such situations.

During the Second World War in the United Kingdom (UK), it was felt that many strategic and tactical problems associated with military operations were too complicated to be solved by some individuals for winning the battle. To overcome the situations, groups of scientists and experts from various disciplines joined together as special units within the armed forces. The team approach effectively improved the military operations. Later on in the USA, these concepts along with mathematical techniques were developed for the analysis of military problems. This approach of systematic and scientific study basically originated from military operations and was named as Operations Research (OR) by Trefthen in the year 1940.

Later on, the economic crisis in the UK resulted in industrial revolution and the scientists who were active in solving military problems diverted their attention to solve complex industrial problems through OR approach.

A decade after the Second World War, OR techniques rapidly developed in the fields of industrial, academic and governmental organisations. The Operations Research Society of America (ORSA) was established in 1952. In 1957 the international federation of OR was founded and the subject was introduced in the curriculum of universities and research institutes. In India, the OR Society was founded in 1959.

1.2 DEFINITION OF OR

According to the OR Society of Great Britain, OR is the application of the methods of science to complex problems arising in the direction and management of large systems of men, machines, materials and money, in industry, business, government and defence. The distinctive approach is to develop a scientific model of the system incorporating measurements of factors such as chance and risk which is used to predict and compare the outcome of alternative decisions or strategies. The purpose is to help management determine its policy and actions scientifically.

1.3 AREAS OF APPLICATION OF OR

Whenever there is any problem, simple or complicated, the OR techniques can be applied to find the best solution. Let us see the application of OR in various fields of everyday life.

In defence operations

The Second World War initiated the use of OR techniques for the study of communication, control and replacement of men by machines in various defence operations. The defence operations are carried out by a number of independent components, namely Airforce, Army and Navy. The activities in each of these components can be further divided into four subcomponents, viz., administration, intelligence, operation, training and supply. The application of modern techniques in each of the components of military organisation requires expertise knowledge in the respective fields. Furthermore, each component works to derive maximum gains from its operations and there is always a possibility that strategy beneficial to one component may have an adverse effect on the other. Thus in defence operation, there is a necessity to co-ordinate the activities of various components which give maximum benefit to the organisation as a whole, having maximum use of the individual components.

The final strategy is formulated by a team of scientists drawn from various disciplines who study the strategies of different components and after appropriate analysis of the various courses of action, the best course of action known as optimum strategy is chosen. Similarly both the parties in war have a number of courses of action and the one who selects the best strategy wins the battle.

In Industry

Post-World War II period marked the application of OR techniques in industrial operation. The system of modern industries is so complex that the optimum point of operation in its various components cannot be judged by an individual. The business environment is always changing and any decision useful at one time may not be good some time later. There is always a need to check the validity of decisions continuously against the situation. The industrial revolution with increased division of labour and introduction of management responsibilities has made each component an independent unit having its own goals. Naturally, there is a need to integrate the different components of an industry and to frame a policy, keeping in view the activities and aims of various components and to make the most effective use of these components so as to serve the best interest of the organisation as a whole.

In an industry, the production department wants to manufacture good quality products. Marketing department aims at maximizing the sale of the product but for a lower price. Finance department tries to minimise capital investment. Personnel department wishes to appoint qualified and highly skilled people. Thus, each department plans its own objectives and all these objectives of various departments, may conflict with each other and may not conform to the overall objectives of the organisation. The application of OR techniques helps in overcoming this difficulty by integrating the diversified activities of various components so as to serve efficiently the interest of the organisation as a whole. The application of OR methodology needs careful and thorough study of various operations of an industry in a systematic and scientific manner to remove their conflicts in the best possible way.

1.4 OR METHODOLOGY

OR is a scientific methodology applied to study the activities of an organisation with a view to access the implementation of various alternative courses of action. The OR approach to solve problems consists of the following steps.

1. Problem definition
2. Model construction

3. Model solution
4. Model validity
5. Implementation

1. Problem definition

Firstly, the OR team should define the problem under investigation, set-up objectives and determine the constraints against which decisions should be adopted.

2. Model construction

The next step is to express all the variables of the problem into a mathematical model such as Linear Programming. The generalised mathematical model is of the form

$$z = f(x_i, y_j)$$

where, z is the objective, x_i is a controllable variable, and y_j is an uncontrollable variable.

3. Model solution

In the next step, the values of decision variables that optimise the given objective function are determined. It deals with the mathematical calculations used to obtain model solution.

4. Model validity

It is the measure of accuracy. A model is valid or accurate if all the objectives, constraints and variables included in the model are relevant and the functional relationships are valid. If the proposed model is representing a new system, no historical data would be available to make the comparison. In such cases, we may use simulation as an independent tool for verifying the output of the mathematical model.

5. Implementation

Finally, the tested results of the model are implemented. Implementation involves the translation of the results into understandable operating instructions to the managers.

1.5 FEATURES AND LIMITATIONS OF OR

Using OR approach, a decision-maker can solve a complex problem involving multiple variables more quickly. A decision-maker can examine a situation from various angles by simulating the model constructed for the real problem. He can change various conditions of the problem under consideration and determine the optimal solution by examining the effect of changes through experiments on the model.

Following are the limitations of OR

1. Construction of OR model for solving certain problems is too expensive.
2. OR model constructed may not represent the real world problem for which the decision has to be made.
3. If the basic data are subject to frequent changes, modification of the OR models becomes costly.
4. Other shortcomings are the magnitude of computation involved, lack of consideration for non-quantifiable factors, etc.

Chapter 2

LINEAR PROGRAMMING

2.1 INTRODUCTION

Linear programming consists of two words namely, Linear and Programming. The term linear implies proportionality and additivity. Proportionality refers to the fact that doubling the production of a product will exactly double the profit and required resources. Additivity implies that the objective function be the direct sum of the individual contribution of different variables. The term program refers to a course of action covering a specified period of time. Since it is always possible to design a number of programs, a manager attempts to identify the programs that are optimal. A program is optimal, if it maximises or minimises some measures such as profit, costs or sales.

The term programming refers to a systematic procedure by which a particular program or plan of action is designed. Programming consists of a series of instruments and rules for solving a problem.

Linear programming is a method of determining an optimal program of independent activities in view of limited resources that are available to the manager during a specified period of time.

2.2 FORMULATION OF LINEAR PROGRAMMING MODELS

Step I Identify the unknown variables to be determined.

Suppose an industrial worker has a wage rate of Rs. 15 per hour and if he works for 40 hours in a week, he gets Rs. 600 per week. During some other week if he works for 30 hours, he earns Rs. 450 per week.

$$\text{Total earnings} = 15 \times \text{Number of hours worked}$$

$$W = 15\,H$$

Similarly, another worker may have a wage rate of Rs. 25 per hour, his earnings is given by

$$W = 25\,H$$

In general, $W = k.H$, where k is a constant

For a particular worker, W and H vary. Since they can assume various values in a given problem, they are called as variables.

Step II Identify all the constraints in the problem and express them as linear equations.

The manager will have to work in a peculiar situation. There will be many factors which restrict him to take a particular course of action. These restrictions are known as constraints.

Step III Identify the objective. An objective is the end result we desire to achieve by selecting a specific course of action. The common objective of money making endeavour is to maximise profit or minimise cost.

2.3 GRAPHICAL SOLUTION OF LINEAR PROGRAMMING PROBLEM (LPP)

A graphical procedure is used to solve Linear programming problems. Though such problems are not encountered in practice, the graphical procedure illustrates some of the basic concepts.

Locating feasible region

For example, $2X + 4Y \leq 8$ represents a constraint where X and Y are variables. Bringing this inequality to equality, we get,

$$2X + 4Y = 8$$

Put $X = 0$ and find values of Y; $Y = 2$; Points are (0, 2) in the co-ordinate system.

Put $Y = 0$ and find values of X; $X = 4$; Points are (4, 0).

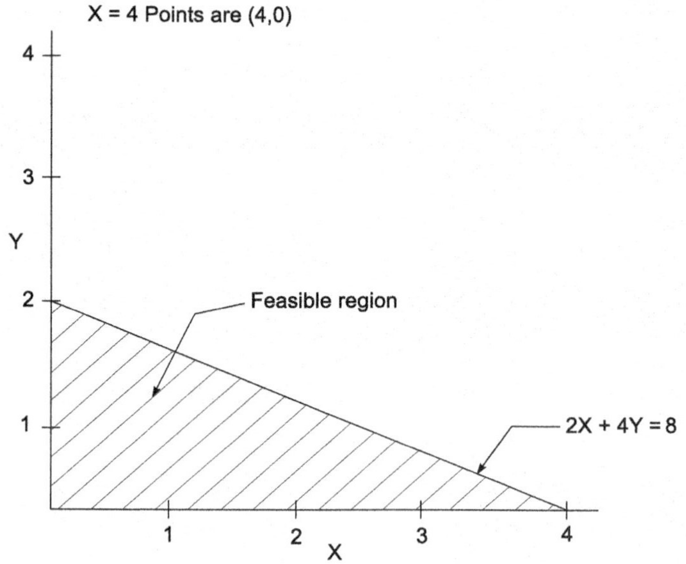

Figure 2.1 Graphical representation of a constraint

The optimum solution always occurs at the corner of the feasible region.

Example 2.1

Nirmala Insecticides Ltd. is a factory which produces two types of agro-based insecticides for whose sale distribution two basic raw materials A and B are used to manufacture the insecticides. The daily requirements of the raw materials per ton of Insecticide I and II are summarised in the following table.

Raw material	Tons of raw material per ton of Insecticides		Maximum available in tons
	Insecticide I	Insecticide II	
A	1	2	6
B	2	1	8

A market survey has established that the daily demand for Insecticide II cannot exceed that of Insecticide I by one ton. The survey also shows that the maximum demand for Insecticide II is limited to two tons only.

The wholesale price per ton is Rs. 3000 for Insecticide I and Rs. 2000 for Insecticide II. How much Insecticide I and Insecticide II should the company produce to maximise the gross income?

Solution

Formulation

Variables The variables are Insecticide I and Insecticide II. These are variables because their volume vary.

Let the company produce X tons of Insecticide I and Y tons of Insecticide II.

Objective function It is to maximise income. By selling one ton of Insecticide I, we get Rs. 3000. Therefore, by selling X tons of Insecticide I, we get Rs. $3000X$. Similarly, by selling Y tons of Insecticide II, we get Rs. $2000Y$. The revenue is $3000X + 2000Y$.

Maximise $Z = 3000X + 2000Y$.

This is an objective function.

Constraints

(a) *Raw material restrictions*

$$X + 2Y \leq 6 \quad (1)$$

$$2X + Y \leq 8 \quad (2)$$

(b) *Marketing restrictions*

Demand for Insecticide II cannot exceed Insecticide I by more than one ton.

$$Y - X \leq 1 \quad (3)$$

The maximum demand for Insecticide II cannot exceed 2 tons.

$$Y \leq 2 \quad (4)$$

(c) *Non-negativity constraints*

As it is not possible to manufacture negative units of Insecticide I and II, we have

$$X \geq 0 \quad (5)$$

$$Y \geq 0 \quad (6)$$

Formulated LPP is given below

$$\text{Maximise } Z = 3000X + 2000Y$$

subject to

$$X + 2Y \leq 6 \quad (1)$$

$$2X + Y \leq 8 \quad (2)$$

$$Y - X \leq 1 \quad (3)$$

$$Y \leq 2 \quad (4)$$

$$X \geq 0 \quad (5)$$

$$Y \geq 0 \quad (6)$$

Graphical solution

Considering Equation (1)

$$X + 2Y \leq 6$$

Bringing inequality to equality, we have

$$X + 2Y = 6$$

Let $X = 0$; then $Y = 3$. Let $Y = 0$; then $X = 6$. The values are (0, 3) and (6, 0).

Considering Equation (2)

$$2X + Y \leq 8$$

Bringing inequality to equality

$$2X + Y = 8$$

When $X = 0$, $Y = 8$. When $Y = 0$, $X = 4$. The values are (0, 8) and (4, 0).

Considering Equation (3)

$$Y - X \leq 1$$

Bringing inequality to equality

$$Y - X = 1$$

When $X = 0$, $Y = 1$. When $Y = 0$, $X = -1$. The values are (–1, 0) and (0, 1).

Considering Equation (4)

$$Y \leq 2$$

Bringing inequality to equality

$$Y = 2$$

Plot the values on a graph by selecting a suitable scale.

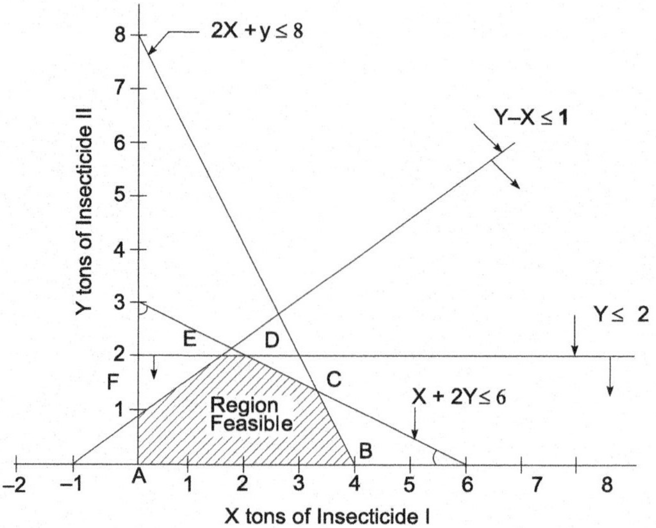

Figure 2.2 Insecticide I vs Insecticide II

Optimum solution

The optimum solution always occurs at the corners of the feasible region. Name these corners. Find the values of Z at all these corners.

$$Z = 3000X + 2000Y$$

At corner A, $X = 0$ and $Y = 0$.

$$Z = 3000(0) + 2000(0) = 0$$

At corner B, $X = 4$ and $Y = 0$.

$$Z = 3000(4) + 2000(0) = \text{Rs. } 12{,}000$$

At corner C, $X = 3.4$ and $Y = 1.35$.

$$Z = 3000(3.4) + 2000(1.35) = \text{Rs. } 12{,}900$$

At corner D, $X = 2$ and $Y = 2$.

$$Z = 3000(2) + 2000(2) = \text{Rs. } 10{,}000$$

At corner E, $X = 1$ and $Y = 2$.

$$Z = 3000(1) + 2000(2) = \text{Rs. } 7000$$

At corner F, $X = 0$ and $Y = 1$.

$$Z = 3000(0) + 2000(1) = 2000(1) = \text{Rs. } 2000$$

The maximum value is at the corner C. Hence $X = 3.4$, $Y = 1.35$, i.e., the company has to produce 3.4 tons of Insecticide I and 1.35 tons of Insecticide II. The revenue is Rs. 12,900.

$$\text{Maximise } Z = \text{Rs. } 12{,}900$$

subject to

$$X = 3.4 \text{ tons}$$
$$Y = 1.35 \text{ tons}$$

Example 2.2

Two products are manufactured by passing sequentially through three machines. Time per machine allocated to the two products is limited to 10 hours per day. The production time and profit per unit of each product is given below

Product	Minutes per unit			Profit (Rs.)
	Machine I	Machine II	Machine III	
I	10	6	8	2
II	5	20	15	3

Find the optimal mix of the products.

Solution

Formulation

Variables The variables are Product I and Product II. Let us produce X units of Product I and Y units of Product II.

Objective function It is to maximise profit.

$$\text{Maximise } Z = 2X + 3Y$$

By selling one unit of Product I we get Rs. 2 profit. By selling X units, Rs. $2X$. Similarly, by selling Y units of Product II, profit gained is Rs. $3Y$

Constraints

(a) *Machining time constraints*

$$10X + 5Y \leq 600 \quad (1)$$
$$6X + 20Y \leq 600 \quad (2)$$
$$8X + 15Y \leq 600 \quad (3)$$

(b) *Non-negativity constraints*

$$X \geq 0 \quad (4)$$
$$Y \geq 0 \quad (5)$$

Formulated LPP is given below

$$\text{Maximise } Z = 2X + 3Y$$

subject to

$$10X + 5Y \leq 600 \quad (1)$$
$$6X + 20Y \leq 600 \quad (2)$$
$$8X + 15Y \leq 600 \quad (3)$$
$$X \geq 0 \quad (4)$$
$$Y \geq 0 \quad (5)$$

Graphical solution

Considering Equation (1)

$$10X + 15Y \leq 600$$

Bringing this inequality to equality, we get

$$10X + 5Y = 600$$

If $X = 0$, $Y = 120$. If $Y = 0$, $X = 60$. The values are (0, 120) and (60, 0).

Considering Equation (2)

$$6X + 20Y \leq 600$$

Bringing inequality to equality

$$6X + 20Y = 600$$

Put $X = 0$, then $Y = 30$. Put $Y = 0$, then $X = 100$.

Considering Equation (3)

$$8X + 15Y \leq 600$$

Bringing inequality to equality, we have

$$8X + 15Y = 600$$

Let $X = 0$, then $Y = 40$. Let $Y = 0$ then $X = 75$. The values are (0, 40) and (75, 0). Plot the values of X and Y on the graph by taking suitable scale.

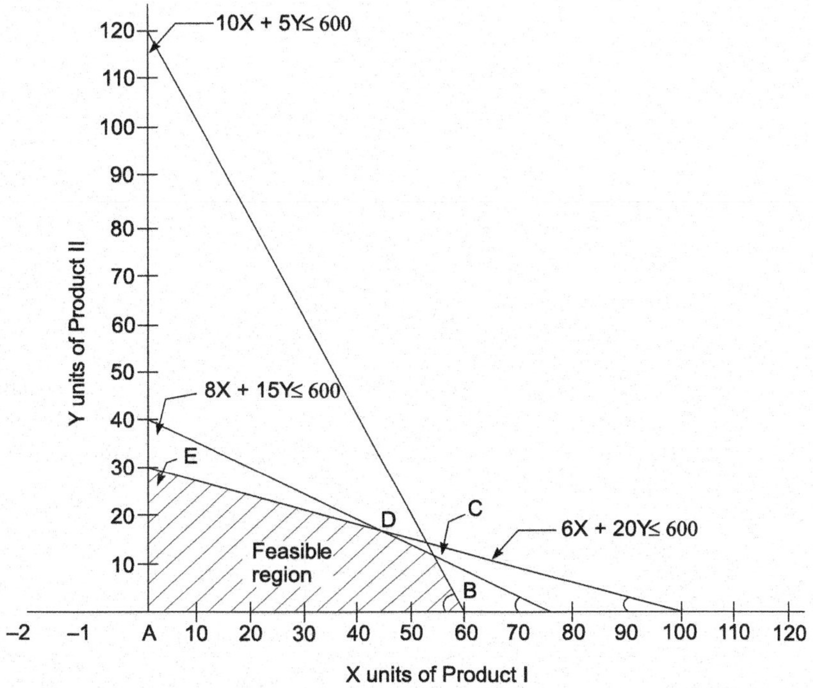

Figure 2.3 Product I vs Product II

Optimum solution

The optimum solution occurs at the corners of the feasible region. Name these corners. Find the values of Z at all these corners.

$$Z = 2X + 3Y$$

At corner A, $X = 0$ and $Y = 0$.

$$Z = 2(0) + 3(0) = 0$$

At corner B, $X = 60$ and $Y = 0$.

$$Z = 2(60) + 3(0) = \text{Rs. } 120$$

At corner C, $X = 55$ and $Y = 11$.

$$Z = 2(55) + 3(11)$$
$$= \text{Rs. } 143$$

At corner D, $X = 43$ and $Y = 17$.

$$Z = 2(43) + 3(17)$$
$$= \text{Rs. } 137$$

At corner E, $X = 0$ and $Y = 30$.

$$Z = 2(0) + 3(30)$$
$$= \text{Rs. } 90$$

The maximum value occurs at corner C. Here, $X = 55$, $Y = 11$. The company has to produce 55 units of Product I and 11 units of Product II. The profit is Rs. 143.

$$\text{Maximise } Z = \text{Rs. } 143$$

subject to

$$X = 55$$
$$Y = 11.$$

Example 2.3

A firm makes two types of furnitures, chairs and tables. The contribution for each product as calculated by the accounting department are Rs. 200 per chair and Rs. 300 per table. Both products are processed on three machines M_1, M_2 and M_3. The time required to produce each product and the total time available per week on each machine are as follows

Machine	Chair	Table	Available hours
M_1	3	3	36
M_2	5	2	50
M_3	2	6	60

(a) Formulate this as LPP.

(b) How should the manufacturer schedule his production in order to maximise the contribution?

Solution

Formulation

Variables The variables are chairs and tables. Let the firm manufacture X number of chairs and Y number of tables.

Objective function By selling one chair the firm earns Rs. 200 as profit. By selling X units of chairs the firm earns Rs. $200X$. Similarly by selling one table the firm earns Rs. 300. For Y units of tables, the firm earns Rs. $300Y$ as the profit.

$$\text{Maximise } Z = 200X + 300Y$$

Constraints

(a) *Time restrictions*

$$3X + 3Y \leq 36 \quad (1)$$
$$5X + 2Y \leq 50 \quad (2)$$
$$2X + 6Y \leq 60 \quad (3)$$

(b) *Non-negativity constraints*

$$X \geq 0 \quad (4)$$
$$Y \geq 0 \quad (5)$$

The formulated LPP is

$$\text{Maximise } Z = 200X + 300Y$$

subject to

$$3X + 3Y \leq 36 \quad (1)$$
$$5X + 2Y \leq 50 \quad (2)$$
$$2X + 6Y \leq 60 \quad (3)$$
$$X \geq 0 \quad (4)$$
$$Y \geq 0 \quad (5)$$

Graphical solution

Considering Equation (1)

$$3X + 3Y \leq 36$$

Bringing inequality to equality, we get

$$3X + 3Y = 36$$

If $X = 0$, $Y = 12$. If $Y = 0$, $X = 12$. The values are (0, 12) and (12, 0).

Considering Equation (2)

$$5X + 2Y \leq 50$$

Bringing inequality to equality

$$5X + 2Y = 50$$

When $X = 0$, $Y = 25$. When $Y = 0$, $X = 10$. The values of X and Y are (0, 25) and (10, 0).

Considering Equation (3)

$$2X + 6Y \leq 60$$

Bringing inequality to equality

$$2X + 6Y = 60$$

If $X = 0$, $Y = 10$. If $Y = 0$, $X = 30$. The values are (0, 10) and (30, 0). Select the suitable scale and plot the graph.

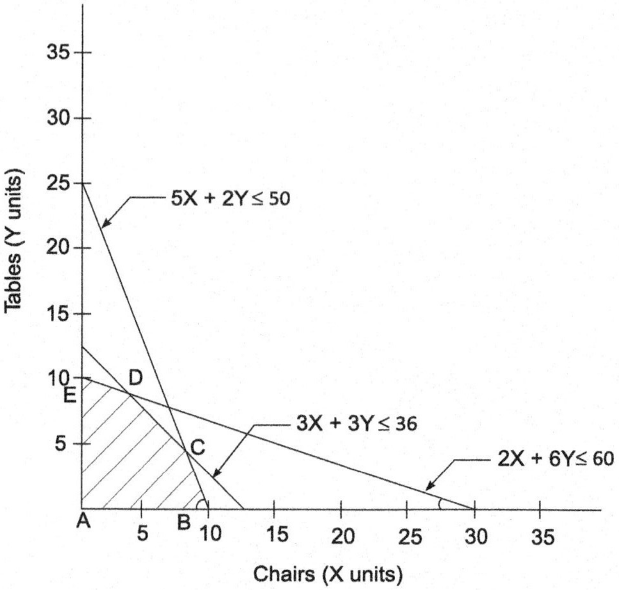

Figure 2.4 Chairs vs Tables

Optimum solution

The optimum solution occurs at the corner of the feasible region. Name these corners. Find the values of Z at all these corners.

$$Z = 200X + 300Y$$

At corner A, $X = 0$ and $Y = 0$.

$$Z = 200(0) + 300(0) = 0$$

At corner B, $X = 10$ and $Y = 10$.

$$Z = 200(10) + 300(0) = \text{Rs. } 2000$$

At corner C, $X = 8.6$ and $Y = 3.4$.

$$Z = 200(8.6) + 300(3.4) = \text{Rs. } 2740$$

At corner D, $X = 3$ and $Y = 9$.

$$Z = 200(3) + 300(9) = \text{Rs. } 3300$$

At corner E, $X = 0$ and $Y = 10$.

$$Z = 200(10) + 300(10) = \text{Rs. } 3000$$

The maximum value occurs at point D. The contribution is Rs. 3300. The firm has to manufacture 3 chairs and 9 tables.

Example 2.4

The manufacturer of a line of patent medicines is preparing a production plan of medicines A and B. There are sufficient ingredients available to make 20,000 bottles of A and 40,000 bottles of B, but there are only 45,000 bottles into which either of the medicines can be put. Furthermore it takes 3 hours to prepare enough material to fill 1000 bottles of B and there are 66 hours available for this operation. The profit is Rs. 8 per bottle of A and Rs. 7 per bottle of B.

(a) Formulate this problem as LPP

(b) Solve this problem.

Solution

Formulation

Variables Medicine A and medicine B are the variables. Let the manufacturer make X bottels of medicine A and Y bottles of medicine B.

Objective function The manufacturer earns Rs. 8 per bottle of medicine A. If he sells X bottles of medicine A, the profit is Rs. $8X$. Similarly profit per bottle of medicine B is Rs. 7. If Y bottles of medicine B are sold, the manufacturer earns the profit of Rs. $7Y$.

$$\text{Maximise } Z = 8X + 7Y$$

This is an objective function.

Constraints

(a) *Capacity restriction*

$$X \leq 20{,}000 \quad (1)$$

$$Y \leq 40{,}000 \quad (2)$$

(b) *Bottling restriction*

$$X + Y \leq 45{,}000 \quad (3)$$

(c) *Time restriction*

To fill 1000 bottles of A, it takes 3 hours. Therefore, to fill one bottle of A, it takes $\frac{3}{1000}$ hours therefore, to fill X bottles of A, it takes $\frac{3}{1000}X$. Similarly to fill 1000 bottles of B, it takes one hour. To fill one bottle of B, $\frac{1}{1000}$ hours is required and to fill Y bottles of B, it takes $\frac{1}{1000}Y$.

$$\frac{3X}{1000} + \frac{Y}{1000} \leq 66$$

Therefore,

$$3X + Y \leq 66{,}000$$

(d) *Non-negativity constraints*

$$X \geq 0$$
$$Y \geq 0$$

Following is the formulated LPP

$$\text{Maximise } Z = 8X + 7Y$$

subject to

$$X \leq 20{,}000 \quad (1)$$
$$Y \leq 40{,}000 \quad (2)$$
$$X + Y \leq 45{,}000 \quad (3)$$
$$3X + Y \leq 66{,}000 \quad (4)$$
$$X \geq 0$$
$$Y \geq 0$$

Graphical solution

Considering Equation (1)

$$X \leq 20{,}000$$

Bringing inequality to equality, we get

$$X = 20{,}000$$

Considering Equation (2)

$$Y \leq 40{,}000$$

Bringing inequality to equality

$$Y = 40{,}000$$

Considering Equation (3)

$$X + Y \leq 45{,}000$$

Bringing inequality to equality, we have

$$X + Y = 45{,}000$$

When $X = 0$, $Y = 45{,}000$. When $Y = 0$, $X = 45{,}000$. The values are (0, 45,000)(45,000, 0).

Considering Equation (4)

$$3X + Y \leq 66{,}000$$

Bringing inequality to equality

$$3X + Y = 66{,}000$$

If $X = 0$, $Y = 66{,}000$. If $Y = 0$, $X = 22{,}000$. The values are (0, 66,000) and (22,000, 0).

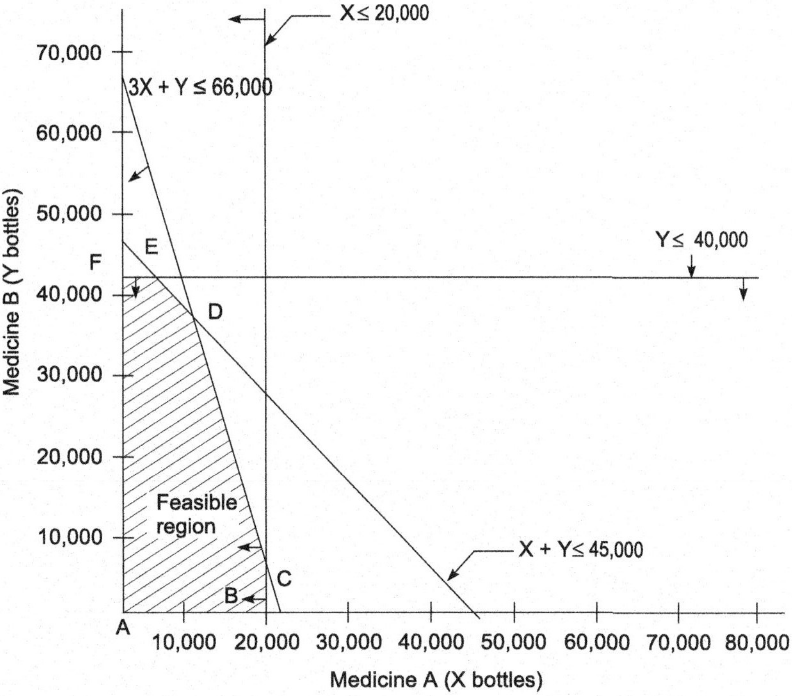

Figure 2.5 Medicine A vs Medicine B

Optimum solution

The optimum solution always occurs at the corners of the feasible region. Name these corners. Find the values of Z at all these corners.

$$Z = 8X + 7Y$$

At corner A, $X = 0$ and $Y = 0$.

$$Z = 8(0) + 7(0)$$
$$= 0$$

At corner B, $X = 20{,}000$ and $Y = 0$.

$$Z = 8(20{,}000) + 7(0)$$
$$= \text{Rs. } 1{,}60{,}000$$

At corner C, $X = 20{,}000$ and $Y = 6000$.

$$Z = 8(20{,}000) + 7(6000)$$
$$= \text{Rs. } 2{,}02{,}000$$

At corner D, $X = 10{,}500$ and $Y = 34{,}000$.

$$Z = 8(10{,}500) + 7(34{,}000)$$
$$= \text{Rs. } 10{,}78{,}000$$

Maximum value occurs at corner D. Manufacturer has to make 10,500 bottles of medicine A and 34,000 bottles of medicine B.

$$\text{Maximise } Z = \text{Rs. } 10{,}78{,}000$$

subject to

$$X = 10{,}500$$
$$Y = 34{,}000$$

Example 2.5

The Broad motor company manufactures two car models, the Pontillac and the Bulkswagon. These are sold to a car dealer at a profit of Rs. 20,000 per Pontillac and Rs. 10,000 per Bulk Swagon. A Pontillac requires 150 man-hours for assembling, 50 man-hours for painting and finishing, and 10 man-hours for checking out and testing. Bulkswagon requires 60 man-hours for assembling, 40 man-hours for painting and finishing, and 20 man-hours for checking and testing. There are 30,000 man-hours available in the assembly shop, 13,000 man-hours in painting and finishing shop, and 5000 man-hours in the checking and testing division. How many Pontillacs and how many Bulkswagons should the company produce so as to gain maximum profit?

Solution

Formulation

Variables Variables are the car models Pontillac and Bulkswagon. Let the Broad motor company manufacture X units of Pontillac and Y units of Bulkswagon.

Objective function It is to maximise the profit. Profit per Pontillac is Rs. 10,000. If the company sells X units of Pontillac, the earnings is Rs. 10,000X. Similarly, if the company sells Y units of Bulkswagon, it earns the profit of Rs. 20,000Y.

Constraints

(a) *Time constraint in the assembly section*

$$150X + 60Y \leq 30{,}000 \quad (1)$$

(b) *Man-hour restriction in painting and finishing section*

$$50X + 40Y \leq 13{,}000 \quad (2)$$

(c) *Man-hour restriction in checking and testing section*

$$10X + 20Y \leq 5000 \quad (3)$$

(d) *Non-negativity constraints*

$$X \geq 0 \quad (4)$$
$$Y \geq 0 \quad (5)$$

Formulated LPP is as follows

$$\text{Maximise } Z = 10{,}000X + 20{,}000Y$$

subject to

$$150X + 60Y \leq 30{,}000 \quad (1)$$
$$50X + 40Y \leq 13{,}000 \quad (2)$$

$$10X + 20Y \leq 5000 \quad (3)$$
$$X \geq 0 \quad (4)$$
$$Y \geq 0 \quad (5)$$

Graphical solution

Considering Equation (1)

$$150X + 60Y \leq 30{,}000$$

Bringing inequality to equality

$$150X + 60Y = 30{,}000$$

If $X = 0$, $Y = 500$. If $Y = 0$, $X = 200$. The values are (0, 500) and (200, 0).

Considering Equation (2)

$$50X + 40Y \leq 13{,}000$$

Bringing inequality to equality, we have

$$50X + 40Y = 13{,}000$$

When $X = 0$, $Y = 325$. When $Y = 0$, $X = 260$. The values are (0, 325), (260, 0).

Considering Equation (3)

$$10X + 20Y \leq 5000$$

Bringing inequality to equality

$$10X + 20Y = 5000$$

If $X = 0$, $Y = 250$. If $Y = 0$, $X = 500$. The values are (0, 250) and (500, 0). Plot a graph.

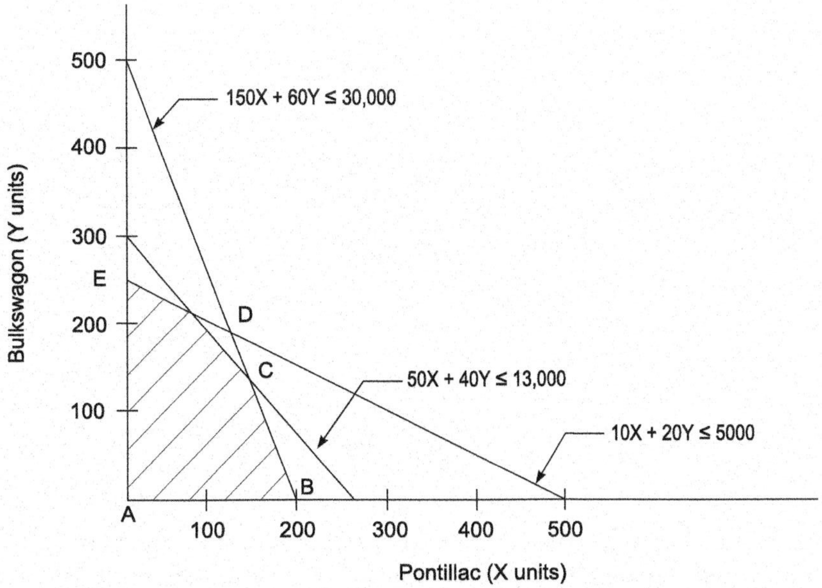

Figure 2.6 Pontillac vs Bulkswagon

Optimum solution

The optimum solution always occurs at the corners of the feasible region. Name these corners. Find the values of Z at all these corners.

$$Z = 10{,}000X + 20{,}000Y$$

At corner A, $X = 0$ and $Y = 0$.

$$Z = 10{,}000(0) + 20{,}000(0) = 0$$

At corner B, $X = 200$ and $Y = 0$.

$$Z = 10{,}000(200) + 20{,}000(0) = \text{Rs. } 20{,}00{,}000$$

At corner C, $X = 140$ and $Y = 145$.

$$Z = 10{,}000(140) + 20{,}000(145) = 14{,}00{,}000 + 29{,}00{,}000 = \text{Rs. } 43{,}00{,}000$$

At corner D, $X = 100$ and $Y = 200$.

$$Z = 10{,}000(100) + 20{,}000(200) = 10{,}00{,}000 + 40{,}00{,}000 = \text{Rs. } 50{,}00{,}000$$

At corner E, $X = 0$ and $Y = 250$.

$$Z = 10{,}000(0) + 20{,}000(250) = \text{Rs. } 50{,}00{,}000$$

Maximum value occurs at D and E. The Broad motor company has to manufacture 100 Pontillac and 200 Bulkswagon models or alternatively, 250 Bulkswagon models.

$$\text{Maximise } Z = \text{Rs. } 50{,}00{,}000$$

subject to

$$X = 100$$
$$Y = 200$$

or

$$\text{Maximise } Z = 50{,}00{,}000$$

subject to

$$X = 0$$
$$Y = 250$$

***Example* 2.6**

A farm is engaged in breeding pigs. The pigs are fed on various products grown in the farm. Because of the need to ensure certain nutrient constituents, it is necessary to buy constituents A and B. The nutrient constituents in each unit of the products are as follows

Product	Nutrient constituents		Minimum required
	A	B	
1	36	6	108
2	3	12	36
3	20	10	100

A costs Rs. 20 per unit and B costs Rs. 40 per unit. How much of constituents A and B should be purchased at the lowest possible cost so as to provide the pigs, nutrient not less than the minimum required as given in the table.

Solution

Formulation

Variables The variables are the nutrient constituents A and B. Let us produce X units of nutrient constituent A and Y units of nutrient constituent B.

Objective function Objective is to minimise the production cost.

$$\text{Minimise } Z = 20X + 40Y$$

Constraints

(a) *Minimum requirement*

$$36X + 6Y \geq 108 \quad (1)$$
$$3X + 12Y \geq 36 \quad (2)$$
$$20X + 10Y \geq 100 \quad (3)$$

(b) *Non-negative constraints*

$$X \geq 0$$
$$Y \geq 0$$

Graphical solution

Considering Equation (1)

$$36X + 6Y \geq 108$$

Bringing inequality to equality, we have

$$36X + 6Y = 108$$

When $X = 0$, $Y = 18$. When $Y = 0$, $X = 3$.

Considering Equation (2)

$$3X + 12Y \geq 36$$

Bringing inequality to equality

$$3X + 12Y = 36$$

If $X = 0$, $Y = 3$. If $Y = 0$, $X = 12$.

Considering Equation (3)

$$20X + 10Y \geq 100$$

Bringing inequality to equality

$$20X + 10Y = 100$$

Let $X = 0$, then $Y = 10$. Let $Y = 0$, then $X = 5$.

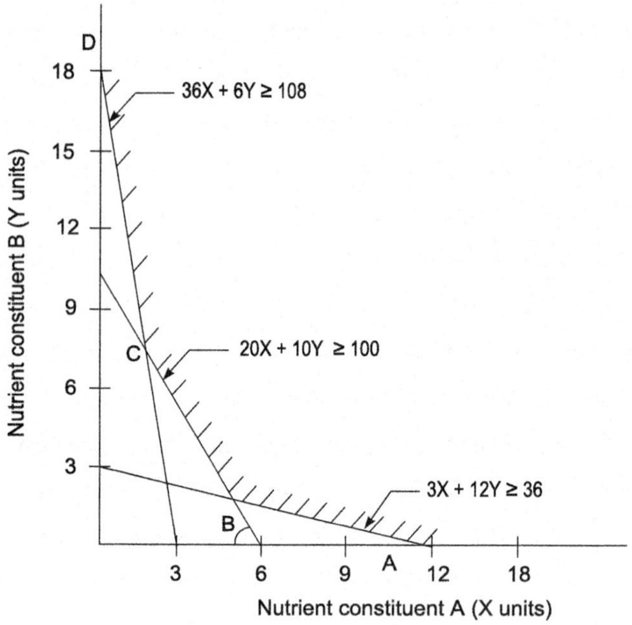

Figure 2.7 Nutrient constituent A vs Nutrient constituent B

The optimum solution occurs at the corners of the feasible region. Name the corners. Find the value of Z at these corners.

$$Z = 20X + 40Y$$

At corner A, $X = 12$ and $Y = 0$.

$$Z = 20(12)+40(0)$$
$$= \text{Rs. } 240$$

At corner B, $X = 4$ and $Y = 2$.

$$Z = 20(4) + 40(2)$$
$$= 80 + 80$$
$$= \text{Rs. } 160$$

At corner C, $X = 2$ and $Y = 6$.

$$Z = 20(2) + 40(6)$$
$$= \text{Rs. } 280$$

At corner D, $X = 0$ and $Y = 18$.

$$Z = 20(0) + 40(18)$$
$$= \text{Rs. } 720$$

Minimum value is at corner B. Here, $X = 4$, $Y = 2$ and $Z = $ Rs. 160.

Example 2.7

The Food and Agricultural Organisation's (FAO) research department has developed two high nutrient foods. One using soya-bean derivatives and the other fishmeal. Each unit of soya-bean has 20 units of nutrient 1 and 60 units of nutrient 2. Each unit of fishmeal food has 30 units of nutrient 1 and 20 units of nutrient 2. The FAO has determined the per capita daily minimum requirements of nutrient constituent 1 and 2 to be 350 and 700 units respectively. The cost of producing soya-bean food is Rs. 20 per unit and Rs. 9 per unit for fishmeal food. The FAO wants to send food packets (separate for each individual to famine ridden area). Determine the units of soya-bean and fishmeal food that should be in each packets in order to minimise the cost. Solve this problem graphically.

Solution

Formulation

Variables Variables are soya-bean food and fishmeal food. Let FAO produce X units of soya-bean food and Y units of fishmeal food.

Objective function It is to minimise the production cost. FAO spends Rs. 20 per unit of soya-bean food and for X units it is Rs. $20X$. For Y units of fishmeal food FAO spends Rs. $9Y$.

$$\text{Minimise } Z = 20X + 9Y$$

Constraints

(a) *Nutrient restriction*

$$20X + 30Y \geq 350 \quad (1)$$
$$60X + 20Y \geq 700 \quad (2)$$

(b) *Non-negative constraints*

$$X \geq 0 \quad (3)$$
$$Y \geq 0 \quad (4)$$

Graphical solution

Considering Equation (1)

$$20X + 30Y \geq 350$$

Bringing inequality to equality, we get

$$20X + 30Y = 350$$

If $X = 0$, $Y = 11.66$. If $Y = 0$, $X = 17.5$. The values are (0, 11.66) (17.5, 0).

Considering Equation (2)

$$60X + 20Y \geq 700$$

Bringing inequality to equality

$$60X + 20Y = 700$$

When $X = 0$, $Y = 35$. When $X = 0$, $Y = 11.66$. The values are (0, 35) and (11.66, 0). Plot the graph with suitable scale.

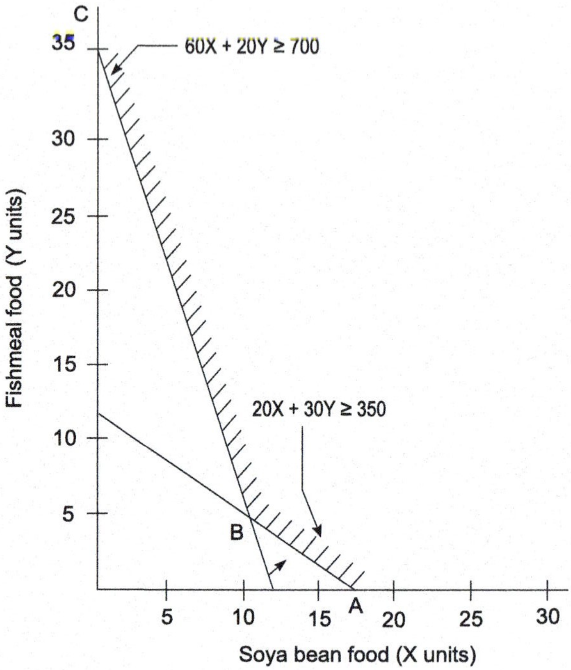

Figure 2.8 Soya bean food vs Fishmeal food

The optimum solution occurs at the corners of the feasible region. Name the corners. Find the value of Z at these corners.

$$Z = 20X + 9Y$$

At corner A, $X = 17.5$ and $Y = 0$

$$Z = 20(17.5) + 9(0)$$
$$= \text{Rs. } 350$$

At corner B, $X = 10$ and $Y = 5$

$$Z = 20(10) + 9(5)$$
$$= \text{Rs. } 245$$

At corner C, $X = 0$ and $Y = 35$

$$Z = 9(35)$$
$$= \text{Rs. } 315$$

Minimum value occurs at corner B.

$$\text{Minimise } Z = \text{Rs. } 245$$

subject to

$$X = 10$$
$$Y = 5$$

Example 2.8

A company manufactures washing machines and driers. The major manufacturing departments and their monthly capacity are as follows

Department	Monthly capacity in units		
	Washing machine		Drier
Stamping	10,000	or	10,000
Motor	16,000	or	7000
Transmission assembly line for washer	9000		–
Assembly line for drier	–		5000

The contribution to profit is Rs. 900 per washer and Rs. 1000 per drier. Determine the optimum product mix and profit.

Solution

Formulation

Variables Washing machines and driers are the variables. Let the company produce X units of washers and Y units of driers.

Objective function It is to maximise profit. Profit is Rs. 900 per unit for washer and Rs. 1000 per unit for drier.

$$\text{Maximise } Z = 900X + 1000Y$$

Constraints

(a) *Stamping restriction*

The stamping time taken for each washer and each drier is the same. Either 10,000 washers or 10,000 driers can be stamped.

Therefore, $$X + Y \leq 10{,}000 \quad (1)$$

(b) *Motor and constraint*

The capacity of this section is restricted to either 16,000 washers or 7000 driers, i.e., we can manufacture either 16,000 washers or 7000 driers.

$$16{,}000W = 7000D$$

$$W = \frac{7}{10}D$$

$$\frac{7}{16}X + Y \leq 7000$$

$$7X + 16Y \leq 1{,}12{,}000 \quad (2)$$

(c) *Transmission assembly line for washer*

$$X \leq 9000 \tag{3}$$

$$Y \leq 5000 \tag{4}$$

(d) *Assembly line for drier*

$$X \geq 0$$

(e) *Non negativity constraints*

$$Y \geq 0$$

The formulated LPP is as follows

$$\text{Maximise } Z = 900X + 1000Y$$

subject to

$$\begin{aligned} X + Y &\leq 10{,}000 \\ 7X + 16Y &\leq 1{,}12{,}000 \\ X &\leq 9000 \\ Y &\leq 5000 \\ X &\geq 0 \\ Y &\geq 0 \end{aligned}$$

Graphical solution

Considering Equation (1)

$$X + Y \leq 10{,}000$$

Bringing inequality to equality, we get

$$X + Y = 10{,}000$$

If $X = 0$, $Y = 10{,}000$. If $Y = 0$, $X = 10{,}000$. The values are (0, 10,000) (10,000, 0).

Considering Equation (2)

$$7X + 16Y \leq 1{,}12{,}000$$

Bringing inequality to equality

$$7X + 16Y = 1{,}12{,}000$$

When $X = 0$, $Y = 7000$. When $Y = 0$, $X = 16{,}000$. The values are (0, 7000) and (16,000, 0).

Considering Equation (3)

$$X \leq 9000$$

Bringing inequality to equality, we have

$$X = 9000$$

Considering Equation (4)

$$Y \leq 5000$$

Bringing inequality to equality

$$Y = 5000$$

Plot the graph with suitable scale.

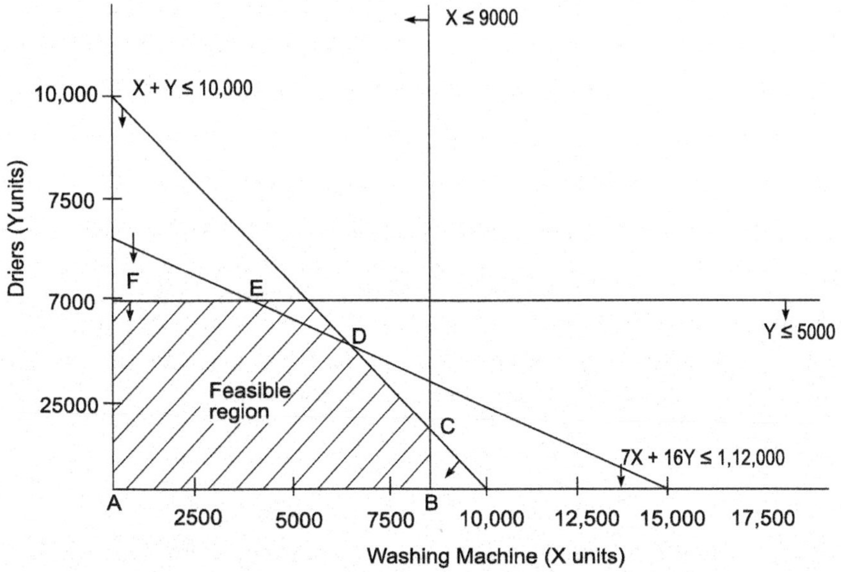

Figure 2.9 Washing machines vs Driers

The optimum solution occurs at the corners of the feasible region. Name the corners. Find the value of Z at each corner.

$$Z = 900X + 1000Y$$

At corner A, $X = 0$ and $Y = 0$.

$$Z = 900(0) + 1000(0)$$
$$= 0$$

At corner B, $X = 9000$ and $Y = 0$.

$$Z = 900(9000) + 1000(0)$$
$$= \text{Rs. } 81,00,000$$

At corner C, $X = 9000$ and $Y = 1000$.

$$Z = 900(9000) + 1000(1000)$$
$$= \text{Rs. } 91,00,000$$

At corner D, $X = 5300$ and $Y = 4700$.

$$Z = 900(5300) + 1000(4700)$$
$$= \text{Rs. } 94,70,000$$

At corner E, $X = 4600$ and $Y = 5000$.

$$Z = 900(4600) + 1000(5000)$$
$$= 41,40,000 + 50,00,000$$
$$= \text{Rs. } 9,40,000$$

Maximum profit occurs at corner D.

$$\text{Maximise } Z = \text{Rs. } 94{,}70{,}000$$

subject to

$$X = 5300$$
$$Y = 4700$$

Example 2.9

A company produces two types of leather belts, type A and type B. Profit on the two types of belts are Rs. 40 and Rs. 30, respectively per belt. Each A type belt requires twice as much time as required by belt B. If all the belts are of type B, the company would produce 1000 belts per day. But the supply of leather is sufficient only to produce 800 belts per day. Belt A requires fancy buckels and 400 fancy buckles are available per day. For belt B 700 buckels are available per day. How should the company manufacture the two types of belts in order to maximize the profits? Solve graphically.

Solution

Formulation

Variables Belts A and B are the variables. Let the company manufacture X units of belt A and Y units of belt B.

Objective It is to maximise profit. Profit per unit of belt A is Rs. 40 and profit per unit of belt B is Rs. 30.

$$\text{Maximise } Z = 40X + 30Y$$

Constraints

(a) *Time for belt*

$$\text{A} = 2 \times \text{Time for belt B}$$

Therefore, quantity produced in unit time

$$\frac{X}{2} = Y$$

$$\frac{X}{2} + Y \le 1000 \qquad (1)$$

(b) *Supply restriction*

$$X + Y \le 800 \qquad (2)$$

(c) *Buckle restriction*

$$X \le 400 \qquad (3)$$
$$Y \le 700 \qquad (4)$$

(d) *Non-negativity constraints*

$$Y \ge 0$$
$$X \ge 0$$

The formulated LPP is as follows

$$\text{Maximise } Z = 40X + 30Y$$

subject to

$$\frac{X}{2} + Y \leq 1000$$

$$X + Y \leq 800$$

$$X \leq 400$$

$$Y \leq 700$$

$$X \geq 0$$

$$Y \geq 0$$

Graphical solution

Considering Equation (1)

$$\frac{X}{2} + Y \leq 1000$$

Bringing inequality to equality, we get

$$\frac{X}{2} + Y = 1000$$

$$X + 2Y = 2000$$

If $X = 0$, $Y = 1000$. If $Y = 0$, $X = 2000$. The values are (0, 1000) (2000, 0).

Considering Equation (2)

$$X + Y \leq 800$$

Bringing inequality to equality

$$X + Y = 800$$

When $X = 0$, $Y = 800$. When $Y = 0$, $X = 800$. The values are (0, 800) and (800, 0).

Considering Equation (3)

$$X \leq 400$$

Bringing inequality to equality

$$X = 400$$

Considering Equation (4)

$$Y \leq 700$$

Bringing inequality to equality

$$Y = 700$$

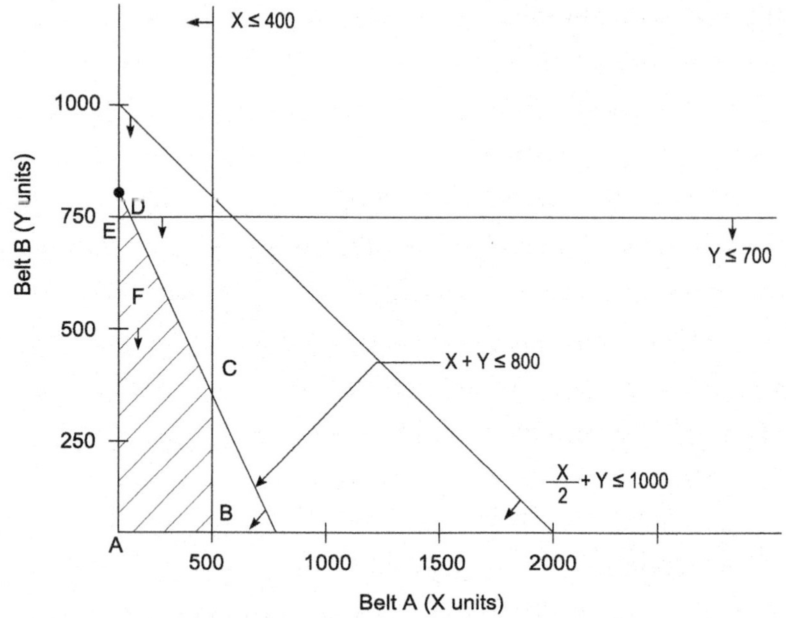

Figure 2.10 Belt A vs Belt B

The optimum solution occurs at the corners of the feasible region. Name the corners and find the value of Z at each corner

$$Z = 40X + 30Y$$

At corner A, $X = 0$ and $Y = 0$.

$$\begin{aligned} Z &= 40(0) + 30(0) \\ &= 0 \end{aligned}$$

At corner B, $X = 400$ and $Y = 0$.

$$\begin{aligned} Z &= 40(40) + 30(0) \\ &= \text{Rs. } 16{,}000 \end{aligned}$$

At corner C, $X = 400$ and $Y = 420$.

$$\begin{aligned} Z &= 40(400) + 30(420) \\ &= 16{,}000 + 12{,}600 \\ &= \text{Rs. } 28{,}600 \end{aligned}$$

At corner D, $X = 120$ $Y = 700$.

$$\begin{aligned} Z &= 40(120) + 30(700) \\ &= 4800 + 21{,}000 \\ &= \text{Rs. } 25{,}800 \end{aligned}$$

At corner E, $X = 0$ $Y = 700$.

$$\begin{aligned} Z &= 40(0) + 30(700) \\ &= \text{Rs. } 21{,}000 \end{aligned}$$

Maximum value occurs at corner C.

$$\text{Maximise } Z = \text{Rs. } 28{,}600$$

subject to

$$X = 400$$
$$Y = 420$$

Example 2.10

A company produces two products A and B. The sales volume for product A is at least 60% of the total sales of the two products. Both products use the same raw material of which the daily availability is limited to 100 lbs. A and B use this raw material at the rate of 2 lbs/unit and 4 lbs/unit respectively. The sales prices for the two products are Rs. 20 and Rs. 40/unit. Determine the optimal allocation of the raw materials to the products.

Solution

Formulation

Variables Products A and B are the variables. Let the company produce X units of product A and Y units of product B.

Objective It is to maximise sales revenue. Sales revenue/unit of product A is Rs. 20. Sales revenue/unit of product B is Rs. 40. Therefore,

$$\text{Maximise } Z = 20X + 40Y$$

Constraints

(a) *Raw material restriction*

$$2X + 4Y \leq 100 \qquad (1)$$

(b) *Market restriction*

$$X = 0.6(X + Y)$$
$$X - 0.6(X + Y) \geq 0$$
$$X - 0.6X - 0.6Y \geq 0$$
$$0.4X - 0.6Y \geq 0 \qquad (2)$$

(c) *Non-negativity constraints*

$$X \geq 0$$
$$Y \geq 0$$

Formulated LPP

$$\text{Maximise } Z = 20X + 40Y$$

subject to

$$2X + 4Y \leq 100$$
$$0.4X - 0.6Y \geq 0$$
$$X \geq 0$$
$$Y \geq 0$$

Graphical solution

Considering Equation (1)

$$2X + 4Y \leq 100$$

Bringing inequality to equality, we get

$$2X + 4Y = 100$$

If $X = 0$, $Y = 25$. If $Y = 0$, $X = 50$. The values are (0, 25) and (50, 0).

Considering Equation (2)

$$0.4X - 0.6Y \geq 0$$

Bringing inequality to equality

$$0.4X - 0.6Y = 0$$

$$4X = 6Y$$

$$2X = 3Y$$

If $X = 30$, $Y = 20$. If $Y = 10$, $X = 15$. The values are (30, 20) and (15, 10). Plot the graph.

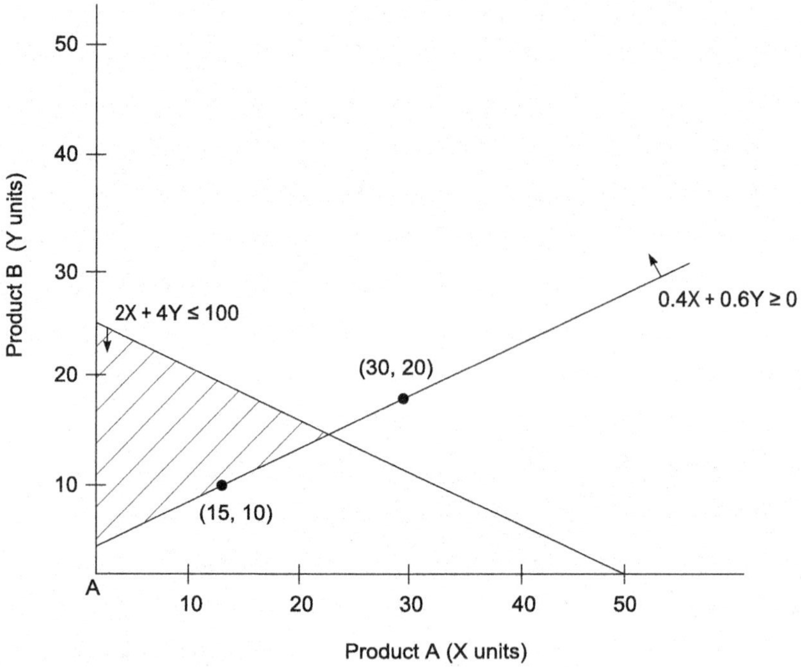

Figure 2.11 Product A vs Product B

Optimum solution always occurs at the corners of the feasible region. Name the corners. Find the value of Z at each corner.

$$Z = 20X + 40Y$$

At corner A, $X = 0$ and $Y = 5$.

$$Z = 20(0) + 40(5)$$

$$= \text{Rs. } 200$$

At corner B, $X = 21$ and $Y = 14$.

$$Z = 20(21) + 40(14)$$
$$= 420 + 560$$
$$Z = \text{Rs. } 980$$

At corner C, $X = 0$ and $Y = 25$.

$$Z = 20(0) + 40(25)$$
$$= \text{Rs. } 1000$$

Maximum value occurs at corner C.

$$\text{Maximise } Z = \text{Rs. } 1000$$

subject to

$$X = 0$$
$$Y = 25$$

Example 2.11

A factory has two bottling machines A and B. A is designed for eight-ounce bottels and B is designed for sixteen-ounce bottles. However each machine can be used on both types of bottels, with some loss of efficiency. The following data is available.

Machine	Eight-ounce bottle		Sixteen-ounce bottle
A	100/minute	or	40/minute
B	60/minute	or	75/minute

The machines can be run 8 hours/day and 5 days/week. Profit on eight-ounce bottle is Rs. 1.5 and sixteen-ounce bottle is Rs. 2.5 and the market can absorb 2,50,000 bottles of eight-ounce and 17,000 bottles of sixteen-ounce/week. Weekly production of the drink cannot exceed three lakh ounces. Formulate this as a LPP and solve graphically.

Solution

Formulation

Variables Variables are eight-ounce bottle and sixteen-ounce bottle. Let the factory produce X eight-ounce bottles and Y sixteen-ounce bottles.

Objective It is to maximise profit. Profit for X eight-ounce bottles is Rs. $1.5X$ and profit for Y sixteen-ounce bottles is Rs. $2.5Y$. Therefore,

$$\text{Maximise } Z = 1.5X + 2.5Y$$

Constraints

(a) *Bottling capacity of machines*

Machine runs 5 days/week and 8 hours/day.

$$\text{Total time} = 5 \times 8 = 40 \text{ hours}$$
$$\text{Total time} = 2400 \text{ minutes}$$
$$100 \text{ eight-ounce bottles} = 40 \text{ sixteen-ounce bottles}$$

$$1 \text{ eight-ounce bottle} = 0.4 \text{ sixteen-ounce bottle}$$

$$X \text{ eight-ounce bottles} = 0.4X \text{ sixteen-ounce bottels}$$

$$0.4X + Y \leq 2400 \quad (1)$$

$$60 \text{ eight-ounce bottles} = 75 \text{ sixteen-ounce bottles}$$

$$1 \text{ eight-ounce bottle} = \frac{75}{60} \text{ sixteen-ounce bottles}$$

$$X \text{ eight-ounce bottles} = \frac{75}{60} X \text{ sixteen-ounce bottles}$$

$$1.25X + Y \leq 2400 \quad (2)$$

(b) *Production restriction*

$$8X + 16Y = 3,00,000 \quad (3)$$

(c) *Market restriction*

$$X \leq 25,000 \quad (4)$$

$$Y \leq 17,000 \quad (5)$$

$$X \geq 0 \quad (6)$$

$$Y \geq 0 \quad (7)$$

Formulated LPP is given below

$$\text{Maximise } Z = 1.5X + 2.5Y$$

subject to

$$0.4X + Y \leq 2400$$

$$1.25X + Y \leq 2400$$

$$8X + 16Y \leq 3,00,000$$

$$X \leq 25,000$$

$$Y \leq 17,000$$

$$X \geq 0$$

$$Y \geq 0$$

Graphical solution

Considering Equation (1)

$$0.4X + Y \leq 2400$$

Bringing inequality to equality, we have

$$0.4X + Y = 2400$$

If $X = 0$, $Y = 2400$. If $Y = 0$, $X = 6000$. The values are (0, 2400) and (6000, 0).

Considering Equation (2)

$$1.25X + Y \leq 2400$$

Bringing inequality to equality

$$1.25X + Y = 2400$$

When $X = 0$, $Y = 2400$. When $Y = 0$, $X = 1920$.

Considering Equation (3)

$$8X + 16Y \leq 3{,}00{,}000$$

Bringing inequality to equality, we have

$$8X + 16Y = 3{,}00{,}000$$

When $X = 0$, $Y = 18{,}750$. When $Y = 0, X = 37{,}500$.

Considering Equation (4)

$$X \leq 25{,}000$$

Bringing inequality to equality, we have

$$X = 25{,}000$$

Considering Equation (5)

$$Y \leq 17{,}000$$

Bringing inequality to equality

$$Y = 17{,}000$$

Plot the graph with suitable scale.

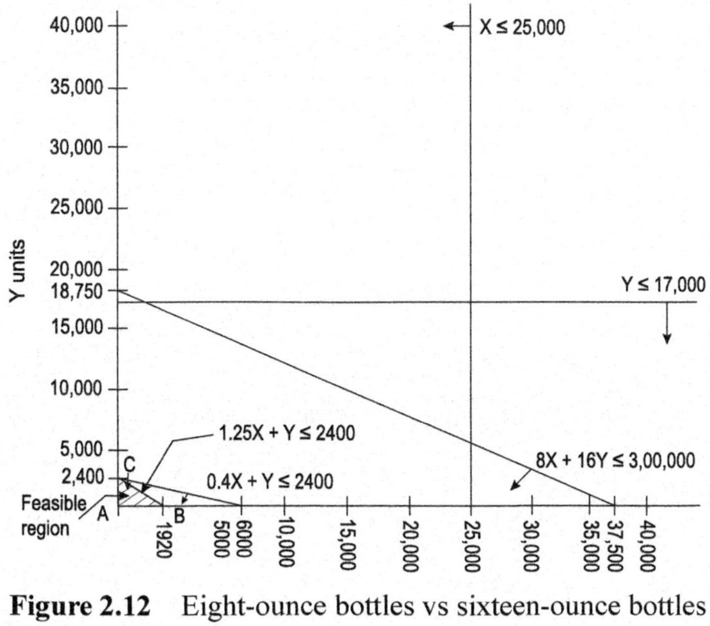

Figure 2.12 Eight-ounce bottles vs sixteen-ounce bottles

Optimum solution occurs at corners of the feasible region.

Find the value of Z at each corner.

$$Z = 1.5X + 2.5Y$$

At corner A, $X = 0$ and $Y = 0$.

$$Z = 1.5\,(0) + 2.5\,(0)$$

$$Z = 0$$

At corner B, $X = 1920$ and $Y = 0$.

$$Z = 1.5\,(1920) + 2.5\,(0)$$
$$= \text{Rs. } 2880$$

At corner C, $X = 0$ and $Y = 2400$.

$$Z = 1.5\,(0) + 2.5\,(2400)$$
$$= \text{Rs. } 6000$$

Maximum Z occurs at corner C.

$$\text{Maximise } Z = 6000$$

subject to

$$X = 0$$
$$Y = 2400$$

Example 2.12

A company makes sweepers and polishers. Each sweeper has a profit of Rs. 100. Each polisher has a profit of Rs. 150. Each sweeper requires one labour hour and one run, on a special machine. Each polisher requires one labour hour and two runs, on the special machine. For the scheduling period the demand is such that at least 10 sweepers must be made and at most 5 polishers.

There are 20 labour hours and 20 machine hours available. Use graphical method and find maximum profit for the scheduling period. An additional 5 runs on the special machine could be rented for Rs. 50 per run. Should the company rent the additional hours?

Solution

Formulation

Variables Variables are the sweepers and polishers. Let the company produce X units of sweepers and Y units of polishers.

Objective It is to maximise profit. By selling X sweepers, the company earns Rs. $100X$. By selling Y polishers, the company earns Rs. $150Y$. Therefore,

$$\text{Maximise } Z = 100X + 150Y$$

Constraints

(a) *Labour hours*

$$X + Y \leq 20 \quad (1)$$

(b) *Machine hours*

$$X + 2Y \leq 20 \quad (2)$$

(c) *Demand restriction*

$$X \geq 10 \quad (3)$$
$$Y \geq 5 \quad (4)$$

(d) *Non-negativity constraints*

$$X \geq 0 \quad (5)$$
$$Y \geq 0 \quad (6)$$

Graphical solution

Considering Equation (1)

$$X + Y \le 20$$

Bringing inequality to equality, we get

$$X + Y = 20$$

If $X = 0$, $Y = 20$. If $Y = 0$, $X = 20$. The values are (0, 20) and (20, 0).

Considering Equation (2)

$$X + 2Y \le 20$$

Bringing inequality to equality, we have

$$X + 2Y = 20$$

When $X = 0$, $Y = 10$. When $Y = 0$, $X = 20$. The values are (0, 10) and (20, 0).

Considering Equation (3)

$$X \ge 10$$

Bringing inequality to equality

$$X = 10$$

Considering Equation (4)

$$Y \le 5$$

Bringing inequality to equality

$$Y = 5$$

Plot the graph with suitable scale.

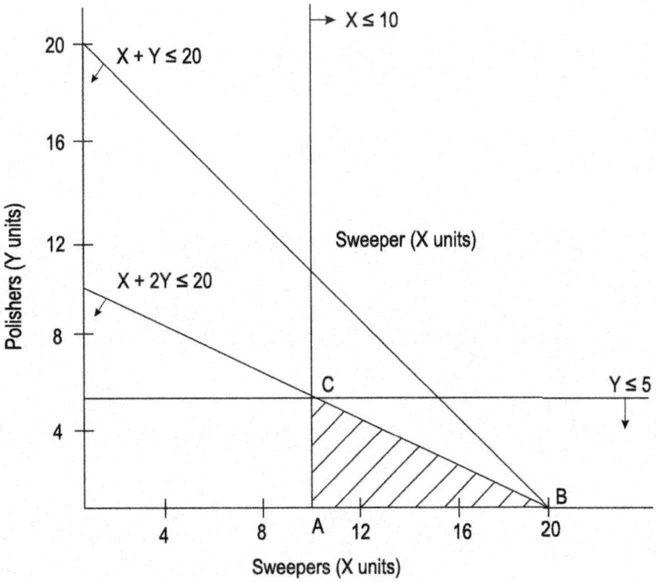

Figure 2.13 Sweepers vs Polishers

Optimum solution occurs at the corners of the feasible region. Name the corners. Find the value of Z at each corner.

$$Z = 100X + 150Y$$

At corner A, $X = 10$ and $Y = 0$.

$$Z = 100(10) + 150(0)$$
$$= \text{Rs. } 1000$$

At corner B, $X = 20$ and $Y = 0$.

$$Z = 100(20) + 150(0)$$
$$= \text{Rs. } 2000$$

At corner C, $X = 10$ and $Y = 5$.

$$Z = 100(10) + 150(5)$$
$$= \text{Rs. } 1750$$

Maximum value occurs at corner B

Therefore, the company has to produce 20 units of sweepers and the profit is Rs. 2000.

$$\text{Maximise } Z = \text{Rs. } 2000$$

subject to

$$X = 20$$
$$Y = 0$$

By renting additional 5 machine runs, the total machine run available will be 20 + 5 = 25 hours.

The machine restriction is

$$X + 2Y \leq 25$$

Bringing inequality to equality

$$X + 2Y = 25$$

When $X = 0$, $Y = 12.5$. When $Y = 0$, $X = 25$. The values are (0, 12.5) and (25, 0).

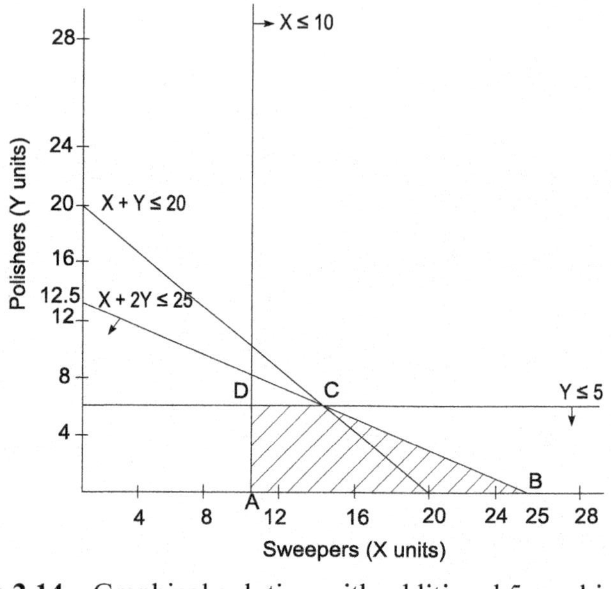

Figure 2.14 Graphical solution with additional 5 machine runs

Optimum solution occurs at the corners of the feasible region. Name the corners and find the value of Z at each corner.

$$Z = 100X + 150Y$$

At corner A, $X = 10$ and $Y = 0$.

$$Z = 100(10) + 150(0) = \text{Rs. } 1000$$

At corner B, $X = 20$ and $Y = 0$.

$$Z = 100(20) + 150(5) = \text{Rs. } 2000$$

At corner C, $X = 15$ and $Y = 5$.

$$Z = 100(15) + 150(5) = \text{Rs. } 2250$$

At corner D, $X = 10$ and $Y = 5$.

$$Z = 100(10) + 150(5) = \text{Rs. } 1750$$

Maximum is at C and the profit is Rs. 2250. The company has to produce 15 units of sweepers and 5 units of polishers. The cost of additional run is Rs. 5/hour. Therefore for five runs, the cost is 5×50 = Rs. 250.

$$\text{The net profit} = 2250 - 250 = \text{Rs. } 2000$$

The original profit is also Rs. 2000. Therefore, it is not profitable to have additional runs.

Example 2.13

Old hens can be bought for Rs. 20 each but, young hens cost Rs. 50 each. The old hens lay 3 eggs per week and young hens lay 5 eggs per week, each being worth Rs. 1.5. The feed cost for old hen is Rs. 5 per week. If I have only Rs. 800 to spend on hens, determine graphically how many of each kind I should buy to get a profit of more than Rs. 30 per week. Assume that I can house only 20 hens.

Solution

Formulation

Variables Variables are old hens and young hens. Let us purchase X number of old hens and Y number of young hens.

Objective It is to maximise profit.

$$\begin{aligned}\text{Profit per week for old hens} &= \text{Cost of eggs/week} - \text{Feed cost/week}\\ &= 3 \text{ eggs/week} \times \text{Rs. } 1.5\text{/egg} - \text{Rs. } 5\\ &= \text{Rs. } 4.5\text{/week} - \text{Rs. } 5\text{/week}\\ &= \text{Rs. } -0.5\text{/week}\end{aligned}$$

$$\text{Profit per week for } X \text{ old hens} = \text{Rs. } -0.5$$

$$\text{Profit per week for } Y \text{ young hens} = 5 \text{ eggs/week} \times \text{Rs. } 1.5\text{/egg} - \text{Rs. } 5\text{/week}$$
$$= \text{Rs. } 7.5\text{/week} - \text{Rs. } 5\text{/week}$$
$$= \text{Rs. } 2.5\text{/week}$$

Profit per week for Y young hens = Rs. 2.5 per week

$$\text{Maximise } Z = -0.5X + 2.5Y$$

Constraints

(a) *Money restriction*

$$20X + 50Y \leq 800 \qquad (1)$$

(b) *Profit restriction*

$$-0.5X + 2.5Y \geq 30 \qquad (2)$$

(c) *Housing restriction*

$$X + Y \leq 20 \qquad (3)$$
$$X \geq 0 \qquad (4)$$
$$Y \geq 0 \qquad (5)$$

Graphical solution

Considering Equation (1)

$$20X + 50Y \leq 800$$

Bringing inequality to equality, we get

$$20X + 50Y = 800$$

If $X = 0$, $Y = 16$. If $Y = 0$, $X = 40$. The values are (0, 16) and (40, 0).

Considering Equation (2)

$$-0.5X + 2.5Y \geq 30$$

Bringing inequality to equality, we have

$$-0.5X + 2.5Y = 30$$

If $X = 0$, $Y = 12$. If $Y = 0$, $X = -60$. The values are (0, 12) and (–60, 0).

Considering Equation (3)

$$X + Y \leq 20$$

Bringing inequality to equality

$$X + Y = 20$$

When $X = 0$, $Y = 20$. When $Y = 0$, $X = 20$. The values are (0, 20) and (20, 0). Plot the graph with suitable scale.

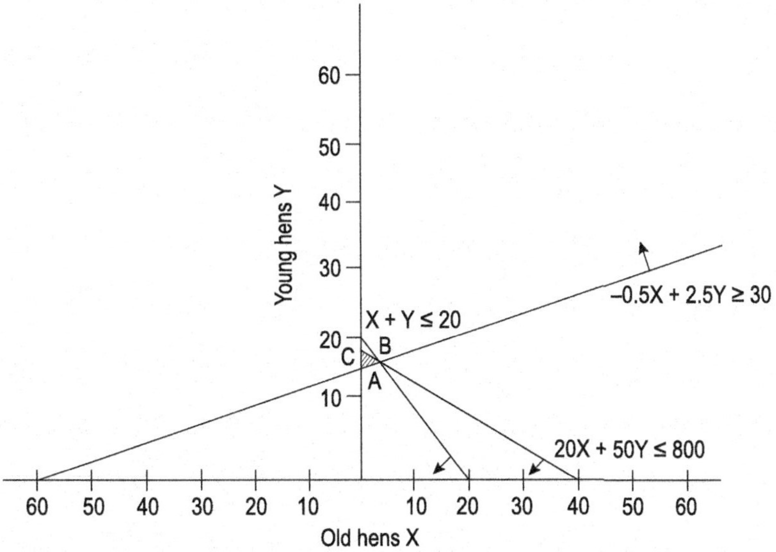

Figure 2.15 Old hens vs Young hens

Optimum solution occurs at the corners of the feasible region. Name the corners and find the value of Z at each corner.

$$Z = -0.5X + 2.5Y$$

At corner A, $X = 0$ and $Y = 12$.

$$\begin{aligned} Z &= -0.5(0) + 2.5(12) \\ &= \text{Rs. } 25 \end{aligned}$$

At corner B, $X = 6$ and $Y = 13$.

$$\begin{aligned} Z &= -0.5(6) + 2.5(13) \\ &= -3 + 32.5 \\ &= \text{Rs. } 29.5 \end{aligned}$$

At corner C, $X = 0$ and $Y = 16$.

$$\begin{aligned} Z &= -0.5(0) + 2.5(16) \\ &= \text{Rs. } 40 \end{aligned}$$

Maximum profit occurs at corner C.

Therefore, we have to purchase 16 young hens. The profit is Rs. 40.

$$\text{Maximise } Z = \text{Rs. } 40$$

subject to

$$\begin{aligned} X &= 0 \\ Y &= 16 \end{aligned}$$

Example 2.14

Ten gram of Alloy A contains 2 gm of Copper, 1 gm of Zinc and 1 gm of Lead. Ten gram of Alloy B contains 1 gm of Copper, 1 gm of Zinc and 3 gm of Lead. It is required to produce a mixture of these alloys which contain at least 10 gm of Copper, 8 gm of Zinc and 12 gm of Lead. Alloy B costs 1.5 times as much per kg as Alloy A. Find the amount of Alloy A and Alloy B which must be mixed in order to satisfy the condition in the cheapest way.

Solution

Formulation

Variables Amount of Alloy A and Alloy B used in the mixture. Let us use X gm of Alloy A and Y gm of Alloy B.

Objective B costs 1.5 times that of A.

$$\text{Minimise } Z = X + 1.5Y$$

Constraints

$$2X + Y \geq 10 \quad (1)$$

$$X + Y \geq 8 \quad (2)$$

$$X + 3Y \geq 12 \quad (3)$$

(a) *Non-negativity constraints*

$$X = 0 \quad (4)$$

$$Y = 0 \quad (5)$$

Graphical solution

Considering Equation (1)

$$2X + Y \geq 10$$

Bringing inequality to equality

$$2X + Y = 10$$

If $X = 0$, $Y = 10$. If $Y = 0$, $Y = 5$. The values are (0, 10) and (5, 0).

Considering Equation (2)

$$X + Y \geq 8$$

Bringing inequality to equality

$$X + Y = 8$$

When $X = 0$, $Y = 8$. When $Y = 0$, $X = 8$. The values are (0, 8) and (8, 0).

Considering Equation (3)

$$X + 3Y \geq 12$$

Bringing inequality to equality

$$X + 3Y = 12$$

Put $X = 0$ and $Y = 4$. Put $Y = 0$ and $X = 12$. The values are (0, 4) and (12, 0). Plot the graph with suitable scale.

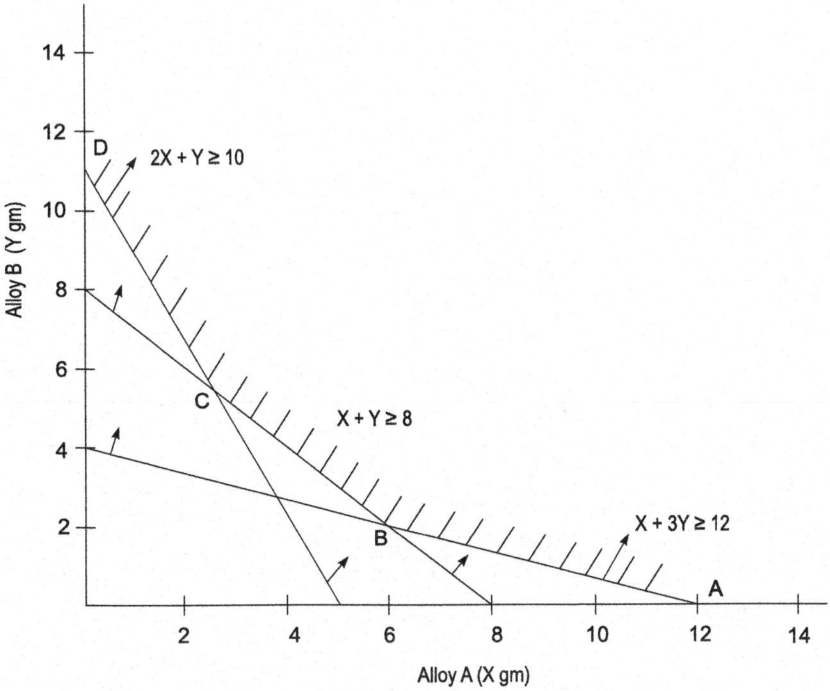

Figure 2.16 Alloy A vs Alloy B

Optimum solution always occurs at the corners of the feasible region. Name the corner and find the value of Z at these corners.

$$Z = X + 1.5Y$$

At corner A, $X = 12$ and $Y = 0$.

$$\begin{aligned} Z &= 12 + 1.5(0) \\ &= 12 \text{ Units} \end{aligned}$$

At corner B, $X = 6$ and $Y = 2$.

$$\begin{aligned} Z &= 6 + 1.5(2) \\ &= 9 \text{ units} \end{aligned}$$

At corner C, $X = 2.2$ and $Y = 58$.

$$\begin{aligned} Z &= 2.2 + 1.5(58) \\ &= 10.9 \text{ units} \end{aligned}$$

At corner D, $X = 0$ and $Y = 10$.

$$\begin{aligned} Z &= 0 + 1.5(10) \\ &= 15 \text{ units} \end{aligned}$$

Minimum value occurs at corner *B*.

$$\text{Minimise } Z = 9 \text{ units}$$

subject to

$$X = 6 \text{ gm}$$
$$Y = 2 \text{ gm}$$

Example 2.15

$$\text{Minimise } Z = 2X_1 + X_2$$

subject to

$$X_1 - X_2 \leq 10 \quad (1)$$
$$2X_1 \leq 40 \quad (2)$$
$$X_1 \geq 0 \quad (3)$$
$$X_2 \geq 0 \quad (4)$$

Solution

Bringing inequality to equality

Equation (1) becomes

$$X_1 - X_2 = 10$$

If $X_1 = 0, X_2 = -10$. If $X_2 = 0, X_1 = -0$. The values are (0, – 10) and (10, 0).

Equation (2) becomes

$$2X_1 = 40$$
$$X_1 = 20$$

Plot the graph.

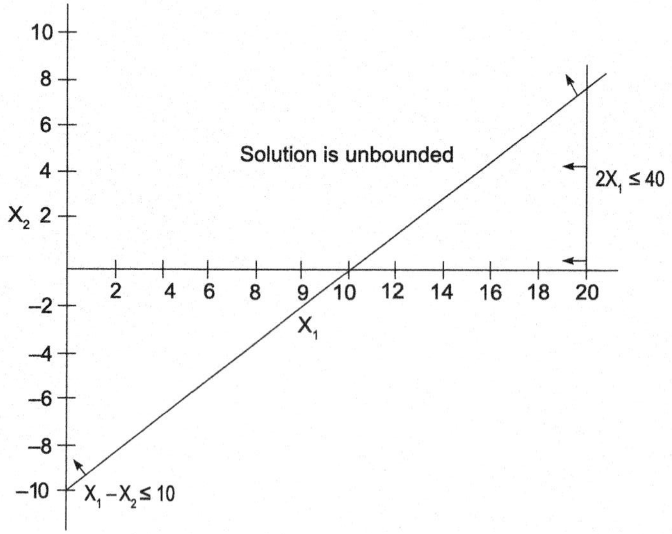

Figure 2.17 Unbounded solution

Equation $X_1 - X_2 \leq 10$ should move towards origin. Therefore, the solution space is the space between the constraint and the origin. Here the solution is unbounded.

It means, we can make as much profit as we want, if it is a maximisation case or we can have as least cost as possible, if it is a minimisation problem. This is not possible in practice.

This has happened for the reason, some constraints might have been omitted or some mistake has been done in taking the constraints.

EXERCISES

1. A furniture manufacturer produces two types of desks, standard and executive. These desks are sold to an office furniture wholesaler and for all practical purposes, there is an unlimited market for any mix of these desks, at least within the manufactures production capacity. Each desk has to go through four basic operations: Cutting of the timber, joining of the pieces, pre-finishing, and final finish. Each unit of the standard desk produced takes 48 minutes for cutting, 2 hours for joining, 40 minutes for pre-finishing and 4 hours for final finishing. Each unit of the executive desk requires 72 minutes for cutting, 3 hours for joining, 2 hours for pre-finishing and five hours and 20 minutes for final finishing. The daily capacity for each operation amounts to 16 hours of cutting, 30 hours of joining 16 hours of pre-finishing and 64 hours of final finishing. The profit per unit produced is Rs. 40 for the standard desk and Rs. 50 for the executive desk. Determine the product mix that maximizes total revenue.
2. A retired person wants to invest upto an amount of Rs. 30,000 in fixed income securities. His broker recommends investing in two bonds: Bond A yielding 7% and Bond B yielding 10%. After some consideration, he decides to invest at most Rs. 12,000 in Bond B and at least Rs. 6,000 in Bond A. He also wants the amount invested in Bond A to be at least equal to the amount invested in Bond B. What should the broker recommend if the investor wants to maximise his return on investment? Solve graphically.
3. A local travel agent is planning a charter trip to a major sea port. The eight day/seven night package includes the fare for round trip, surface transportation, boarding and lodging and selected tour options. The charter trip is restricted to 200 persons and the past experience indicates that there will not be any problem for getting 200 clients. The problem for the travel agent is to determine the number of deluxe, standard and economy tour packages to offer for this charter. These three plans differs, according to seating and service for the flight, quality of accommodation, meal plans and tour options. The following table summarises the estimated prices for the three packages and the corresponding expenses for the travel agent. The travel agent has hired an aircraft for a flat fee of Rs. 2,00,000 for the entire trip. In planning the trip, the following consideration must be taken into account.
   i. At least 10% of the packages must be of the deluxe type.
   ii. At least 35% but not more than 70% must be of the standard type.
   iii. At least 30% should be of the economy type.
   iv. At maximum number of deluxe packages available in any air craft is restricted to 60.
   v. The hotel desires that at least 120 of the tourists should be on the deluxe and standard packages taken together.

Price and costs for tour packages per person

Tour plan	Price	Hotel costs	Meal and other expenses
Deluxe	10,000	3000	4750
Standard	7000	2200	2500
Economy	6500	1900	2200

The travel agent wants to determine the number of packages to offer in each type so as to maximise the total profit.

(a) Formulate this as a LPP.

(b) Relate the above LPP in terms of two decision variables, taking advantage of the fact those 200 packages will be sold.

(c) Find the optimal solution using graphical method in the restated problem and interpret your results.

4. A farmer has a 100 acre farm. He can sell all the tomatoes, lettuce or radishes he can raise. The price he can obtain is Re. 1 per kg for tomatoes Rs. 0.75 per kg for lettuce and Rs. 2 per kg for radishes. The average yield per acre is 2000 kg of tomatoes, 3000 heads of lettuce and 1000 kg of radishes.

   Fertilizer is available at Rs. 0.05 per kg and the amount required per acre is 100 kg each, for tomatoes and lettuces and 50 kg for radishes. Labour required for sowing, cultivating and harvesting per acre is 5 man-days for tomatoes and radishes and 6 man-days for lettuce. A total of 400 man-days of labour are available at Rs. 20 per man-day. Formulate this problem as a linear programming model. Do not solve.

5. The standard weight of a special purpose brick is 5 kg and it contains two basic ingredients B_1 and B_2. B_1 costs Rs. 5 per kg and B_2 costs Rs. 8 per kg. Strength considerations state that the brick contains not more than 4 kg of B_1 and minimum of 2 kg of B_2. Since the demand for the product is likely to be related to satisfying the above conditions.

6. Semicond is an electronics company manufacturing tape recorders and radios. Its per unit labour costs, raw material costs and selling price are given in Table 1. An extract from its balance sheet as on 31.3.1994 is shown in Table 2. Its current asset/current liability ratio is 2.

Table 1 Cost information

For Products	Selling price (Rs.)	Labour cost (Rs.)	Raw material cost (Rs.)
Tape Recorder	1000	500	300
Radio	900	350	400

Table 2 Information

Particulars	Current liabilities (Rs.)	Current assets (Rs.)
Cash	–	1,00,000
Accounts Receivable	–	30,000
Inventory	–	70,000
Short term bank borrowings	1,00,000	–

* Accounts receivable is the amount due from customers.
* 100 units of raw material is used for tape recorder.
* 100 units of raw material is used for radio.

Semicond must determine how many tape recorders and radios should be produced during April 1994. Demand is large enough to ensure that all goods produced will be sold. All sales are on credit and payment for goods sold in April 1994 will not be received until May 31, 1994. During April 1994, it will collect Rs. 20,000 in accounts receivable and it must pay-off Rs. 10,000 of the outstanding short term bank borrowing and a monthly rent of Rs. 10,000. On April 30, 1994 it will receive a shipment of raw material worth Rs. 20,000 which will be paid on May 31, 1994. The management has decided that the cash balance on April 30, 1994 must be at least Rs. 40,000. Also its banker requires that the current ratio as on April 30, 1994 be at least 2. In order to maximise the contribution of profit for April 94 production it has to find the product mix for April 1994. Assume that labour costs (wages) are paid in the month in which they are incurred. Formulate this as a linear programming problem and graphically solve it.

7. A firm plans to purchase at least 200 quintals of scrap containing high quality metal X and low quality metal Y. It decides that the scrap to be purchased must contain at least 100 quintals of X metal and not more than 35 quintals of Y metal. The firm can purchase the scrap from two suppliers (A and B) in unlimited quantities. The percentage of X and Y metals in terms of weight in the scraps supplied by A and B is given below

Metals	Supplier A	Supplier B
X	25%	75%
Y	10%	20%

The price of A's scrap is Rs. 200 per quintal and that of B's scrap is Rs. 400 per quintal. Formulate this problem as LP model and solve it to determine the quantities that the firm should buy from the two suppliers so as to minimise the total purchase cost.

8. A television company has three major departments for manufacturing of its two models A and B. Monthly capacities are given as follows

Department	Per unit time requirements (hours)		Hours available this month
	Model A	Model B	
I	4.0	2.0	1600
II	2.5	1.0	1020
III	4.5	1.5	1600

The marginal profit of model A is Rs. 400 each and that of model B is Rs. 100 each. Assuming that the comapny can sell any quanity of either product due to favourable market conditions. Determine the optimum output for both the models, the highest possible profit for this month and the slack time in three departments.

9. A poultry producer must feed his stock daily at least 124 units of nutritional element A and 60 units of nutritional element B. He has two feeds available. One kg of the first feed costs Rs. 16 and contains 10 units of A and 3 units of B. One kg of the second feed costs Rs. 14 and contains 4 units of A and 5 units of B. Determine least expensive adequate diet, by graphical method.

10. Upon completing the construction of his house Dr. Sharma discovers that 100 sq. ft of plywood scrap and 80 sq. ft of white pine scrap are in usable form for the construction of tables and book cases. It takes 16 sq. ft of plywood and 8 sq. ft of white pine to make a table; 12 sq. ft of plywood and 16 sq. ft of white pine are required to construct a book case. By selling the finished products to a local furniture store, Dr. Sharma can realise a profit of Rs. 25 on each table and Rs. 290 on each book case. How may he most profitably use the left-over wood? Use graphical method to solve the problems.

11. An automobile manufacturer makes automobiles and trucks in a factory that is divided into two shops. Shop A, which perform the basic assembly operation must work 5 man-days on each truck but only 2 man-days on each automobile. Shop B, which performs finishing operations must work 3 man-days for each automobile or truck that it produces. Because of men and machine limitation, shop A has 180 man-days per week available, while shop B has 135 man-days per week. If the manufacturer makes a profit of Rs. 300 on each truck and Rs. 200 on each automobile, how many of each should be produced to maximise his profit?

12. A company can advertise its product by using local radio or TV station. The budget limits the advertisement expenditure to Rs. 50,000 a month. Each minute of radio advertisement costs Rs. 150 and each minute of TV advertisement costs Rs. 3000. The company would like to use the radio at least twice as much as the TV. Past experience shows that each minute of TV advertisement will usually generate 25 times as many sales as each minute of radio advertisement. Determine the optimum allocation of monthly budget to TV and radio advertisements.

13. The owner of metro sports wishes to determine how many advertisements to place in the two selected monthly magazines A and B. Percentage of readers for each magazine are known. Exposure in any particular magazine is the number of advertisements placed multiplied by the number of principal buyers. The following data may be used.

	Magazine		Availability (Rs.)
	A (Rs.)	B (Rs.)	
Readers	1 lakh	0.6 lakh	–
Principal buyer	10%	15%	–
Cost/advertisement	5000	4500	1,00,000

The budgeted amount is at most Rs. 1 lakh for the advertisements. The owner has already decided that magazine A should have no more than 6 advertisements and that B should have at least 10 advertisements to be placed in each magazine.

14. A company machines and drills two castings X and Y. The time required to machine and drill one casting including machine set-up time is as follows

Casting	Machining hours	Drilling hours
X	4	2
Y	2	5

There are two lathes and 3 drilling machines. The working week is of 40 hours; there is no lost time and overtime. Variable costs for both castings are Rs. 120 per unit while the total fixed costs amount to Rs. 1000 per week. The selling price of casting X is Rs. 300 per unit and Y is Rs. 360 per unit. There are no limitations on the number X and Y casting that can be sold. The company wishes to maximise its profits. (i) Formulate LP model for the problem and (ii) solve the problem graphically.

15. A student has two final exams to prepare. Each hour of study he devotes to course A is expected to return him Rs. 600 in terms of long range benefits and each hour of study he devotes to course B is expected to return him Rs. 300 in terms of long range benefits. The shops are all closed; the student has only 15 chocolates remaining. He feels he needs one chocolate every 20 minutes while studying for course B and one chocolate every 12 minutes while studying for course A. Time is running short and only 4 hours remain to prepare for the exam. It is necessary that the student must devote at least 2 hours for studying. The student would like to maximise his return for the effort expended. Solve this problem as an LPP and determine the optimum policy for the student by solving the problem.

16. An oil company has 2 units A and B which produce 3 different grades of oil, superfine, medium and low grade oil. The company has to supply 12, 8 and 24 barrels of superfine, medium and low grade oils respectively per week. It costs the company Rs. 100 and Rs. 80 per day to run the units A and B respectively. On a day, unit A produces 6, 2 and 4 barrels and unit B produces 2, 2 and 12 barrels of superfine, medium and low grade oil per day. The manager has to decide on how many days per week should each unit be operated in order to meet the requirement at minimum cost. Formulate the LPP and solve it graphically.

17. Consider a small plant which makes two types of automobile parts say A and B. The plant buys castings from a supplier. Casting for part A costs Rs. 2 each and casting for part B costs Rs. 3 each. The castings are machined, bored and polished. The details of machinery are as follows

Capacity in pieces per hour

Machine	Part A	Part B	Machining cost (Rs. per hour)
Lathe	25	40	20
Boring	28	35	14
Polishing	35	25	17.5
Selling price per unit	Rs. 5	Rs. 6	–

The above table also indicates the selling price for two parts. Assuming that any combination of part A and part B can be sold in the market. Determine the optimal product mix (formulate the above as an LPP and solve graphically).

18. A company has two grades of inspectors 1 and 2, who are to be assigned for a quality-control inspection. It is required that at least 1800 pieces be inspected in an 8 hour day. Grade 1 inspectors can check pieces at the rate of 25 per hour, with an accuracy of 98%. Grade 2 inspectors check at the rate of 15 pieces per hour with an accuracy of 95%.

The wage rate of grade 1 inspectors is Rs. 200/hour and that of grade 2 inspectors is Rs. 150 per hour. Each time an error is made by an inspector, the cost to the company is Rs. 100. The company has eight grade 1 inspectors and ten grade 2 inspectors, for the inspection job. The company wants to determine the optimal assignment of inspectors, which will minimise the total cost of the inspection.

Chapter 3

SIMPLEX METHOD

3.1 INTRODUCTION

Graphical method of solving LPP has a drawback in the sense it cannot be used for those problems that have more than two variables. Simplex method is an algebraic iterative process used to determine the optimal solution, when the variables are more than two. The simplex method checks whether it is possible to find a better basic feasible solution.

3.2 SLACK AND SURPLUS VARIABLES

Consider a constraint $2X_1 + 4X_2 \leq 20$. This inequality can be brought to equality by adding a variable to the left-hand side $2X_1 + 4X_2 + S_1 = 20$. S_1 is called the slack variable.

In general, let the constraints of an LPP be

$$a_{i1}X_1 + a_{i2}X_2 + a_{i3}X_3 + \cdots + a_{in}X_n \leq b_i, \qquad \text{where } i = 1, 2, 3, \ldots, m.$$

Then the inequality can be brought to equality by adding a non-negative variable to the left-hand side. Let these variables be X_{n+i}.

$$a_{i1}X_1 + a_{i2}X_2 + a_{i3}X_3 + \cdots + a_{in}X_n + X_{n+i} = b_i.$$

X_{n+i} is called the slack variable.

Consider the equation $5X_1 + 3X_2 \geq 15$. This inequality can be brought to equality by deducting S_1 from the left-hand side.

$5X_1 + 3X_2 - S_1 = 15$. S_1 is called a surplus variables. In general,

$$a_{i1}X_1 + a_{i2}X_2 + a_{i3}X_3 + \cdots + a_{in}X_n \geq b_i \qquad \text{where } i = 1, 2, 3, \ldots, m$$

This inequality can be brought to equality by deducting X_{n+i} from the left-hand side.

Then $$a_{i1}X_1 + a_{i2}X_2 + a_{i3}X_3 + \cdots + a_{in}X_n - X_{n+i} = b_i$$

X_{n+i} is called the surplus variable.

3.3 BASIC SOLUTION

Consider the system of equations.

$$X_1 + 2X_2 + X_3 = 4$$
$$2X_1 + X_2 + 5X_3 = 5$$

Here there are three variables and two constraints. We keep (3 – 2) variables at zero and solve the remaining variables.

Let us put $X_3 = 0$ and solve X_1 and X_2. X_3 which is kept at zero is called a non-basic variable. X_1 and X_2 are called basic variables.

The solution obtained is the basic solution.

In general, let us consider a set of constraints.

$$a_{i1}X_1 + a_{i2}X_2 + a_{i3}X_3 + \dots a_{in}X_n = b_i \qquad \text{where } i = 1, 2, 3, \dots, m.$$

This has got 'n' variables. There are m such constraints. Let us keep $(n - m)$ variables at zero. Solve for the remaining variables. The solution obtained is known as basic solution $(n - m)$, variables which are kept at zero level are non-basic variables. The remaining variables are basic variables.

3.4 BASIC FEASIBLE SOLUTION

A solution which satisfies all the constraints and non-negativity restriction is called a feasible solution. A feasible solution which is also basic is called a basic feasible solution.

3.5 OPTIMUM BASIC FEASIBLE SOLUTION

The basic feasible solution to LPP is called Optimum basic feasible solution if $Z_0 \geq Z^*$. Where Z^* is the value of the objective function for any feasible solution and Z_0 is the maximum value of the objective function.

3.6 SIMPLEX PROBLEM IN TABLE FORM

Basic	X_1	X_2	X_3	S_1	S_2	Solution
Z	–5	–2	–3	0	0	0
S_1	1	2	2	1	0	8
S_2	3	4	1	0	1	7

A simplex problem can be expressed in a detached co-efficient form. The table above illustrates the initial basic feasible solution. Basic refers to the basic variables in the basic feasible solution.

From the table above, the basic feasible solution is written as $S_1 = 8$, $S_2 = 7$.

Example 3.1

Solve by the simplex method.

$$\text{Maximise } Z = 5X_1 + 2X_2$$

subject to

$$\begin{aligned} 10X_1 + 2X_2 &\leq 2100 \\ X_1 + 0.5X_2 &\leq 600 \\ X_2 &\leq 800 \\ X_1 &\geq 0 \\ X_2 &\geq 0 \end{aligned}$$

Solution

Add slack variables and bring inequalities to equality

$$10X_1 + 2X_2 + S_1 = 2100$$
$$X_1 + 0.5X_2 + S_2 = 600$$
$$X_2 + S_3 = 800$$

Here, there are five variables and three constraints. Therefore, keep 5 – 3 = 2 variables at zero. Let us keep X_1 and X_2 at zero, then,

$$S_1 = 2100$$
$$S_2 = 600$$
$$S_3 = 800$$

Objective function

$$Z = 5X_1 + 2X_2$$
$$Z - 5X_1 - 2X_2 = 0$$

Write the simplex problem in table form.

↓PC

Basic	X_1	X_2	S_1	S_2	S_3	Solution		
Z	–5	–2	0	0	0	0		
S_1	10*	2	1	0	0	2100	$\frac{2100}{10} = 210$	→PR
S_2	1	0.5	0	1	0	600	$\frac{600}{1} = 600$	
S_3	0	1	0	0	1	800		

Look for the maximum value with a negative sign in the Z row. It is – 5. The variable corresponding to this value is X_1. X_1 tries to enter the basic column. This X_1 column is called a pivot column.

Which leaves the basic column?

Ignore all zeroes and negative values in the pivot column. Divide the elements in the solution column by the corresponding elements in the pivot column, i.e.,

$$\frac{2100}{10} = 210$$

and

$$\frac{600}{1} = 600$$

Select the minimum of these two. It is 210. This row is called a pivot row. Now X_1 enters the basic column and S_1 leaves. An element which is common to both pivot row and pivot column is called pivot element.

In the aforementioned table 10 is the pivot element. It is common to both pivot row S_1 and pivot column X_1. '*' represents pivot element.

The solution is optimum when all the values in the 'Z' row are either positive or zero. Here, there are two variables with negative value. Therefore, the solution is not optimum.

Iteration 1

$$\text{New pivot equation} = \frac{\text{Old pivot equation}}{\text{Pivot elements}}$$

New pivot equation (NPE) : $\frac{10}{10}\ \frac{2}{10}\ \frac{1}{10}\ \frac{0}{10}\ \frac{0}{10}\ \frac{2100}{10}$

NPE : 1 0.2 0.1 0 0 210

For all other equations,

New equation = Old equation – Corresponding pivot × NPE column element

Old Z	–5	–2	0	0	0	0
–(–5) × NPE	5	1	0.5	0	0	1050
New Z	0	–1	0.5	0	0	1050
Old S_2	1	0.5	0	1	0	600
–(–1) × NPE	–1	–0.2	–0.1	0	0	–210
New S_2	0	0.3	–0.1	1	0	390

Formulate the new basic table.

↓PC

Basic	X_1	X_2	S_1	S_2	S_3	Solution	
Z	0	–1	0.5	0	0	1050	
S_1	1	0.2	0.1	0	0	210	$\frac{210}{0.2} = 1050$
S_2	0	0.3	–0.1	1	0	390	$\frac{390}{0.3} = 1300$
S_3	0	1*	0	0	1	800	$\frac{800}{1} = 800$ →PR

Z row has a negative value –1. Therefore, the solution is not optimum.

Iteration 2

X_2 enters the basic column. X_2 is the pivot column. Ignore –ve value in the pivot column. Divide the elements in the solution column by the corresponding element in the pivot column, i.e.,

$$\frac{210}{0.2} = 1050$$

$$\frac{390}{0.3} = 1300$$

$$\frac{800}{1} = 800$$

Select the minimum of these three. It is 800. Row S_3 is called the pivot row.

NPE	0	1	0	0	1	800
Old Z	0	–1	0.5	0	0	1050
–(–1)NPE	0	1	0	0	1	800
New Z	0	0	0.5	0	1	1850
Old X_1	1	0.2	0.1	0	0	210
–(0.2). NPE	0	–0.2	0	0	–0.2	–160
New X_1	1	0	0.1	0	–0.2	50
Old S_2	0	0.3	–0.1	1	0	390
–(0.3). NPE	0	–0.3	0	0	–0.3	–240
New S_2	0	0	–0.1	1	–0.3	150

Formulate the basic table.

Basic	X_1	X_2	S_1	S_2	S_3	Solution
Z	0	0	0.5	0	1	1850
X_1	1	0	0.1	0	–0.2	50
S_2	0	0	–0.1	1	–0.3	150
X_2	0	1	0	0	1	800

All the values in the Z row are either positive or zero. Therefore, the solution is optimum.

$$\text{Maximise } Z = 1850$$

subject to

$$X_1 = 50$$
$$X_2 = 800$$

Example 3.2

Use the simplex method to maximise

$$Z = 22x - 30y + 25z$$

where x, y and z are non-negative variables,

subject to

$$2x + 2y \leq 100$$
$$2x + y + z \leq 100$$

$$\begin{aligned} x + y + 2z &\leq 100 \\ x &\geq 0 \\ y &\geq 0 \\ z &\geq 0 \end{aligned}$$

Solution

Add slack variables and bring inequalities to equality.

$$\begin{aligned} 2x + 2y + S_1 &= 100 \\ 2x + y + z + S_2 &= 100 \\ x + y + 2z + S_3 &= 100 \end{aligned}$$

Here, there are six variables and three constraints. Therefore, keep 6 – 3 = 3 variables at zero. Let us keep x, y and z at zero, then,

$$\begin{aligned} S_1 &= 100 \\ S_2 &= 100 \\ S_3 &= 100 \end{aligned}$$

Objective function

$$Z = 22x - 30y + 25z$$

Therefore,

$$Z - 22x + 30y - 25z = 0$$

Write the simplex problem in table form.

↓PC

Basic	x	y	z	S_1	S_2	S_3	Solution	
Z	–22	30	–25	0	0	0	0	
S_1	2	2	0	1	0	0	100	
S_2	2	1	1	0	1	0	100	$\frac{100}{1} = 100$
S_3	1	1	2*	0	0	1	100	$\frac{100}{2} = 50$ →PR

Look for the maximum value with a negative sign in the Z row. It is –25. The variable corresponding to this value is z. z tries to enter the basic column. This z column is called as a pivot column.

Which leaves the basic column?

Ignore all zeroes and negative values in the pivot column. Divide the elements in the solution column by the corresponding elements in the pivot column, i.e.,

$$\frac{100}{1} = 100$$

and

$$\frac{100}{2} = 50$$

Select the minimum of these two. It is 50. Row S_3 leaves. This is called a pivot row.

Now z enters the basic column and S_1 leaves. In the aforementioned table Z is the pivot element. It is common to both pivot row S_3 and pivot column z.

The solution is optimum when all the values in the z row are either positive or zero. Here there are two variables with negative sign. Therefore the solution is not optimum.

Iteration 1

$$\text{New pivot equation} = \frac{\text{Old pivot equation}}{\text{Pivot elements}}$$

New pivot equation	$\frac{1}{2}$	$\frac{1}{2}$	1	0	0	$\frac{1}{2}$	50
Old Z	–22	30	–25	0	0	0	0
–(–25)NPE	$\frac{25}{2}$	$\frac{25}{2}$	25	0	0	$\frac{25}{2}$	1250
New Z	$-\frac{19}{2}$	$\frac{85}{2}$	0	0	0	$\frac{25}{2}$	1250
Old S_1 = New S_1	2	2	0	1	0	0	100
Old S_2	2	1	1	0	1	0	100
–(1)NPE	$-\frac{1}{2}$	$-\frac{1}{2}$	–1	0	0	$-\frac{1}{2}$	–50
New S_2	$\frac{3}{2}$	$\frac{1}{2}$	0	0	1	$-\frac{1}{2}$	50

Formulate the basic table.

Basic	x ↓PC	Y	Z	S_1	S_2	S_3	Solution		
Z	$-\frac{19}{2}$	$\frac{85}{2}$	0	0	0	$\frac{25}{2}$	1250		
S_1	2	2	0	1	0	0	100	$\frac{100}{2} = 50$	
S_2	$\frac{3}{2}$	$\frac{1}{2}$	0	0	1	$-\frac{1}{2}$	50	$\frac{100}{3} = 38.33$	→ PR
Z	$\frac{1}{2}$	$\frac{1}{2}$	1	0	0	$\frac{1}{2}$	50	$\frac{100}{1} = 100$	

In the table, z row has a negative value. Therefore, the solution is not optimum.

Look for the negative value in the 'z' row. It is $-\frac{19}{2}$. The variable corresponding to this value is x. x tries to enter the basic column. This x column is called a pivot column.

Which leaves the basic column?

Ignore all zeroes and negative values in the pivot column. Divide the elements in the solution column by the corresponding elements in the pivot column, i.e.,

$$\frac{100}{2} = 50$$

$$50 \times \frac{2}{3} = 33.33$$

and

$$50 \times \frac{2}{1} = 100$$

Select the minimum of these three. It is 33.33. Row S_2 leaves. This is called a pivot row.

Now x enters the basic column and S_2 leaves. In the aforementioned table, $\frac{3}{2}$ is the pivot element. It is common to both pivot row S_3 and pivot column x.

Iteration 2

New pivot Equation	1	$\frac{1}{3}$	0	0	$\frac{2}{3}$	$-\frac{1}{3}$	$\frac{100}{3}$
Old Z	$-\frac{19}{2}$	$\frac{85}{2}$	0	0	0	$\frac{25}{2}$	1250
$-\left(-\frac{19}{2}\right) NPE$	$\frac{19}{2}$	$\frac{19}{6}$	0	0	$\frac{19}{3}$	$-\frac{19}{6}$	$\frac{1900}{6}$
New Z	0	$\frac{137}{3}$	0	0	$\frac{19}{3}$	$\frac{28}{3}$	1566.67
Old S_1	2	2	0	1	0	0	100
–2 NPE	–2	$-\frac{2}{3}$	0	0	$-\frac{4}{3}$	$\frac{2}{3}$	$-\frac{200}{3}$
New S_1	0	$\frac{4}{3}$	0	1	$-\frac{4}{3}$	$\frac{2}{3}$	$\frac{100}{3}$
Old Z	$\frac{1}{2}$	$\frac{1}{2}$	1	0	0	$\frac{1}{2}$	50
$-\frac{1}{2}$ NPE	$-\frac{1}{2}$	$-\frac{1}{6}$	0	0	$-\frac{1}{3}$	$\frac{1}{6}$	$-\frac{50}{3}$
New Z	0	$\frac{1}{3}$	1	0	$-\frac{1}{3}$	$\frac{2}{3}$	$\frac{100}{3}$

Basic	x	y	z	S_1	S_2	S_3	Solution
Z	0	$\frac{137}{3}$	0	0	$\frac{19}{3}$	$\frac{28}{3}$	1566.67
S_1	0	$\frac{4}{3}$	0	1	$-\frac{4}{3}$	$\frac{2}{3}$	$\frac{100}{3}$
X	1	$\frac{1}{3}$	0	0	$\frac{2}{3}$	$-\frac{1}{3}$	$\frac{100}{3}$
Z	0	$\frac{1}{3}$	1	0	$-\frac{1}{3}$	$\frac{2}{3}$	$\frac{100}{3}$

In this table, all the values in Z row are either positive or zero. Therefore the solution is optimum.

$$\text{Maximise } Z = 1566.67$$

subject to

$$x = \frac{100}{3}$$
$$y = 0$$
$$z = \frac{100}{3}$$

Example 3.3

Solve the following LPP using simplex method.

$$\text{Maximise } Z = 30X_1 + 23X_2 + 2\,X_3$$

subject to

$$6X_1 + 5X_2 + 3X_3 + 3 \leq 26$$
$$4X_1 + 2X_2 + 5X_3 \leq 7$$
$$X_1, X_2, X_3 \geq 0$$

Solution

Add slack variables and bring inequalities to equality.

$$6X_1 + 5X_2 + 3X_3 + S_1 = 26$$
$$4X_1 + 2X_2 + 5X_3 + S_2 = 7$$

Here, there are 5 variables and 2 constraints. Therefore, keep $5 - 2 = 3$ variables at zero. Let us keep X_1, X_2 and X_3 at zero,

Then $$S_1 = 26$$
$$S_2 = 7$$

Objective function

$$Z = 30X_1 + 23X_2 + 29X_3$$

Therefore,

$$Z = 30X_1 - 23X_2 - 29X_3 = 0$$

Write the simplex X problem in table form.

↓PC

Basic	x_1	x_2	x_3	S_1	S_2	Solution		
Z	–30	–23	–29	0	0	0		
S_1	6	5	3	1	0	26	$\frac{26}{6} = \frac{13}{3}$	
S_2	4*	2	5	0	1	7	$\frac{7}{4}$	→PR

z row has negative values. Therefore, solution is not optimum.

Look for the maximum value with a negative sign in the Z row. It is –30. The variable corresponding to this value is x_1. x_1 tries to enter the basic column. This x_1 column is called a pivot column.

Which leaves the basic column?

Ignore all zeroes and negative values in the pivot column. Divide the elements in the solution column by the corresponding elements in the pivot column, i.e.,

$$\frac{26}{6} = \frac{13}{3} \text{ and } \frac{7}{4}$$

Select the minimum of these two. It is $\frac{7}{4}$. Row S_2 leaves. This is called pivot row.

Now X_1 enters the basic column and S_2 leaves. In the aforementioned table, 4 is the pivot element.

Iteration 1

$\text{NPE} = \frac{\text{OPE}}{\text{PE}}$	1	$\frac{1}{2}$	$\frac{5}{4}$	0	$\frac{1}{4}$	$\frac{7}{4}$
Old Z	–30	–23	–29	0	0	0
–(–30) NPE	30	15	$+\frac{75}{2}$	0	$+\frac{30}{4}$	$\frac{210}{4}$
New Z	0	–8	$\frac{17}{2}$	0	$\frac{15}{2}$	$\frac{105}{2}$
Old S_1	6	5	3	1	0	26
–6 NPE	–6	–3	$\frac{-15}{2}$	0	$-\frac{3}{2}$	$-\frac{21}{2}$
New S_1	0	2	$-\frac{9}{2}$	1	$-\frac{3}{2}$	$+\frac{31}{2}$

Formulate the new table.

Basic	x_1	x_2 ↓PC	x_3	S_1	S_2	Solution		
Z	0	–8	$\frac{17}{2}$	0	$\frac{15}{2}$	$\frac{105}{2}$		
S_1	0	2	$-\frac{9}{2}$	1	$\frac{3}{2}$	$+\frac{31}{2}$	$\frac{31}{4}$	
X_1	1	$\frac{1}{2}$ *	$\frac{5}{4}$	0	$\frac{1}{4}$	$\frac{7}{4}$	$\frac{7}{4}\times 2$	→ PR

Look for the negative value in the z row. It is –8. The variable corresponding to this value is X_2. X_2 tries to enter the basic column. This X_2 column is called a pivot column.

Which leaves the basic column?

Ignore all zeroes and negative values in the pivot column. Divide the elements in the solution column by the corresponding elements in the pivot column.

Select the minimum of the ratios from $\frac{31}{4}$ and $\frac{14}{4}$. The minimum is $\frac{14}{4}$. Row X_1 leaves. This is called pivot row. Now X_2 enters and X_1 leaves the basic column. In the aforementioned table, $\frac{1}{2}$ is the pivot element.

Iteration 2

NPE	2	1	$\frac{5}{2}$	0	$\frac{1}{2}$	$\frac{7}{2}$
Old Z	0	–8	$\frac{17}{2}$	0	$\frac{15}{2}$	$\frac{105}{2}$
–(–8) NPE	16	8	2	0	+4	28
New Z	16	0	$\frac{57}{2}$	0	$\frac{23}{2}$	$\frac{161}{2}$
Old S_1	0	2	$-\frac{9}{2}$	1	$-\frac{3}{2}$	$\frac{3}{2}$
–2 NPE	–4	–2	–5	0	–1	–7
New S_1	–4	0	$-\frac{19}{2}$	1	$-\frac{5}{2}$	$\frac{17}{2}$

Basic	x_1	x_2	x_3	S_1	S_2	Solution
Z	+16	0	$\frac{57}{2}$	0	$\frac{23}{2}$	$\frac{161}{2}$
S_1	–4	0	$-\frac{19}{2}$	1	$-\frac{5}{2}$	$\frac{17}{2}$
X_2	2	1	$\frac{5}{2}$	0	$\frac{1}{2}$	$\frac{7}{2}$

All the values in Z row are either positive or zero. Hence the solution in optimum.

$$\text{Maximise } Z = \frac{161}{2}$$

subject to

$$X_1 = 0$$
$$X_2 = \frac{7}{2}$$

Example 3.4

$$\text{Maximise } Z = X_1 - X_2 + 3X_3$$

subject to

$$X_1 + X_2 + X_3 \leq 10$$
$$2X_1 - X_3 \leq 2$$
$$2X_1 - 2X_2 + 3X_3 \leq 0$$
$$X_1, X_2, X_3 \geq 0$$

Solve this LPP by simplex method.

Solution

Add slack variables and bring inequalities to equality.

$$X_1 + X_2 + X_3 + S_1 = 10$$
$$2X_1 - X_3 + S_2 = 2$$
$$2X_1 - 2X_2 + 3X_3 + S_3 = 0$$

Here, there are 6 variables and 3 constraints. Therefore, keep 6 – 3 = 3 variables at zero. Let us keep X_1, X_2 and X_3 at zero, then,

$$S_1 = 10$$
$$S_2 = 2$$
$$S_3 = 0$$

Objective function

$$Z = X_1 - X_2 + 3X_3$$

Therefore,

$$Z - X_1 + X_2 - 3X_3 = 0$$

Write the simplex problem in table form.

↓PC

Basic	x_1	x_2	x_3	S_1	S_2	S_3	Solution	
Z	–1	–1	–3	0	0	0	0	
S_1	1	1	1	1	0	0	10	$\frac{10}{1} = 10$
S_2	2	0	–3	0	1	0	2	
S_3	2	–2	3*	0	0	1	0	$\frac{0}{3} = 0$ → PR

Z row has negative values. Therefore, solution is not optimum.

Look for the maximum value with negative sign in the z row. It is –3. The variable corresponding to this value is X_3. X_3 tries to enter the basic column. This X_3 column is called a pivot column.

Which leaves the basic column?

Ignore all the zeroes and negative value in the pivot column. Divide the elements in the solution column by these corresponding elements in the pivot column, i.e.,

$$\frac{10}{1} = 10$$

and

$$\frac{0}{3} = 0$$

Select the minimum of these two. It is zero. Row S_3 leaves. This is called pivot row. Now X_3 enters and S_3 leaves the basic column. In the aforementioned table, 3 is the pivot element.

Iteration 1

$NPE = \frac{OPE}{PE}$	$\frac{2}{3}$	$-\frac{2}{3}$	1	0	0	$\frac{1}{3}$	0
Old Z	–1	1	–3	0	0	0	0
–(–3)NPE	2	–2	3	0	0	1	0
New Z	1	–1	0	0	0	1	0
Old S_1	1	1	1	1	0	0	10
–1(NPE)	$-\frac{2}{3}$	$+\frac{2}{3}$	–1	0	0	$-\frac{1}{3}$	0
New S_1	$\frac{1}{3}$	$\frac{5}{3}$	0	1	0	$-\frac{1}{3}$	10
Old S_2	2	0	–3	0	1	0	2
–(–3)NPE	2	–2	3	0	0	1	0
New S_2	4	–2	0	0	1	1	2

Write the simplex problem in table form.

Basic	x_1	x_2 ↓PC	x_3	S_1	S_2	S_3	Solution	
Z	1	−1	0	0	0	1	0	
S_1	$\frac{1}{3}$	$\frac{5}{3}$ *	0	1	0	$-\frac{1}{3}$	10	→ PR
S_2	4	−2	0	0	1	1	2	
X_3	$\frac{2}{3}$	$-\frac{2}{3}$	1	0	0	$\frac{1}{3}$	0	

Z row has negative value. Therefore, solution is not optimum.

Look for the negative value in the z row. It is −1. The variable corresponding to this value is X_2. X_2 tries to enter the basic column. This X_2 column is called a pivot column.

Which leaves the basic column?

Ignore all zero and negative values in the pivot column. Therefore, S_1 leaves the basic column. In the aforementioned table, $\frac{5}{3}$ is the pivot element.

Iteration 2

$NPE = \frac{OPE}{PE}$	$\frac{1}{5}$	1	0	$\frac{3}{5}$	0	$-\frac{1}{5}$	6
old Z	1	−1	0	0	0	1	0
−(−1)NPE	$\frac{1}{5}$	1	0	$\frac{3}{5}$	0	$-\frac{1}{5}$	6
New Z	$\frac{6}{5}$	0	0	$\frac{3}{5}$	0	$\frac{4}{5}$	6
Old S_2	4	−2	0	0	1	1	2
−1(−2)(NPE)	$\frac{2}{5}$	2	0	$\frac{6}{5}$	0	$-\frac{2}{5}$	12
New S_2	$\frac{22}{5}$	0	0	$\frac{6}{5}$	1	$\frac{3}{5}$	14
old X_3	$\frac{2}{3}$	$-\frac{2}{3}$	1	0	0	$\frac{1}{3}$	0
$-\left(-\frac{2}{3}\right)$ NPE	$\frac{2}{15}$	$\frac{2}{3}$	0	$\frac{2}{5}$	0	$-\frac{2}{15}$	$\frac{12}{3}$
New X_3	$\frac{4}{5}$	0	1	$\frac{2}{5}$	0	$-\frac{3}{5}$	$\frac{12}{3}$

Formulate the basic table.

Basic	X_1	X_2	X_3	S_1	S_2	S_3	Solution
Z	$\frac{6}{5}$	0	0	$\frac{3}{5}$	0	$\frac{4}{5}$	6
X_2	$\frac{1}{5}$	1	0	$\frac{3}{5}$	0	$-\frac{1}{5}$	6
S_2	$\frac{22}{5}$	0	0	$\frac{6}{5}$	1	$\frac{3}{5}$	14
X_3	$\frac{4}{5}$	0	1	$\frac{2}{5}$	0	$-\frac{3}{5}$	$\frac{12}{3}$

All the values in Z row are positive and zero. Therefore, solution is optimum.

$$\text{Maximise } Z = 6$$
$$X_1 = 0$$
$$X_2 = 6$$
$$X_3 = \frac{12}{3}$$

Example 3.5

Solve by simplex method

$$\text{Maximise } Z = 3X_1 + 5X_2 + 4X_3$$

subject to

$$2X_1 + 3X_2 \leq 8$$
$$3X_1 + 2X_2 + 4X_3 \leq 15$$
$$2X_2 + 5X_3 \leq 10$$
$$X_1, X_2, X_3 \geq 0$$

Solution

Add slack variables and bring inequalities to equality.

$$2X_1 + 3X_2 + S_1 = 8$$
$$3X_1 + 2X_2 + 4X_3 + S_2 = 15$$
$$2X_2 + 5X_3 + S_3 = 10$$

Here, there are 6 variables and 3 constraints. Therefore, keep 6 – 3 = 3 variables at zero. Let us keep X_1, X_2 and X_3 at zero, then,

$$S_1 = 8$$
$$S_2 = 15$$
$$S_3 = 10$$

Objective function

$$Z = 3X_1 + 5X_2 + 4X_3 = 0$$

Therefore,

$$Z - 3X_1 - 5X_2 - 4X_3 = 0$$

Write the simplex problem in table form.

↓PC

Basic	X_1	X_2	X_3	S_1	S_2	S_3	Solution	
Z	–3	–5	–4	0	0	0	0	
S_1	2	3*	0	1	0	0	8	$\frac{8}{3} = 2.66$ → PR
S_2	3	2	4	0	1	0	15	$\frac{15}{2} = 7.5$
S_3	0	–2	5	0	0	1	10	$\frac{10}{2} = 5$

Z row has negative values. Therefore, solution is not optimum.

Look for the maximum value with negative sign in the Z row. It is –5. The variable corresponding to this value is X_2. X_2 tries to enter the basic column. This X_2 column is called a pivot column.

Which leaves the basic column?

Ignore all the zeroes and negative values in the pivot column. Divide the element in the solution column by the corresponding elements in the pivot column, i.e.,

$$\frac{8}{3} = 2.66$$

$$\frac{15}{2} = 7.5$$

and

$$\frac{10}{2} = 5$$

Select the minimum of these three. It is 3.66. Row S_1leaves. This is called pivot row. Now X_2 enters and S_1 leaves the basic column. In the aforementioned table, 3 is the pivot element.

Iteration 1

$NPE = \frac{OPE}{PE}$	$\frac{2}{3}$	1	0	$\frac{1}{3}$	0	0	$\frac{8}{3}$
Old Z	–3	–5	–4	0	0	0	0
–(–5)NPE	$\frac{10}{3}$	5	0	$\frac{5}{3}$	0	0	$\frac{40}{3}$
New Z	$\frac{1}{3}$	0	–4	$\frac{5}{3}$	0	0	$\frac{40}{3}$

Old S_2	3	2	4	0	1	0	15
–(2)NPE	$-\frac{4}{3}$	–2	0	$-\frac{2}{3}$	0	0	$-\frac{16}{3}$
New S_2	$\frac{5}{3}$	0	4	$-\frac{2}{3}$	1	0	$\frac{29}{3}$
Old S_3	0	2	5	0	0	1	10
–(2)NPE	$-\frac{4}{3}$	–2	0	$-\frac{2}{3}$	0	0	$-\frac{16}{3}$
New S_3	$-\frac{4}{3}$	0	5	$-\frac{2}{3}$	0	1	$\frac{14}{3}$

Formulate the basic table.

Basic	X_1	X_2	↓PC X_3	S_1	S_2	S_3	Solution		
Z	$\frac{1}{3}$	0	–4	$\frac{5}{3}$	0	0	$\frac{40}{3}$		
X_2	$\frac{2}{3}$	1	0	$\frac{1}{3}$	0	0	$\frac{8}{3}$		
S_2	$\frac{5}{3}$	0	4	$-\frac{2}{3}$	1	0	$\frac{29}{3}$	$\frac{29}{12}$	
S_3	$-\frac{4}{3}$	0	5_*	$-\frac{2}{3}$	0	1	$\frac{14}{3}$	$\frac{14}{15}$	→ PR

Z row has negative value. Therefore, solution is not optimum.

Look for the negative value in the Z row. It is –4. The variable corresponding to this value is X_3. X_3 tries to enter the basic column. This X_3 column is called a pivot column.

Which leaves the basic column?

Ignore all zeroes and negative values in the pivot column. Divide the elements in the solution column by the corresponding elements in the pivot column, i.e.,

$$\frac{29}{12} \text{ and } \frac{14}{15}$$

Select the minimum of these two. It is $\frac{14}{15}$. Row S_3 leaves. This is called pivot row. Now X_3 enters and S_3 leaves the basic column. In the above table, 5 is the pivot element.

Iteration 2

$NPE = \frac{OPE}{PE}$	$\frac{-4}{15}$	0	1	$-\frac{2}{15}$	0	$\frac{1}{5}$	$\frac{14}{15}$
Old Z	$\frac{1}{3}$	0	–4	$\frac{5}{3}$	0	0	$\frac{40}{3}$
–(–4)NPE	$-\frac{16}{5}$	0	4	$-\frac{8}{15}$	0	$\frac{4}{5}$	$\frac{56}{15}$
New Z	$-\frac{11}{15}$	0	0	$\frac{17}{15}$	0	$\frac{4}{5}$	$\frac{256}{15}$
Old X_2	$\frac{2}{3}$	1	0	$\frac{1}{3}$	0	0	$\frac{8}{3}$
–(0)NPE	0	0	0	0	0	0	0
New X_2	$\frac{2}{3}$	1	0	$\frac{1}{3}$	0	0	$\frac{8}{3}$
Old S_2	$\frac{5}{3}$	0	4	$-\frac{2}{3}$	1	0	$\frac{29}{3}$
–(4)NPE	$\frac{16}{15}$	0	–4	$\frac{8}{15}$	0	$-\frac{4}{5}$	$-\frac{56}{15}$
New S_2	$\frac{41}{15}$	0	0	$-\frac{2}{15}$	1	$-\frac{4}{5}$	$\frac{89}{15}$

Formulate the basic table.

Basic	X_1 ↓PC	X_2	X_3	S_1	S_2	S_3	Solution	
Z	$-\frac{11}{15}$	0	0	$\frac{17}{15}$	0	$\frac{4}{5}$	$\frac{256}{15}$	
X_2	$\frac{2}{3}$	1	0	$\frac{1}{3}$	0	0	$\frac{8}{3}$	$\frac{8}{3} \times \frac{3}{2} = 4$
S_2	$\frac{41}{15}$ *	0	0	$-\frac{2}{15}$	1	$-\frac{4}{5}$	$\frac{89}{15}$	$\frac{89}{15} \times \frac{15}{41} = \frac{89}{41}$ → PR
X_3	$\frac{-4}{15}$	0	1	$-\frac{2}{15}$	0	$\frac{1}{5}$	$\frac{14}{15}$	

Z row has a negative value. Therefore, solution is not optimum.

Look for the negative value in Z row. It is $-\frac{11}{15}$. The variable corresponding to this value is X_1. X_1 tries to enter the basic column. This X_1 column is called a pivot column.

Which leaves the basic column?

Ignore all zeroes and negative values in the pivot column. Divide the elements in the solution column by the corresponding elements in the pivot column, i.e.,

$$\frac{8}{3} \times \frac{3}{2} = 4$$

and

$$\frac{89}{15} \times \frac{15}{41} = \frac{89}{41}$$

Select the minimum of these two. It is $\frac{89}{41}$. Row S_2 leaves. This is called pivot row. Now X_1 enters and S_2 leaves the basic column. In the aforementioned table, $\frac{41}{15}$ is the pivot element.

Iteration 3

$\text{NPE} = \frac{\text{OPE}}{\text{PE}}$	1	0	0	$-\frac{2}{41}$	$\frac{15}{41}$	$-\frac{12}{41}$	$\frac{89}{41}$
Old Z	$-\frac{11}{15}$	0	0	$\frac{17}{15}$	0	$\frac{4}{5}$	$\frac{256}{15}$
$-\left(-\frac{11}{15}\right)$ NPE	$\frac{11}{15}$	0	0	$-\frac{22}{615}$	$\frac{11}{41}$	$-\frac{132}{615}$	$\frac{979}{615}$
New Z	0	0	0	$\frac{676}{615}$	$\frac{11}{41}$	$\frac{360}{615}$	$\frac{11475}{615}\left(\frac{765}{41}\right)$
Old X_2	$\frac{2}{3}$	1	0	$\frac{1}{3}$	0	0	$\frac{8}{3}$
$-\frac{2}{3}$ NPE	$-\frac{2}{3}$	0	0	$+\frac{4}{123}$	$-\frac{10}{41}$	$\frac{8}{41}$	$-\frac{178}{123}$
New X_2	0	1	0	$\frac{15}{41}$	$-\frac{10}{41}$	$\frac{8}{41}$	$\frac{50}{41}$
Old X_3	$-\frac{4}{15}$	0	1	$-\frac{2}{15}$	0	$\frac{1}{5}$	$\frac{14}{15}$
$-\left(-\frac{4}{15}\right)$ NPE	$\frac{4}{5}$	0	0	$-\frac{8}{615}$	$\frac{4}{41}$	$-\frac{48}{615}$	$\frac{356}{615}$
New X_3	0	0	1	$\frac{6}{41}$	$\frac{4}{41}$	$\frac{5}{41}$	$\frac{62}{41}$

Formulate the basic table.

Basic	X_1	X_2	X_3	S_1	S_2	S_3	Solution
Z	0	0	0	$\frac{676}{615}$	$\frac{11}{41}$	$\frac{360}{615}$	$\frac{765}{41}$
X_2	0	1	0	$\frac{15}{41}$	$-\frac{10}{41}$	$\frac{8}{41}$	$\frac{50}{41}$
X_1	1	0	0	$-\frac{2}{41}$	$\frac{15}{41}$	$-\frac{12}{41}$	$\frac{89}{41}$
X_3	0	0	1	$\frac{6}{41}$	$\frac{4}{41}$	$\frac{5}{41}$	$\frac{62}{41}$

All the values in Z row are either positive or zero. Therefore, solution is optimum.

$$\text{Maximise } Z = \frac{765}{41}$$

subject to

$$X_1 = \frac{89}{41}$$

$$X_2 = \frac{50}{41}$$

$$X_3 = \frac{62}{41}$$

3.7 MINIMISATION PROBLEM

Maximise profit is same as minimise loss. Therefore, an objective function of minimisation can be converted into objective function of maximisation.

For example, Minimize $Z = 3X + 2Y$ is same as Maximise $Z = -3X - 2Y$. The procedure of simplex is for maximisation. Therefore, convert the minimisation problem to a maximisation problem.

Example 3.6

Solve by simplex method.

$$\text{Minimise } Z = X_1 - 3X_2 + 2X_3$$

subject to

$$\begin{aligned} 3X_1 - X_2 + 2X_3 &\leq 7 \\ -2X_1 + 4X_2 &\leq 12 \\ -4X_1 + 3X_2 + 8X_3 &\leq 10 \\ X_1, X_2, X_3 &\geq 0 \end{aligned}$$

Solution

Convert this problem into maximisation problem.

$$\text{Maximise } Z = -X_1 + 3X_2 - 2X_3$$

subject to

$$3X_1 - X_2 + 2X_3 \leq 7$$
$$-2X_1 + 4X_2 \leq 12$$
$$-4X_1 + 3X_2 + 8X_3 \leq 10$$
$$X_1, X_2, X_3 \geq 0$$

Add slack variables and bring inequalities to equality.

$$3X_1 - X_2 + 2X_3 + S_1 = 7$$
$$-2X_1 + 4X_2 + S_2 = 12$$
$$-4X_1 + 3X_2 + 8X_3 + S_3 = 10$$

Here, there are 6 variables and 3 constraints. Therefore, keep 6 – 3 = 3 variables at zero. Let us keep X_1, X_2 and X_3 at zero, then

$$S_1 = 7$$
$$S_2 = 12$$
$$S_3 = 10$$

Objective function

$$Z = -X_1 + 3X_2 - 2X_3$$

Therefore,

$$Z + X_1 - 3X_2 + 2X_3 = 0$$

Write the simplex problems in table form.

↓PC (on X_2 column)

Basic	X_1	X_2	X_3	S_1	S_2	S_3	Solution		
Z	1	–3	2	0	0	0	0		
S_1	3	–1	2	1	0	0	7		
S_2	–2	4*	0	0	1	0	12	$\frac{12}{4} = 3$	→ PR
S_3	–4	3	8	0	0	1	10	$\frac{10}{3} = 3.33$	

Z row has a negative value. Therefore, solution is not optimum.

Look for the negative value in Z row. It is –3. The variable corresponding to this value is X_2. X_2 tries to enter the basic column. This X_2 column is called as pivot column.

Which leaves the basic column?

Ignore all zeroes and negative values in the pivot column. Divide the elements in the solution column by the corresponding elements in the pivot column, i.e.,

$$\frac{12}{4} = 3$$

and

$$\frac{10}{3} = 3.33$$

Select the minimum of these two. It is 3. Row S_2 leaves. This is called pivot row.

Now X_2 enters and S_2 leaves the basic column. In the aforementioned table, 4 is the pivot element.

Iteration 1

$NPE = \frac{OPE}{PE}$	$-\frac{1}{2}$	1	0	0	$\frac{1}{4}$	0	3
Old Z	1	–3	2	0	0	0	0
–(–3)NPE	$-\frac{3}{2}$	3	0	0	$\frac{3}{4}$	0	9
New Z	$-\frac{1}{2}$	0	2	0	$\frac{3}{4}$	0	9
Old S_1	3	–1	2	1	0	0	7
–(–1)NPE	$-\frac{1}{2}$	1	0	0	$\frac{1}{4}$	0	3
New S_1	$\frac{5}{2}$	0	2	1	$\frac{1}{4}$	0	10
Old S_3	–4	3	8	0	0	1	10
–(3)NPE	$\frac{3}{2}$	–3	0	0	$-\frac{3}{4}$	0	–9
New S_3	$-\frac{5}{2}$	0	8	0	$-\frac{3}{4}$	1	1

Formulate the table.

Basic	X_1	X_2	X_3	S_1	S_2	S_3	Solution	
Z	$-\frac{1}{2}$	0	2	0	$\frac{3}{4}$	0	9	
S_1	$\frac{5}{2}$ *	0	2	1	$\frac{1}{4}$	0	10	→ PR
X_2	$-\frac{1}{2}$	1	0	0	$\frac{1}{4}$	0	3	
S_3	$-\frac{5}{2}$	0	8	0	$-\frac{3}{4}$	1	1	

Z row has a negative value. Therefore, solution is not optimum.

Look for the negative value in Z row. It is $-\frac{1}{2}$. The variable corresponding to this value is X_1. X_1 tries to enter the basic column. This X_1 column is called a pivot column.

Which leaves the basic column?

Ignore all zeroes and negative values in the pivot column. S_1 leaves the basic column. In the previous table, $\frac{5}{2}$ is the pivot element.

Iteration 2

$\text{NPE} = \frac{\text{OPE}}{\text{PE}}$	1	0	$\frac{4}{5}$	$\frac{2}{5}$	$\frac{1}{10}$	0	4
Old Z	$-\frac{1}{2}$	0	2	0	$\frac{3}{4}$	0	9
$-\left(-\frac{1}{2}\right)$ NPE	$\frac{1}{2}$	0	$\frac{2}{5}$	$\frac{1}{5}$	$\frac{1}{20}$	0	2
New Z	0	0	$\frac{12}{15}$	$\frac{1}{5}$	$\frac{4}{5}$	0	11
Old X_2	$-\frac{1}{2}$	1	0	0	$\frac{1}{4}$	0	3
$-\left(-\frac{1}{2}\right)$ NPE	$\frac{1}{2}$	0	$\frac{2}{5}$	$\frac{1}{5}$	$\frac{1}{20}$	0	2
New X_2	0	1	$\frac{2}{5}$	$\frac{1}{5}$	$\frac{3}{10}$	0	5
Old S_3	$-\frac{5}{2}$	0	8	0	$-\frac{3}{4}$	1	1
$-\left(-\frac{5}{2}\right)$ NPE	$\frac{5}{2}$	0	2	1	$\frac{1}{4}$	0	10
New S_3	0	0	10	1	$-\frac{1}{2}$	1	11

Formulate the table.

Basic	X_1	X_2	X_3	S_1	S_2	S_3	Solution
Z	0	0	$\frac{12}{15}$	$\frac{1}{5}$	$\frac{4}{5}$	0	11
X_1	1	0	$\frac{4}{5}$	$\frac{2}{5}$	$\frac{1}{10}$	0	4
X_2	0	1	$\frac{2}{5}$	$\frac{1}{5}$	$\frac{3}{10}$	0	5
S_3	0	0	10	1	$-\frac{1}{2}$	1	11

Solution is optimum.

$$\text{Maximise } Z = 11$$

Therefore,

$$\text{Minimise } Z = -11$$

subject to

$$X_1 = 4$$
$$X_2 = 5$$

3.8 ARTIFICIAL VARIABLE

Consider the equation $2X_1 + 3X_2 \geq 20$. To bring this inequality to equality, we use a surplus variable. Then, $2X_1 + 3X_2 - S_1 = 20$. When X_1 and X_2 are zero,

$$S_1 = -20$$

This violates the non-negativity restriction. Therefore, we use another variable R_1 which has a big value M. Thus, $2X_1 + 3X_2 - S_1 + R_1 = 20$

This R_1 is known as an artificial variable. This is supposed to have a very big value 'M' in the objective function. As the number with the highest negative value leaves the non-basic column, we attach to 'R' '$-M$'. The procedure of solving simplex problem by this method is known as Big-M method.

Example 3.7

$$\text{Minimise } Z = 5X_1 + 6X_2$$

subject to

$$2X_1 + 5X_2 \geq 1500$$
$$3X_1 + X_2 \geq 1200$$
$$X_1, X_2 \geq 0$$

Solution

Convert the problem into maximisation problem.

$$\text{Maximise } Z = -5X_1 - 6X_2$$

subject to

$$2X_1 + 5X_2 \geq 1500$$
$$3X_1 + X_2 \geq 1200$$
$$X_1, X_2 \geq 0$$

Bringing inequality to equality, we get,

$$2X_1 + 5X_2 - S_1 + R_1 = 1500$$
$$3X_1 + X_2 - S_2 + R_2 = 1200$$

There are 6 variables and 2 constraints. Therefore, keep $6 - 2 = 4$ variables at zero. Let us keep X_1, X_2, S_1 and S_2 at zero, then,

$$R_1 = 1500$$
$$R_2 = 1200$$
$$R_1 = 1500 - 2X_1 - 5X_2 + S_1$$
$$R_2 = 1200 - 3X_1 - X_2 + S_2$$
$$Z = -5X_1 - 6X_2 - MR_1 - MR_2$$
$$= -5X_1 - 6X_2 - M(1500 - 2X_1 - 5X_2 + S_1) - M(1200 - 3X_1 - X_2 + S_2)$$
$$= -5X_1 - 6X_2 - 1500M + 2MX_1 + 5MX_2 - MS_1 - 1200M + 3MX_1 + MX_2 - MS_2$$
$$= X_1(-5 + 5M) + X_2(-6 + 6M) - MS_2 - MS_2 - 2700M$$
$$Z + X_1(5 - 5M) + X_2(6 - 6M) + MS_1 + MS_2 = -2700M$$

Formulate the basic table.

Basic	X_1	↓PC X_2	S_1	S_2	R_1	R_2	Solution	
Z	$5 - 5M$	$6 - 6M$	M	M	0	0	$-2700M$	
R_1	2	5*	–1	0	1	0	1500	$\frac{1500}{5} = 300$ → PR
R_2	3	1	0	–1	0	1	1200	$\frac{1200}{1} = 1200$

Z row has negative values. Therefore, solution is not optimum. Look for the highest negative value in Z row. Consider $5 - 5M$ and $6 - 6M$.

Put,

$$M = 2$$
$$5 - 5(2) \quad \text{and } 6 - 6(2)$$

we get

$$-5 \text{ and } -6$$

Therefore, $6 - 6M$ is the highest negative value. Therefore, X_2 enters the basic column. This X_2 column is called pivot column.

Which leaves the basic column?

Ignore all zeroes and negative values in the pivot column. Divide the elements in the solution column by the corresponding elements in the pivot column, i.e.,

$$\frac{1500}{5} = 300$$

and

$$\frac{1200}{1} = 1200$$

Select the minimum of these two. It is 300. Row 'R_1' leaves. This is called pivot row. Delete column R_1.

Now X_2 enters and R_2 leaves the basic column. In the above table, 5 is the pivot element.

Iteration 1

$NPE = \frac{OPE}{PE}$	$\frac{2}{5}$	1	$-\frac{1}{5}$	0	0	300
Old Z	$5-5M$	$6-6M$	M	M	0	$-2700M$
$-(6-6M)$NPE	$-\frac{12}{5}+\frac{12}{5}M$	$-6+6M$	$-\frac{6}{5}M+\frac{6}{5}$	0	0	$1800M-1800$
New Z	$\frac{13}{5}-\frac{13}{5}M$	0	$-\frac{1}{5}M+\frac{6}{5}M$	M	0	$-900M-1800$
Old R_2	3	1	0	–1	1	1200
–(1)NPE	$\frac{-2}{5}$	–1	$\frac{1}{5}$	0	0	–300
New R_2	$\frac{13}{5}$	0	$\frac{1}{5}$	–1	1	900

Formulate the basic table.

↓PC

Basic	X_1	X_2	S_1	S_2	R_2	Solution		
Z	$\frac{13}{5}-\frac{13}{5}M$	0	$-\frac{1}{5}M+\frac{6}{5}$	M	0	$-900M-1800$		
X_2	$\frac{2}{5}$	1	$-\frac{1}{5}$	0	0	300	$300\times\frac{5}{2}=750$	
R_2	$\frac{13}{5}$ *	0	$\frac{1}{5}$	–1	1	900	$900\times\frac{5}{13}=\frac{4500}{13}$	→PR

Z row has negative values. Therefore, solution is not optimum.

Look for the highest negative value in Z row. Consider $\frac{13}{5}-\frac{13}{5}M$ and $-\frac{1}{5}M+\frac{6}{5}$

Put

$$X = 2$$

$$\frac{13}{5}-\frac{13}{5}(2) \quad \text{and} \quad -\frac{1}{5}(2)+\frac{6}{5}$$

We get

$$-\frac{13}{5} \text{ and } \frac{4}{5}.$$

Therefore, $\frac{13}{5}-\frac{13}{5}M$ is the highest negative value. X_1 enters the basic column.

Which leaves the basic column?

Ignore all zeroes and negative values in the pivot column. Divide the elements in the solution column by the corresponding elements in the pivot column, i.e.,

$$300 \times \frac{5}{2} = 750$$

$$900 \times \frac{5}{13} = \frac{4500}{13}$$

Select the minimum of these two. It is $\frac{4500}{13}$ Row R_2 leaves. This is called pivot row. Delete row R_2.

Now X_1 enters and R_2 leaves the basic column, in the aforementioned table, $\frac{13}{5}$ is the pivot element.

Iteration 2

$NPE = \frac{OPE}{PE}$	1	0	$\frac{1}{13}$	$-\frac{5}{13}$	$\frac{4500}{13}$
Old Z	$\frac{13}{5}-\frac{13}{5}M$	0	$-\frac{M}{5}+\frac{6}{5}$	M	$-900M-1800$
$-\left(-\frac{13}{5}-\frac{13}{5}M\right)$ NPE	$-\frac{13}{5}+\frac{13}{5}M$	0	$\frac{M}{5}-\frac{1}{5}$	$-M+1$	$900M-900$
New Z	0	0	1	1	-2700
Old X_2	$\frac{2}{5}$	1	$-\frac{1}{5}$	0	300
$-\left(\frac{2}{5}\right)$ NPE	$-\frac{2}{5}$	0	$-\frac{2}{65}$	$+\frac{2}{13}$	$-\frac{1800}{13}$
New X_2	0	1	$-\frac{3}{13}$	$+\frac{2}{13}$	$\frac{2100}{13}$

Formulate the basic table.

Basic	X_1	X_2	S_1	S_2	Solution
Z	0	0	1	1	-2700
X_2	0	1	$-\frac{3}{13}$	$\frac{2}{13}$	$\frac{2100}{13}$
X_1	1	0	$\frac{1}{5}$	$-\frac{5}{13}$	$\frac{4500}{13}$

Solution is optimum.

Therefore,

$$\text{Maximise } Z = -2700$$
$$\text{Min } Z = 2700$$

subject to

$$X_1 = \frac{4500}{13}$$
$$X_2 = \frac{2100}{13}$$

Example 3.8

Solve the following LPP by simplex method.

$$\text{Minimise } Z = 2X_1 + 9X_2 + X_3$$

subject to

$$\begin{aligned} X_1 + 4X_2 + 2X_3 &\geq 5 \\ 3X_1 + X_2 + 2X_3 &\geq 4 \\ X_1, X_2, X_3 &\geq 0 \end{aligned}$$

Solution

Convert minimisation problem to maximisation problem.

$$\begin{aligned} \text{Maximise } Z &= -2X_1 - 9X_2 - X_3 \\ X_1 + 4X_2 + 2X_3 &\geq 5 \\ 3X_1 + X_2 + 2X_3 &\geq 4 \\ X_1, X_2, X_3 &\geq 0 \end{aligned}$$

Bringing inequality to equality, we have,

$$\begin{aligned} X_1 + 4X_2 + 2X_3 - S_1 + R_1 &= 5 \\ 3X_1 + X_2 + 2X_3 - S_2 + R_2 &= 4 \end{aligned}$$

where S_1 and S_2 are surplus variables, R_1 and R_2 are artificial variables.

$$\begin{aligned} R_1 &= 5 - X_1 - 4X_2 - 2X_3 + S_1 \\ R_2 &= 4 - 3X_1 - X_2 - 2X_3 + S_2 \end{aligned}$$

The objective function is

$$\begin{aligned} Z &= -2X_1 - 9X_2 - X_3 - MR_1 - MR_2 \\ Z &= -2X_1 - 9X_2 - X_3 - M(5 - X_1 - 4X_2 - 2X_3 + S_1) \\ &\quad - M(4 - 3X_1 - X_2 - 2X_3 + S_2) \\ Z &= -2X_1 - 9X_2 - X_3 - 5M + MX_1 + 4MX_2 + 2MX_3 - MS_1 - 4M \\ &\quad + 3MX_1 + MX_2 + 2MX_3 - MS_2 \end{aligned}$$

$$Z + 2X_1 + 9X_2 + X_3 + 5M - MX_1 - 4MX_2 - 2MX_3 + MS_1 + 4M - 3MX_1 - MX_2 - 2MX_3 + MS_2 = 0$$
$$Z + X_1(-4M + 2) + X_2(-5M + 9) + X_3(1 - 4M) + MS_1 + MS_2 = -9M$$

Fomulate the basic table.

↓PC

Basic	X_1	X_2	X_3	S_1	S_2	R_1	R_2	Solution	
Z	$2-4M$	$9-5M$	$1-4M$	M	M	0	0	$-9M$	
R_1	1	4	2	–1	0	1	0	5	$\frac{5}{2}$
R_2	3	+1	2_*	0	–1	0	1	4	$\frac{4}{2}$ → PR

Z row has negative values. Therefore, solution is not optimum. Look for the highest negative value in Z row. Consider,

$$2 - 4M$$
$$9 - 5M$$

and

$$1 - 4M$$

Put

$$M = 2$$
$$2 - 4(2) = -6$$
$$9 - 5(2) = -1$$
$$1 - 4(2) = -7$$

Therefore, $(1 - 4M)$ is the highest negative value. X_3 enters the basic column. This X_3 is called a pivot column.

Which leaves the basic column?

Ignore all zeroes and negative values in pivot column. Divide the elements in the solution column by the corresponding elements in the pivot column, i.e.,

$$\frac{5}{2} \text{ and } \frac{4}{2}$$

Select the minimum of these two. It is $\frac{4}{2}$. Row R_2 leaves the basic column. This is called pivot row. Delete column R_2.

Now X_3 enters and R_2 leaves the basic column. In the above table, 2 is the pivot element.

Iteration 1

NPE	$\frac{3}{2}$	$\frac{1}{2}$	1	0	$-\frac{1}{2}$	0	2
Old Z	$2-4M$	$9-5M$	$1-4M$	M	M	0	$-9M$
$-(1-4M)$NPE	$-\frac{3}{2}+6M$	$-\frac{1}{2}+2M$	$-1+4M$	0	$+\frac{1}{2}-2M$	0	$+8M-2$
New Z	$\frac{1}{2}+2M$	$\frac{17}{2}-3M$	0	M	$-M+\frac{1}{2}$	0	$-M-2$
Old R_1	1	4	2	–1	0	1	5
(2)NPE	–3	–1	–2	0	1	0	–4
New R_1	–2	3	0	–1	1	1	1

Formulate the basic table.

↓PC

Basic	X_1	X_2	X_3	S_1	S_2	R_1	Solution	
Z	$\frac{1}{2}+2M$	$\frac{17}{2}-3M$	0	M	$-M+\frac{1}{2}$	0	$-M-2$	
R_1	–2	3	0	–1	1_*	1	1	$\frac{1}{1}$ → PR
X_3	$\frac{3}{2}$	$\frac{1}{2}$	1	0	$-\frac{1}{2}$	0	2	

Z row has negative values. Therefore, solution is not optimum. Look for the highest negative value in Z row. Consider $\frac{17}{2}-3M$ and $-M+\frac{1}{2}$

Put

$$X = 2$$

$$\frac{17}{2}-6=\frac{5}{2}$$

$$-2+\frac{1}{2}=-\frac{3}{2}$$

$-M+\frac{1}{2}$ is the highest negative value. S_2 enters the basic column.

Which leaves the basic column?

Ignore all zeroes and negative values in pivot column. R_1 leaves the basic column. Delete R_1 column.

Iteration 2

NPE	-2	3	0	-1	1	1
Old Z	$\frac{1}{2}+2M$	$\frac{17}{2}-3M$	0	M	$-M+\frac{1}{2}$	$-M-2$
$\left(-M+\frac{1}{2}\right)$NPE	$1-2M$	$-\frac{3}{2}+3M$	0	$+\frac{1}{2}-M$	$+M-\frac{1}{2}$	$M-\frac{1}{2}$
New Z	$\frac{3}{2}$	7	0	$\frac{1}{2}$	0	$-\frac{5}{2}$
Old X_3	$\frac{3}{2}$	$\frac{1}{2}$	1	0	$-\frac{1}{2}$	2
$-\left(-\frac{1}{2}\right)$NPE	-1	$\frac{3}{2}$	0	$-\frac{1}{2}$	$\frac{1}{2}$	$\frac{1}{2}$
New X_3	$\frac{1}{2}$	2	1	$-\frac{1}{2}$	0	$\frac{3}{2}$

Formulate the basic table.

Basic	X_1	X_2	X_3	S_1	S_2	Solution
Z	$\frac{3}{2}$	7	0	$\frac{1}{2}$	0	$-\frac{5}{2}$
S_2	-2	3	0	-1	1	1
X_3	$\frac{1}{2}$	2	1	$-\frac{1}{2}$	0	$\frac{3}{2}$

Solution is optimum

$$\text{Maximise } Z = -\frac{5}{2}$$

$$\text{Min } Z = \frac{5}{2}$$

subject to

$$X_1 = 0$$
$$X_2 = 0$$
$$X_3 = \frac{3}{2}$$

***Example* 3.9**

Solve by simplex method.

$$\text{Maximise } Z = 3X_1 + 2X_2 + X_3$$

subject to,

$$\begin{aligned} -3X_1 + 2X_2 + 2X_3 &= 8 \\ -3X_1 + 4X_2 + X_3 &= 7 \\ X_1, X_2, X_3 &\geq 0 \end{aligned}$$

Solution

Wherever equality is existing, add artificial variables.

$$\begin{aligned} -3X_1 + 2X_2 + 2X_3 + R_1 &= 8 \\ -3X_1 + 4X_2 + X_3 + R_2 &= 7 \\ R_1 &= 8 + 3X_1 - 2X_2 - 2X_3 \\ R_2 &= 7 + 3X_1 - 4X_2 - X_3 \end{aligned}$$

The objective function is,

$$\begin{aligned} Z &= 3X_1 + 2X_2 + X_3 - MR_1 - MR_2 \\ &= 3X_1 + 2X_2 + X_3 - M(8 + 3X_1 - 2X_2 - 2X_3) \\ &\quad - M(7 + 3X_1 - 4X_2 - X_3) \\ &= 3X_1 + 2X_2 + X_3 - 8M - 3MX_1 + 2MX_2 + \\ &\quad 2MX_3 - 7M - 3MX_1 + 4MX_2 + MX_3 \\ &\quad Z + X_1(6M - 3) + X_2(-2-6M) + X_3(1-3M) = -15M \end{aligned}$$

Formulate the basic table.

↓PC (column X_2)

Basic	X_1	X_2	X_3	R_1	R_2	Solution	
Z	$6M - 3$	$-2 - 6M$	$-1 - 3M$	0	0	$-15M$	
R_1	−3	2	2	1	0	8	$\frac{8}{2} = 4$
R_2	−3	4*	1	0	1	7	$\frac{7}{4} = 1.75$ → PR

Z row has a negative value. Therefore, solution is not optimum.

Look for the highest negative value in Z row. Consider,

$$-2 - 6M$$

and

$$-1 - 3M$$

Put

$$\begin{aligned} M &= 2 \\ -2 - 6(2) &= (-14) \\ -1 - 3(2) &= -7(-5) \end{aligned}$$

X_2 enters the basic column. This X_2 column is called a pivot column.

Which leaves the basic column?

Ignore all zeroes and negative values in Z row. Divide the solution values by the corresponding elements in the pivot column, i.e.,

$$\frac{8}{2} = 4 \text{ and } \frac{7}{4} = 1.75$$

Select the minimum of these two. It is 1.75. R_2 leaves the basic column. Delete R_2 column. In the aforementioned table, 4 is the pivot element.

Iteration 1

$\text{NPE} = \frac{\text{OPE}}{\text{PE}}$	$-\frac{3}{4}$	1	$\frac{1}{4}$	0	$\frac{7}{4}$
Old Z	$6M-3$	$-2-6M$	$-1-3M$	0	$-15M$
$-(-2-6M)$NPE	$-\frac{9}{2}M-\frac{3}{2}$	$2+6M$	$\frac{1}{2}+\frac{3}{2}M$	0	$\frac{21}{2}M+\frac{7}{2}$
New Z	$\frac{3}{2} \quad -\frac{9}{2}$	0	$-\frac{1}{2}-\frac{3}{2}M$	0	$-\frac{9}{2}M+\frac{7}{2}$
Old R_1	-3	2	2	1	8
$-(2)$NPE	$\frac{3}{2}$	-2	$-\frac{1}{2}$	0	$-\frac{7}{2}$
New R_1	$-\frac{3}{2}$	0	$\frac{3}{2}$	1	$\frac{9}{2}$

Formulate the basic table.

Basic	X_1	X_2	↓PC X_3	R_1	Solution		
Z	$\frac{3}{2}M-\frac{9}{2}$	0	$-\frac{3}{2}M-\frac{1}{2}$	0	$-\frac{9}{2}M+\frac{7}{2}$		
R_1	$-\frac{3}{2}$	0	$\frac{3}{2}$	1	$\frac{9}{2}$	$\frac{9}{2}\times\frac{2}{3}=3$	→ PR
X_3	$-\frac{3}{4}$	1	$\frac{1}{4}$	0	$\frac{7}{4}$	$\frac{7}{4}\times\frac{4}{1}=\frac{28}{4}=7$	

Z row has a negative value. Therefore, solution is not optimum. Look for the negative value in Z row. It is $-\frac{3}{2}M-\frac{1}{2}$. The variable corresponding to this value is X_3. X_3 enters the basic column.

Which leaves the basic column?

Ignore all zeroes and negative values in Z row. Divide the solution elements by the corresponding element in the pivot column, i.e.,

$$\frac{9}{2} \times \frac{2}{3} = 3$$

$$\frac{7}{4} \times \frac{4}{1} = \frac{28}{4} = 7$$

Select the minimum of these two. It is 3. Row R_1 leaves basic column. Delete column R_2. $\frac{3}{2}$ is the pivot element.

Iteration 2

$\text{NPE} = \frac{\text{OPE}}{\text{PE}}$	-1	0	1	3
Old Z	$\frac{3}{2}M - \frac{9}{2}$	0	$-\frac{3}{2}M - \frac{1}{2}$	$-\frac{9}{2}M + \frac{7}{2}$
$-\left(-\frac{3}{2}M - \frac{1}{2}\right)$ NPE	$-\frac{3}{2}M - \frac{1}{2}$	0	$\frac{3}{2}M + \frac{1}{2}$	$\frac{9}{2}M + \frac{3}{2}$
New Z	-5	0	0	5
Old X_1	$-\frac{3}{4}$	1	$\frac{1}{4}$	$\frac{7}{4}$
$-\left(\frac{1}{4}\right)$ NPE	$\frac{1}{4}$	0	$-\frac{1}{4}$	$-\frac{3}{4}$
New X_2	$-\frac{1}{2}$	1	0	1

Formulate the basic table.

Basic	X_1	X_2	X_3	Solution
Z	-5	0	0	5
X_3	-1	0	1	3
X_2	$\frac{-1}{2}$	1	0	1

Column X_1 enters. But no variable leaves the basic, as they have negative values in the pivot column. Hence the solution is optimum.

$$\text{Maximise } Z = 5$$

subject to

$$X_1 = 0$$
$$X_2 = 1$$
$$X_3 = 3$$

3.9 TWO-PHASE METHOD

Two-phase method is an alternative method to Big-M method used to solve simplex problems with artificial variables. As the name implies, this method consists of two phases.

Phase 1

In phase 1 the given problem is converted to the form,

$$\text{Maximise } Z = 0X_1 + 0X_2 + 0X_n + \cdots + 0X_n + 0S_1 + 0S_2 + \cdots + 0S_n$$
$$-R_1 - R_2 \cdots - R_n$$

subject to

$$a_{i1}X_{i1} + a_{i2}X_{i2} + \cdots + a_{in}X_n - S_i + R_i = b_i$$

where $i = 1$ to M, and this problem is solved by simplex method and maximise Z value subject to $X_1, X_2, \cdots, X_n$ variables is determined. R_n artificial variables are taken in basic table and all other variables are kept at zero.

Phase 2

Basic variables obtained in phase 1 are taken in initial basic table in phase 2. The constraint equations are equated to get solution obtained in phase 1, on the right-hand side of the equation. With reference to this equation, the basic table is constructed and solved by simplex method and maximise Z value subject to X, $X_1, X_2, X_3, \cdots, X_n$ variables is determined.

Solve the given LPP by two-phase method.

***Example* 3.10**

$$\text{Maximise } Z = -4X_1 - 3X_2 - 9X_3$$

subject to

$$2X_1 + 4X_2 + 6X_3 \geq 15$$
$$6X_1 + X_2 + 6X_3 \geq 12$$
$$X_1, X_2, X_3 \geq 0$$

Solution

Phase 1

The objective function is,

$$\text{Maximise } Z = 0X_1 + 0X_2 + 0X_3 + 0S_1 + 0S_2 - 0R_1 - 0R_2$$

Bringing inequality to equality

$$2X_1 + 4X_2 + 6X_3 - S_1 + R_1 = 15$$
$$6X_1 + X_2 + 6X_3 - S_2 + R_2 = 12$$
$$R_1 = (15 - 2X_1 - 4X_2 - 6X_3 + S_1)$$
$$R_2 = (12 - 6X_1 - X_2 - 6X_3 + S_2)$$
$$Z = -R_1 - R_2$$

Substituting for R_1 and R_2 in Z equation, we get,

$$Z = -15 + 2X_1 + 4X_2 + 6X_3 - S_1 - 12 + 6X_1 + X_2 + 6X_3 - S_2$$

$$Z = -27 + 8X_1 + 5X_2 + 12X_3 - S_1 - S_2$$

$$Z - 8X_1 - 5X_2 - 12X_3 + S_1 + S_2 = -27$$

Formulate the basic table.

↓PC (column X_3)

Basic	X_1	X_2	X_3	S_1	S_2	R_1	R_2	Solution	
Z	–8	–5	–12	1	1	0	0	–27	
R_1	2	4	6	–1	0	1	0	15	$\frac{15}{6} = 2.5$
R_2	6	1	6*	0	–1	0	1	12 =	$\frac{12}{6} = 2$ → PR

Z row has negative values. Therefore, solution is not optimum. Look for the highest negative value in Z row. X_3 with –12 enters the basic column.

It is the pivot column. Ignore zeroes and negative values in pivot column. Divide the elements in the solution column by the corresponding elements in the pivot column, i.e.,

$$\frac{15}{6} = 2.5$$

and

$$\frac{12}{6} = 2$$

Select the minimum of these two. It is 2. Row R_2 leaves. This is called pivot row. Delete column R_2. New X_3 enters and R_2 leaves basic table.

Iteration 1

NPE = $\frac{\text{OPE}}{\text{PE}}$	1	$\frac{1}{6}$	1	0	$-\frac{1}{6}$	0	2
Old Z	–8	–5	–12	1	1	0	–27
–(–12)NPE	12	2	12	0	–2	0	24
New Z	4	–3	0	1	–1	0	–3
Old R_1	2	4	6	–1	0	1	15
–(+6)NPE	–6	–1	–6	0	1	0	–12
New R_1	–4	3	0	–1	1	1	3

Formulate the basic table.

Basic	X_1	X_2 ↓PC	X_3	S_1	S_2	R_1	Solution		
Z	4	–3	0	1	–1	0	–3		
R_1	–4	3*	0	–1	1	1	3	$\frac{3}{3}=1$	→ PR
X_3	1	$\frac{1}{6}$	1	0	$-\frac{1}{6}$	0	2	$\frac{2}{\frac{1}{6}}=12$	

Z row has negative value. Therfore, solution is not optimum. X_2 column with negative value in Z row enters basic column. It is the pivot column.

Ignore zeroes and negative values in pivot column. Divide solution value by corresponding element in pivot column, i.e.,

$$\frac{3}{3}=1 \text{ and } \frac{2}{\frac{1}{6}}=12$$

Select the minimum of these two. It is 1. Row R_1 leaves. Delete R_1 column. Value 3 is common to both pivot row and pivot column, it is the pivot element.

Iteration 2

$NPE = \frac{OPE}{PE}$	$-\frac{4}{3}$	1	0	$-\frac{1}{3}$	$\frac{1}{3}$	1
Old Z	4	–3	0	1	–1	–3
–(–3)NPE	–4	3	0	–1	1	3
New Z	0	0	0	0	0	0
Old X_3	1	$\frac{1}{6}$	1	0	$-\frac{1}{6}$	2
$-\left(\frac{1}{6}\right)$ NPE	$\frac{2}{9}$	$-\frac{1}{6}$	0	$\frac{1}{18}$	$-\frac{1}{18}$	$-\frac{1}{6}$
New X_3	$\frac{11}{9}$	0	1	$\frac{1}{18}$	$-\frac{2}{9}$	$\frac{11}{6}$

Formulate the basic table.

Basic	X_1	X_2	X_3	S_1	S_2	Solution
Z	0	0	0	0	0	0
X_2	$-\frac{4}{3}$	1	0	$-\frac{1}{3}$	$\frac{1}{3}$	1
X_3	$\frac{11}{9}$	0	1	$\frac{1}{18}$	$-\frac{2}{9}$	$\frac{11}{6}$

The solutions of the variables are,

$$X_2 = 1$$
$$X_3 = \frac{11}{6}$$

Phase 2

From phase1 $X_2 = 1$ and $X_3 = \frac{11}{6}$. To determine objective function Z, consider the given problem.

$$\text{Maximise } Z = -4X_1 - 3X_2 - 9X_3$$

Bringing constraints to equality, we get,

$$2X_1 + 4X_2 + 6X_3 - S_1 = 15$$
$$6X_1 + X_2 + 6X_3 - S_2 = 12$$

where S_1 and S_2 are surplus variables.

Equate the above equation to get the solution values on the right-hand side, and $X_3 = 0$ for X_2 equation.

$$2X_1 + 4X_2 + 6X_3 - S_1 = 15$$
$$(-)\quad(-)\quad(-)\quad(+)\quad(-)$$
$$6X_1 + X_2 + 6X_3 - S_2 = 12$$
$$\overline{-4X_1 + 3X_2 - S_1 + S_2 = 3}$$

Divide the above equation by 3 to get solution value 1. X_2 equation is,

$$-\frac{4}{3}X_1 + X_2 - \frac{1}{3}S_1 + S_2 = 1$$

Similarly, equate the same equation to get the solution value $\frac{11}{6}$ on the right-hand side and $X_2 = 0$ for X_3 equation.

$$2X_1 + 4X_2 + 6X_3 - S_1 = 15 \qquad (1)$$

Multiply by 4,

$$6X_1 + X_2 + 6X_3 - S_2 = 12$$
$$24X_1 + 4X_2 + 24X_3 - 4S_2 = 48 \qquad (2)$$

Subtracting Equation (2) from (1), we get,

$$2X_1 + 4X_2 + 6X_3 - S_1 = 15$$
$$(-)\quad(-)\quad(-)\quad(+)\quad(-)$$
$$24X_1 + 4X_2 + 24X_3 - 4S_2 = 48$$
$$\overline{-22X_1 - 18X_3 - S_1 + 4S_2 = -33} \qquad (3)$$

Multiply by $-\frac{1}{18}$ to get $\frac{11}{6}$ on the right-hand side, X_3 equation is,

$$\frac{11}{9}X_1 + X_3 + \frac{1}{18}S_1 - \frac{2}{9}S_2 = \frac{11}{6}$$

From X_2 equation,

$$X_2 = \left[1+\frac{4}{3}X_1+\frac{1}{3}S_1-\frac{1}{3}S_2\right]$$

From X_3 equation,

$$X_3 = \left[\frac{11}{6}-\frac{11}{9}X_1-\frac{1}{18}S_1+\frac{2}{9}S_2\right]$$

Substituting in,

$$Z = -4X_1 - 3X_2 - 9X_3$$

we get,

$$Z = -4X_1 - 3\left[1+\frac{4}{3}X_1+\frac{1}{3}S_1-\frac{1}{3}S_2\right]$$

$$-9\left[\frac{11}{6}-\frac{11}{9}X_1-\frac{1}{18}S_1+\frac{2}{9}S_2\right]$$

$$= -4X_1 - 3 - 4X_1 - S_1 + S_2 - \frac{33}{2} + 11X_1 + \frac{1}{2}S_1 - 2S_2$$

$$= -8X_1 + 11X_1 - \frac{1}{2}S_1 - S_2 - \frac{39}{2}$$

$$= 3X_1 - S_1 - S_2 - \frac{39}{2}$$

$$Z - 3X_1 - \frac{1}{2}S_1 + S_2 = \frac{39}{2}$$

Formulate the basic table.

↓PC

Basic	X_1	X_2	X_3	S_1	S_2	Solution	
Z	-3	0	0	$\frac{1}{2}$	1	$-\frac{39}{2}$	
X_2	$-\frac{4}{3}$	1	0	$-\frac{1}{3}$	$\frac{1}{3}$	1	
X_3	$\frac{11}{9}$ *	0	1	$\frac{1}{18}$	$-\frac{2}{9}$	$\frac{11}{6}$	→ PR

Z row has negative value. Therefore, solution is not optimum. X_1 with negative value enters the basic column. It is the pivot column. Ignore negative values in pivot column. X_3 leaves. It is the pivot row. The value $\frac{11}{9}$ is the pivot element.

Iteration 1

$NPE = \frac{OPE}{PE}$	1	0	$\frac{9}{11}$	$\frac{1}{22}$	$-\frac{2}{11}$	$\frac{3}{2}$
Old Z	–3	0	0	$\frac{1}{2}$	1	$-\frac{39}{2}$
–(–3)NPE	3	0	$\frac{27}{11}$	$\frac{3}{22}$	$-\frac{6}{11}$	$\frac{9}{2}$
New Z	0	0	$\frac{27}{11}$	$\frac{7}{11}$	$\frac{27}{11}$	–15
Old X_2	$-\frac{4}{3}$	1	0	$-\frac{1}{3}$	$\frac{1}{3}$	1
$-\left(-\frac{4}{3}\right)$NPE	$\frac{4}{3}$	0	$\frac{12}{11}$	$\frac{2}{33}$	$\frac{8}{33}$	2
New X_2	0	1	$\frac{12}{11}$	$-\frac{3}{11}$	$\frac{19}{33}$	3

Formulate the basic table.

Basic	X_1	X_2	X_3	S_1	S_2	Solution
Z	0	0	$\frac{27}{11}$	$\frac{7}{11}$	$\frac{27}{11}$	–15
X_2	0	1	$\frac{12}{11}$	$-\frac{3}{11}$	$\frac{19}{33}$	3
X_1	1	0	$\frac{9}{11}$	$\frac{1}{22}$	$-\frac{2}{11}$	$\frac{3}{2}$

Solution is optimum.

$$\text{Maximise } Z = -15$$

subject to

$$X_1 = \frac{3}{2}$$
$$X_2 = 3$$
$$X_3 = 0$$

3.10 DEGENERACY IN LPP

Simplex method is the algebraic procedure. The selection of non-basic variable which enters the basic solution is easy. The variable having most negative co-efficient value in Z row enters the basic column. Variable which leaves the basic solution is selected by dividing the values in the solution column by the corresponding values in the pivot column and select the minimum of these two values. The difficulty arises when the ratio of these variables are the same, then the problem is said to degenerate. The solution of this problem is known as degenerate basic feasible solution.

***Example* 3.11**

$$\text{Maximise } Z = 22X + 30Y + 25Z$$

subject to

$$2X + 2Y \leq 100$$
$$-2X - Y - Z \geq -100$$
$$X + 2Y + 2Z \leq 100$$
$$X, Y, Z \geq 0$$

Solution

The values on the RHS of the constraints should not be negative, consider the equation,

$$-2X - Y - Z \geq -100$$

Multiply this equation by negative sign, when it is multiplied by a negative sign ≥ becomes ≤.

$$2X + Y + Z \leq 100$$

Bringing inequalities to equality, we have,

$$2X + 2Y + S_1 = 100$$
$$2X + Y + Z + S_2 = 100$$
$$X + 2Y + 2Z + S_3 = 100$$

The objective function is,

$$Z = 22X + 30Y + 25Z$$
$$Z - 22X - 30Y - 25Z = 0$$

Formulate the basic table.

↓PC

Basic	x	y	z	S_1	S_2	S_3	Solution		
Z	–22	–30	–25	0	0	0	0		
S_1	2	2	0	1	0	0	100	$\frac{100}{2} = 50$	
S_2	2	1	1	0	1	0	100	$\frac{100}{1} = 100$	
S_3	1	2*	2	0	0	1	100	$\frac{100}{2} = 50$	→PR

Z row has negative values. Therefore, solution is not optimum.

Look for the highest negative value in '*Z*' row. It is –30. The variable corresponding to this value is *y*. *y* enters the basic column. This *y* column is called a pivot column.

Which leaves the basic column?

Divide the values in the solution column by the corresponding values in the pivot column, i.e.,

$$\frac{100}{2} = 50$$

$$\frac{100}{1} = 100$$

and

$$\frac{100}{2} = 50$$

Select the minimum of these three ratios. Here, a tie occurs between S_1 and S_3. The problem degenerates.

To remove degeneracy

1. Consider the elements in the pivot column.
2. Divide the values in the S_1 column by the corresponding elements of the tied row in the pivot column.
3. The minimum of these ratios is the pivot row.

If the solution is not obtained, divide the values in the S_2 column and so on.

In this problem for basic variables S_1, ratio of S_1 value to the pivot column element is $\frac{1}{2}$

Similarly, for S_3, the ratio is $\frac{0}{2} = 0$ The ratio is minimum for S_3. Therefore, S_3 leaves the basic column. It is called the pivot row.

Iteration 1

$NPE = \frac{OPE}{PE}$	$\frac{1}{2}$	1	1	0	0	$\frac{1}{2}$	50
Old Z	–22	–30	–25	0	0	0	0
–(–30)NPE	15	30	30	0	0	15	1500
New Z	–7	0	5	0	0	15	1500
Old S_1	2	2	0	1	0	0	100
–(–2)NPE	–1	–2	–2	0	0	–1	–100
New S_1	1	0	–2	1	0	–1	0
Old S_2	2	1	1	0	1	0	100
–(1)NPE	$-\frac{1}{2}$	–1	–1	0	0	$-\frac{1}{2}$	–50
New S_2	$\frac{3}{2}$	0	0	0	1	$-\frac{1}{2}$	50

Formulate the basic table.

Basic	↓PC x	y	z	S_1	S_2	S_3	Solution		
Z	–7	0	5	0	0	15	1500		
S_1	1_*	0	–2	1	0	–1	0	$\frac{0}{1}$	→ PR
S_2	$\frac{3}{2}$	0	1	0	1	$-\frac{1}{2}$	50	$\frac{100}{3}$	
y	$\frac{1}{2}$	1	1	0	0	$\frac{1}{2}$	50	$50 \times \frac{2}{1} = 100$	

Z row has a negative value. Therefore, solution is not optimum. Look for the negative value in Z row. It is –7. The variable corresponding to this value is x. x enters the basic column.

Which leaves the basic column?

Ignore all zeroes and negative values in the pivot column. Divide the solution values by the corresponding value in the pivot column, i.e.,

$$\frac{0}{1}$$

$$\frac{100}{3}$$

and

$$50 \times \frac{2}{1} = 100$$

Select the minimum of these three. It is zero. The variable which leaves the basic column is S_1. In the aforementioned table, 1 is the pivot element.

Iteration 2

$\text{NPE} = \frac{\text{OPE}}{\text{PE}}$	1	0	–2	1	0	–1	0
Old Z	–7	0	5	0	0	15	1500
–(–7)NPE	7	0	–14	7	0	–7	0
New Z	0	0	–9	7	0	8	1500
Old S_2	$\frac{3}{2}$	0	0	0	1	$-\frac{1}{2}$	50
$-\left(\frac{3}{2}\right)$NPE	$-\frac{3}{2}$	0	3	$-\frac{3}{2}$	0	$\frac{3}{2}$	0
New S_2	0	0	3	$-\frac{3}{2}$	1	1	50

Old y	$\frac{1}{2}$	1	1	0	0	$\frac{1}{2}$	50
$-\left(\frac{1}{2}\right)$NPE	$-\frac{1}{2}$	0	1	$-\frac{1}{2}$	0	$\frac{1}{2}$	0
New y	0	1	2	$-\frac{1}{2}$	0	1	50

Formulate the basic table.

↓PC

Basic	x	y	z	S_1	S_2	S_3	Solution		
Z	0	0	–9	7	0	8	1500		
x	1	0	–2	1	0	–1	0		
S_2	0	0	3*	$-\frac{3}{2}$	1	1	50	$\frac{50}{3}$	→ PR
y	0	1	2	$-\frac{1}{2}$	0	1	50	$\frac{50}{2}$	

Z row has a negative value. Therefore, solution is not optimum. z with –9 enters the basic column.

Which leaves the basic column?

Ignore negative values in the pivot column. Divide the solution value by the corresponding element in the pivot column, i.e.,

$$\frac{50}{3} \text{ and } \frac{50}{2}$$

Select the minimum of these two. S_2 is the pivot row.

Iteration 3

NPE = $\frac{\text{OPE}}{\text{PE}}$	0	0	1	$-\frac{1}{2}$	$\frac{1}{3}$	$\frac{1}{3}$	$\frac{50}{3}$
Old Z	0	0	–9	7	0	8	1500
–(–9)NPE	0	0	9	$-\frac{9}{2}$	+3	+3	+150
New Z	0	0	0	$\frac{5}{2}$	3	11	1650
Old X	1	0	–2	1	0	–1	0
–(–2)NPE	0	0	2	–1	$\frac{2}{3}$	$\frac{2}{3}$	$\frac{100}{3}$
New X	1	0	0	0	$\frac{2}{3}$	$-\frac{1}{3}$	$\frac{100}{3}$

Old y	0	1	2	$-\frac{1}{2}$	0	1	50
$-(2)$NPE	0	0	-2	1	$-\frac{2}{3}$	$-\frac{2}{3}$	$-\frac{100}{3}$
New y	0	1	0	$\frac{1}{2}$	$-\frac{2}{3}$	$-\frac{2}{3}$	$\frac{50}{3}$

Formulate the basic table.

Basic	x	y	z	S_1	S_2	S_3	Solution
Z	0	0	0	$\frac{5}{2}$	3	11	1650
x	1	0	0	0	$\frac{2}{3}$	$-\frac{1}{3}$	$\frac{100}{3}$
z	0	0	1	$-\frac{1}{2}$	$\frac{1}{3}$	$\frac{1}{3}$	$\frac{50}{3}$
y	0	1	0	$\frac{1}{2}$	$-\frac{2}{3}$	$-\frac{2}{3}$	$\frac{50}{3}$

The solution is optimum.

$$Z = \text{Rs. } 1650$$

subject to,

$$x = \frac{100}{3}$$
$$y = \frac{50}{3}$$
$$z = \frac{50}{3}$$

EXERCISES

1. Solve the LPP by simplex method.

$$\text{Minimise } Z = 5X_1 - 6X_2 - 7X_3$$

subject to

$$X_1 + 5X_2 - 3X_3 \geq 15$$
$$5X_1 - 6X_2 + 10X_3 \leq 20$$
$$X_1 + X_2 + X_3 = 5$$
$$X_1, X_2, X_3 \geq 0$$

2. Solve the LPP by simplex method.

$$\text{Maximise } Z = \frac{3}{4}X_1 - 150X_2 + \frac{1}{50}X_3 - 6X_4$$

subject to

$$\frac{1}{4}X_1 - 60X_2 - \frac{1}{25}X_3 + 9X_4 \leq 0$$

$$\frac{1}{2}X_1 - 90X_2 - \frac{1}{50}X_3 + 3X_4 \leq 0$$

$$X_3 \leq 1$$

$$X_1, X_2, X_3, X_4 \geq 0$$

3. Solve by simplex method.

$$\text{Maximise } Z = 3X_1 + 2X_2$$

subject to

$$2X_1 - X_2 \geq 2$$

$$X_1 + 2X_2 \leq 8$$

$$X_1, X_2 \geq 0$$

4. Solve by simplex method.

$$\text{Maximise } Z = X_1 + 2X_2 + X_3$$

subject to

$$3X_1 + X_2 + 3X_3 \geq 2$$

$$3X_1 + 2X_2 = 8$$

$$X_1 - X_2 \leq 1$$

$$X_1, X_2, X_3 \geq 0$$

5. Use the simplex method to solve this LPP.

$$\text{Maximise } Z = 4X_1 + 10X_2$$

subject to

$$2X_1 + X_2 \leq 50$$

$$2X_1 + 5X_2 \leq 100$$

$$2X_1 + 3X_2 \leq 90$$

$$X_1 \geq 0$$

$$X_2 \geq 0$$

6. Use simplex method to solve the LPP.

$$\text{Maximise } Z = 3X_1 + 2X_2$$

subject to

$$X_1 + X_2 \leq 4$$

$$X_1 - X_2 \leq 2$$

$$X_1 \geq 0$$
$$X_2 \geq 0$$

7. Use simplex method to solve the LPP.

$$\text{Maximise } Z = 4X_1 + 3X_2 + 4X_3 + 6X_4$$

subject to

$$X_1 + 2X_2 + 2X_3 - 4X_4 \leq 80$$
$$2X_1 + 2X_3 + X_4 \leq 60$$
$$3X_1 + 3X_2 + X_3 + X_4 \leq 80$$
$$X_1, X_2, X_3, X_4 \geq 0$$

8. Use simplex method to solve the LPP.

$$\text{Maximise } Z = 4X_1 + 5X_2 + 9X_3 + 11X_4$$

subject to

$$X_1 + X_2 + X_3 + X_4 \leq 15$$
$$7X_1 + 5X_2 + 3X_3 + 2X_4 \leq 120$$
$$3X_1 + 5X_2 + 10X_3 + 15X_4 \leq 100$$
$$X_1, X_2, X_3, X_4 \geq 0$$

9. Solve the following problem using simplex method.

$$\text{Maximise } Z = 7X_1 + 14X_2$$

subject to

$$-X_1 - 2X_2 \geq -6$$
$$4X_1 + 3X_2 \leq 12$$
$$X_1, X_2 \geq 0$$

10. Use the Big-M simplex method to find a basic feasible solution to the following linear inequalities.

$$\text{Minimise } Z = 2Y_1 + 4Y_2$$

subject to

$$2Y_1 - 3Y_2 \geq 2$$
$$-Y_1 + Y_1 \geq 3$$
$$Y_1 \geq 0$$
$$Y_2 \geq 0$$

11. Consider the following set of constraints.

$$X_1 + 2X_2 - 2X_3 + 4X_4 \leq 40$$
$$2X_1 - X_2 + X_3 + 2X_4 \leq 8$$
$$4X_1 - 2X_2 + X_3 - X_4 \leq 10$$
$$X_1, X_2, X_3, X_4 \geq 0$$

Solve the problem for each of the following objective functions.

(a) Maximise $Z = 2X_1 + X_2 - 3X_3 + 5X_4$

(b) Maximise $Z = 8X_1 + 6X_2 + 3X_3 - 2X_4$

(c) Maximise $Z = 3X_1 - X_2 + 3X_3 + 4X_4$

(d) Minimise $Z = 5X_1 - 4X_2 + 6X_3 - 8X_4$

(c) Minimise $Z = -4X_1 + 6X_2 - 2X_3 + 4X_4$

12. Solve by simplex method.

$$\text{Maximise } Z = -2X_1 + 3X_2 - 2X_3$$

subject to

$$\begin{aligned} 4X_1 - X_2 - 5X_3 &= 10 \\ 2X_1 + 3X_2 + 2X_3 &= 12 \\ X_1 &\geq 0 \\ X_2 &\geq 0 \\ X_3 &\geq 0 \end{aligned}$$

13. Use simplex method to solve this problem.

$$\text{Maximise } Z = 4X_1 + X_2$$

subject to

$$\begin{aligned} 3X_1 + X_2 &= 3 \\ 4X_1 + 3X_2 &\geq 6 \\ X_1 + 2X_2 &\leq 4 \\ X_1, X_2 &\leq 0 \end{aligned}$$

14. Consider the following set of constraints.

$$\begin{aligned} -2X_1 + 3X_2 &= 3 && (1) \\ 4X_1 + 5X_2 &\geq 10 && (2) \\ X_1 + 2X_2 &\leq 5 && (3) \\ 6X_1 + 7X_2 &\leq 3 && (4) \\ 4X_1 + 8X_2 &\geq 5 && (5) \\ X_1, X_2 &\geq 0 \end{aligned}$$

For each of the following problems use simplex method and find solutions.

(a) Maximize $Z = 5X_1 + 6X_2$, subject to (1) (3) and (4).

(b) Maximize $Z = 2X_1 - 7X_2$, subject to (1) (2) (4) and (5).

Chapter 4

DUAL SIMPLEX METHOD

4.1 THE DUAL PROBLEM

Associated with every linear problem, there exists another linear programming problem which is based on the same data having the same solution. The original problem is called the primal problem while the associated one is called its dual problem. It is important to note that either of the LPP can be treated as primal and the other as its dual. The two problems thus, constitute a primal–dual pair.

4.2 GENERAL PRIMAL–DUAL PAIR

Based on the standard form of primal, there are two important primal–dual pairs

Standard primal problem

$$\text{Maximise } Z = C_1X_1 + C_2X_2 + C_3X_3 + \cdots + C_nX_n$$

subject to the constraints,

$$a_{i1}X_1 + a_{i2}X_2 + \cdots + a_{in}X_n = b_i \ (i = 1, 2, 3, \ldots, m)$$
$$X_j \geq 0 \ (j = 1, 2, 3, \ldots, n)$$

Dual problem

$$\text{Minimise } Z = b_1Y_1 + b_2Y_2 + \ldots + b_mY_m$$

subject to the constraints,

$$a_{1j}Y_1 + a_{2j}Y_2 + \ldots + a_{mj}Y_m \geq C_j$$
$$Y_i \geq 0 \ (i = 1, 2, 3, \ldots, m)$$

4.3 FORMULATION OF A DUAL PROBLEM

Step 1 Write the given LPP in its standard form. Consider it as the primal problem.

Step 2 Identify the variables for the dual problem. The number of variables in the dual problem is equal to the number of constraints in the primal problem.

Step 3 Using the right-hand side constants, write down the objective function of the dual problem. If the primal problem is of maximisation type, the dual will be a minimisation problem and vice-versa.

Step 4 Using the dual variables, write the constraints for the dual problem.

If the primal problem is a maximisation problem, the dual constraints must be of $\geq$ type. If the primal problem is a minimisation problem the dual constraints must be of $\leq$ type.

The column co-efficient of constraints in the primal problem becomes the row co-efficient of constraints in the dual problem.

The co-efficients of the objective function of the primal problem becomes the right-hand side of the dual constraints.

The dual variables are defined to be unrestricted in sign.

Example 4.1

Primal problem

$$\text{Maximise } Z = 5X_1 + 12X_2 + 4X_3$$

subject to

$$\begin{aligned} X_1 + 2X_2 + X_3 &\leq 10 \\ 2X_1 - X_2 + 3X_3 &= 8 \\ X_1, X_2, X_3 &\geq 0 \end{aligned}$$

Bringing inequality to equality, we have,

$$\text{Maximise } Z = 5X_1 + 12X_2 + 4X_3 + 0S_1$$

subject to

$$\begin{aligned} X_1 + 2X_2 + X_3 + S_1 &= 10 \\ 2X_1 - X_2 + 3X_3 + 0S_1 &= 8 \\ X_1, X_2, X_3, S_1 &\geq 0 \end{aligned}$$

where S_1 is the slack variable.

Dual problem

$$\text{Minimise } W = 10Y_1 + 8Y_2$$

subject to

$$\begin{aligned} Y_1 + 2Y_2 &\geq 5 \\ 2Y_1 - Y_2 &\geq 12 \\ Y_1 + 3Y_2 &\geq 4 \\ Y_1 + 0Y_2 &\geq 0 \\ Y_1 &\geq 0 \end{aligned}$$

Y_2 is unrestricted.

Example 4.2

Primal problem

$$\text{Minimise } Z = 15X_1 + 12X_2$$

subject to

$$\begin{aligned} X_1 + 2X_2 &\geq 3 \\ 2X_1 - 4X_2 &\leq 5 \\ X_1, X_2 &\geq 0 \end{aligned}$$

Bringing inequality to equality, we get,

$$\text{Minimise } Z = 15X_1 + 12X_2 + 0S_1 + 0S_2$$

subject to

$$X_1 + 2X_2 - S_1 + 0S_2 = 3$$
$$2X_1 - 4X_2 + 0S_1 + S_2 = 5$$
$$X_1, X_2, S_1, S_2 \geq 0$$

where S_2 is the slack variable and S_1 is the surplus variable.

Dual problem

$$\text{Maximise } W = 3Y_1 + 5Y_2$$

subject to

$$Y_1 + 2Y_2 \leq 15$$
$$2Y_1 - 4Y_2 \leq 12$$
$$-Y_1 + 0Y_2 \leq 0;\ Y_1 \geq 0$$
$$0Y_1 + Y_2 \leq 0;\ Y_2 \leq 0$$

***Example* 4.3**

Primal problem

$$\text{Maximise } Z = 3X_1 + 5X_2$$

subject to

$$3X_1 + 5X_2 \leq 15$$
$$5X_1 + 2X_2 \leq 10$$
$$X_1, X_2 \geq 0$$

Bringing inequality to equality, we get,

$$\text{Maximise } Z = 3X_1 + 5X_2 + 0S_1$$

subject to

$$3X_1 + 5X_2 + S_1 = 15$$
$$5X_1 + 2X_2 + S_1 = 10$$

where S_1 is the slack variable.

Dual problem

$$\text{Minimise } W = 15Y_1 + 10Y_2$$

subject to

$$3Y_1 + 5Y_2 \geq 3$$
$$5Y_1 + 2Y_2 \geq 5$$
$$Y_1 + Y_2 \geq 0$$

Y_1, Y_2 are unrestricted.

Example 4.4

Primal problem

$$\text{Maximise } Z = 4X_1 + 6X_2 + 18X_3$$

subject to

$$X_1 + 3X_2 \geq 3$$
$$X_2 + 2X_3 \geq 5$$
$$X_1, X_2, X_3 \geq 0$$

Bringing inequality to equality,

$$\text{Minimise } Z = 4X_1 + 6X_2 + 18X_3 + 0S_1$$

subject to

$$X_1 + 3X_2 + 0X_3 - S_1 = 3$$
$$0X_1 + X_2 + 2X_3 - S_1 = 5$$
$$X_1, X_2, X_3, S_1 \geq 0$$

where S_1 is the surplus variable.

Dual Problem

$$\text{Minimise } W = 3Y_1 + 5Y_2$$

subject to

$$Y_1 \leq 4$$
$$3Y_1 + Y_2 \leq 6$$
$$2Y_2 \leq 18$$
$$-Y_1 - Y_2 \leq 0$$
$$Y_1 + Y_2 \geq 0$$

Y_1, Y_2 are unrestricted.

Example 4.5

Primal problem

$$\text{Maximise } Z = 5X_1 + 6X_2$$

subject to

$$X_1 + 2X_2 = 5$$
$$-X_1 + 5X_2 \geq 3$$
$$4X_1 + 7X_2 \leq 8$$

X_1 is unrestricted

$$X_2 \geq 0$$

X_1 is unrestricted in sign. Therefore, substitute,

$$X_1^+ - X_1^-$$

Bringing inequalities to equality,

$$\text{Maximise } Z = 5X_1^+ - 5X_1^- + 6X_2 + 0S_1 + 0S_2$$

subject to

$$\begin{aligned} X_1^+ - X_1^- + 2X_2 + 0S_1 + 0S_2 &= 5 \\ -X_1^+ + X_1^- + 6X_2 - S_1 + 0S_2 &= 3 \\ 4X_1^+ - 4X_1^- + 7X_2 + 0S_1 + S_2 &= 8 \\ X_1^+, X_1^-, X_2, S_1, S_2 &\geq 0 \end{aligned}$$

where, S_1 is the surplus variable and S_2 is the slack variable.

Dual Problem

$$\text{Minimise } W = 5Y_1 + 3Y_2 + 8Y_3$$

subject to

$$\begin{aligned} Y_1 - Y_2 + 4Y_3 &\geq 5 \\ -Y_1 + Y_2 - 4Y_3 &\geq -5 \\ 6Y_1 + 2Y_2 + 5Y_3 &\geq 6 \\ 0Y_1 + Y_2 + 5Y_3 &\geq 0 \\ -Y_2 &\geq 0 \text{ or } Y_2 \leq 0 \\ 0Y_1 + 0Y_2 + Y_3 &\geq 0 \text{ or } Y_3 \geq 0 \\ Y_1 &\text{ is unrestricted.} \end{aligned}$$

***Example* 4.6**

Primal problem

$$\text{Minimise } Z = 2X_1 + X_2 - X_3$$

subject to

$$\begin{aligned} X_1 - X_2 - X_3 &= 1 \\ X_1 - X_2 + X_3 &\geq 2 \\ X_2 + X_3 &\leq 3 \\ X_1 &\geq 0 \text{ or } X_2 \leq 0 \\ X_3 &\text{ is unrestricted.} \end{aligned}$$

X_3 is unrestricted in sign. Therefore, substitute,

$$\begin{aligned} X_3 &= X_3^+ - X_3^-, X_2 \leq 0 \\ -X_2 &\leq 0 \end{aligned}$$

Therefore,

$$-X_2 \geq 0$$

Bringing inequalities to equality,

$$\begin{aligned} \text{Minimise } Z &= 2X_1 - X_2 - X_3^+ + X_3^- + 0S_1 + 0S_2 \\ X_2 &\leq 0 \end{aligned}$$

Therefore,

$$-X_2 \geq 0$$

Bringing inequality to equality,

$$\text{Minimise } Z = 2X_1 - X_2 - X_3^+ + X_3^-$$

subjeclt to

$$\begin{aligned} X_1 + X_2 - X_3^+ + X_3^- + 0S_1 + 0S_2 &= 1 \\ X_1 - X_2 + X_3^+ - X_3^- - S_1 + 0S_2 &= 2 \\ X_2 + X_3^+ - X_3^- + 0S_1 + S_2 &= 3 \\ S_1, S_2, X_1, X_2, X_3^+, X_1^- &\geq 0 \end{aligned}$$

where, S_1 is the surplus variable and S_2 is the slack variable.

Dual Problem

$$\text{Minimise } W = Y_1 + 2Y_2 + 3Y_3$$

subject to

$$\begin{aligned} Y_1 + Y_2 &\leq 2 \\ Y_1 - Y_2 + Y_3 &\leq -1 \\ -Y_1 + Y_2 + Y_3 &\leq -1 \\ Y_1 - Y_2 - Y_3 &\leq 1 \\ -Y_2 &\geq 0;\ Y_2 \leq 0 \\ Y_3 &\geq 0; Y_3 \geq 0 \end{aligned}$$

Y_1 is unrestricted.

***Example* 4.7**

Primal problem

$$\text{Maximise } Z = X_1 + 4X_2 + 3X_3$$

subject to

$$\begin{aligned} 2X_1 + 3X_2 - 5X_3 &\leq 2 \\ 3X_1 - X_2 + 6X_3 &\geq 1 \\ X_1 + X_2 + X_3 &\leq 3 \\ X_1 &\geq 0 \text{ or } X_2 \leq 0 \end{aligned}$$

X_3 is unrestricted.

X_3 is unrestricted in sign. Therefore, substitute

$$X_3 = X_3^+ - X_3^-$$

Bringing inequalities to equality,

$$\text{Maximise } Z = X_1 - 4X_2 + 3X_3^+ - 3X_3^- + 0S_1 + 0S_2$$

subject to

$$\begin{aligned} 2X_1 + 3X_2 - 5X_3^+ + 5X_3^- + S_1 + 0S_2 &= 2 \\ 3X_1 - X_2 + 6X_3^+ - 6X_3^- + 0S_1 - S_2 &= 1 \end{aligned}$$

$$X_1 + X_2 + X_3^+ - X_3^- + 0S_1 + 0S_2 = 4$$
$$X_1, X_2, X_3^+, X_3^-, S_1, S_2 \geq 0$$

where S_1 is the slack variable and S_2 is the surplus variable.

Dual Problem

$$\text{Minimise } W = 2Y_1 + Y_2 + 4Y_3$$

subject to

$$2Y_1 + 3Y_2 + Y_3 \geq 1$$
$$3Y_1 - Y_2 + Y_3 \geq -4$$
$$-5Y_1 + 6Y_2 + Y_3 \geq 3$$
$$5Y_1 - 6Y_2 - Y_3 \geq -3$$
$$Y_1 \geq 0$$
$$-Y_2 \geq 0 \text{ or } Y_2 \geq 0$$
$$Y_3 \text{ is unrestricted.}$$

4.4 ECONOMIC INTERPRETATION OF DUAL PROBLEM

Consider the maximisation problem

$$\text{Maximise } Z = 80X + 70Y$$

subject to

$$2X_1 + 3X_2 \leq 60 \qquad (1)$$
$$4X_1 + 3X_2 \leq 96 \qquad (2)$$
$$X_1, X_2 \geq 0$$

Let us say, Equation (1) represents the labour hours and Equation (2) represents the raw material. The optimum solution obtained is as follows:

$$\text{Maximise } Z = 2000$$
$$X_1 = 18$$
$$X_2 = 8$$

It shows that producing 18 units of product 1 and 8 units of product 2 would yield the maximum profit of Rs. 2000. If the firm gives the facilities on rent to a person, the rental rates of the facilities should be at least equal to the profit obtained by producing products 1 and 2. The firm's assets are as follows

Labour hours – 18

Raw material – 8 kg

Let

Y_1 be the rental rate per labour hour, and

Y_2 be the rental rate per kg of raw material

The minimum rental offer must be acceptable to the firm,

i.e., $$\text{Minimum } W = 60Y_1 + 96Y_2$$

We know that 1 unit of product 1 requires 2 labour hours and 4 kg of raw material. Therefore, total rental should be greater than or equal to the profit per unit of product 1, i.e., Rs. 80.

$$2Y_1 + 4Y_2 \geq 80$$

Similarly for product 2, 3 labour hours and 3 kg of raw material are required. Total rental should at least be equal to Rs. 70.

$$3Y_1 + 3Y_2 \geq 70$$

The problem can be formulated as,

$$\text{Minimise } W = 60Y_1 + 96Y_2$$

subject to

$$2Y_1 + 4Y_2 \geq 80$$
$$3Y_1 + 3Y_2 \geq 70$$
$$Y_1, Y_2 \geq 0$$

This problem is dual of the above problem. Y_1 and Y_2 can be determined by solving this LPP by simplex method. The solution obtained by Big-*M* method is as follows

$$\text{Minimise } W = 2000$$

subject to

$$Y_1 = \text{rental rate/labour hour in rupees}$$
$$Y_2 = \text{rental rate/kg of raw material in rupees}$$

Minimum rental rate for product 1 should be at least equal to Rs. 80/unit. The profit that can be obtained by producing product 1 is

$$2Y_1 + 4Y_2 \geq \text{Rs. } 80$$
$$2\left(\frac{20}{3}\right) + 4\left(\frac{50}{3}\right) = \text{Rs. } 80$$

Minimum rental rate for product 2 should be at least equal to Rs. 70/unit. The profit that can be obtained by producing product 2 is,

$$3Y_1 + 4Y_2 \geq \text{Rs. } 70$$
$$3\left(\frac{20}{3}\right) + 3\left(\frac{50}{3}\right) = \text{Rs. } 70$$

An addition of 1 labour hour would add Rs. $\frac{20}{3}$ to the profit and reduction of 1 labour hour would reduce Rs. $\frac{20}{3}$ of Profit. Similarly an addition of 1 kg of raw material would add Rs. $\frac{50}{3}$ to the profit and reduction of 1 kg of raw material would reduce Rs. $\frac{50}{3}$ of profit.

Primal Problem

$$\text{Maximise } Z = 80X_1 + 70X_2$$

Rs. 80 is the profit per unit of product 1.

Rs. 70 is the profit per unit of product 2.

Let

Y_1 be the rental rate/labour hour and

Y_2 be the rental rate/kg of raw material.

$$2Y_1 + 4Y_2 \geq 80$$

$$2\left(\frac{20}{3}\right) + 4\left(\frac{50}{3}\right) = \text{Rs. } 80$$

The total rental rate per unit of product 1 is Rs. 80.

$$3Y_1 + 3Y_2 \geq 70$$

$$3\left(\frac{20}{3}\right) + 3\left(\frac{50}{3}\right) = \text{Rs. } 70$$

The total rental rate per unit of product 2 is Rs. 70.

Thus,

Cost of resources used per unit of product 1 or 2 < Profit per unit of product 1 or 2

or

Rental rate of resources used per unit of product 1 or 2 > Profit per unit of product 1 or 2

4.5 DUAL SIMPLEX METHOD

Every iteration in dual simplex method is associated with a basic solution. The leaving variable is the basic variable having the most negative value. In an iteration, if all the variables take non-negative values, the solution is said to be feasible.

In dual simplex method, the optimality condition of objective function is the same as simplex method. All the constraints must be $\leq$ type, and objective function must be of minimisation type.

If any constraint has $\geq$ sign, convert it to $\leq$ sign by multiplying both sides by –1.

If any constraint has = sign, the equation can be replaced by two inequalities.

For example, $2X + 3Y = 10$ can be written as

$$2X + 3Y \leq 10 \text{ or } 2X + 3Y_2 \geq 10$$

and

$$2X + 3Y \leq 10 \text{ or } -2X - 3Y_2 \leq -10$$

When all the constraints are converted to $\leq$ sign, the LPP will have infeasible solution if any one of the constraints have negative sign after $\leq$ sign.

In case, all the values on the right-hand side of the constraints are non-negative, there is no need to apply the dual simplex method.

Example 4.8

Use dual simplex method to solve the following LPP.

$$\text{Minimize } Z = 3X_1 + X_2$$

subject to

$$X_1 + X_2 \geq 1$$
$$2X_1 + 3X_2 \geq 2$$
$$X_1, X_2 \geq 0$$

Solution

In this example, both the constraints have ≥ sign. Therefore, multiply both sides of the equation by –1 and convert them to have ≤ sign.

$$\text{Minimize } Z = 3X_1 + X_2$$

subject to

$$-X_1 - X_2 \leq -1$$
$$-2X_1 - 3X_2 \leq -2$$
$$X_1, X_2 \geq 0$$

The objective function is,

$$Z = 3X_1 + X_2$$

Therefore,

$$Z - 3X_1 - X_2 = 0$$

Bringing inequalities to equality by adding slack variables, we get,

$$-X_1 - X_2 + S_1 \leq -1$$
$$-2X_1 - 3X_2 + S_2 \leq -2$$

Formulate the basic table.

↓PC

Basic	X_1	X_2	S_1	S_2	Solution	
Z	–3	–1	0	0	0	
S_1	–1	–1	1	0	–1	
S_2	–2	–3*	0	1	–2	→ PR

From the table, $S_1 = -1$, $S_2 = -2$. Therefore, the starting basic solution is infeasible. Basic variable S_2 with highest negative solution value is the leaving variable. Therefore, S_2 is pivot row.

Which variable enters the basic column?

Divide the negative elements in the Z row by the corresponding elements in the pivot row. Ratios are $\frac{3}{1}$ and $\frac{1}{1}$.

The variable corresponding to the minimum ratio is X_2. X_2 enters the basic column. X_2 is pivot column. In the aforementioned table 3 is the pivot element.

Iteration 1

$NPE = \dfrac{OPE}{PE}$	$\dfrac{2}{3}$	1	0	$-\dfrac{1}{3}$	$\dfrac{2}{3}$
Old Z	–3	–1	0	0	0
–(–1)NPE	$\dfrac{2}{3}$	1	0	$-\dfrac{1}{3}$	$\dfrac{2}{3}$
New Z	$-\dfrac{7}{3}$	0	0	$-\dfrac{1}{3}$	$\dfrac{2}{3}$
Old S_1	–1	–1	1	0	–1
–(–1)NPE	$\dfrac{2}{3}$	1	0	$-\dfrac{1}{3}$	$\dfrac{2}{3}$
New S_1	$-\dfrac{1}{3}$	0	1	$-\dfrac{1}{3}$	$-\dfrac{1}{3}$

Formulate the basic table.

Basic	X_1	X_2	S_1	S_2 ↓PC	Solution	
Z	$-\dfrac{7}{3}$	0	0	$-\dfrac{1}{3}$	$\dfrac{2}{3}$	
S_1	$-\dfrac{1}{3}$	0	1	$-\dfrac{1}{3}$*	$-\dfrac{1}{3}$	→ PR
X_2	$\dfrac{2}{3}$	1	0	$-\dfrac{1}{3}$	$\dfrac{2}{3}$	

From the table, $S_1 = -\dfrac{1}{3}$, $S_2 = \dfrac{2}{3}$. Therefore, the basic solution is infeasible. Basic variable S_1 with negative sign is the leaving variable.

Which variable enters the basic column?

Divide the negative elements in the Z row by the corresponding elements in the pivot row.

The ratios are $\dfrac{\frac{-7}{3}}{\frac{-1}{3}}, \dfrac{\frac{-1}{3}}{\frac{-1}{3}}$.

We get 7 and 1.

The variable S_2 with minimum ratio value enters the basic column. S_2 is pivot column. In the above table, $-\dfrac{1}{3}$ is the pivot element.

Iteration 2

$NPE = \dfrac{OPE}{PE}$	1	0	–3	1	1
Old Z	$-\dfrac{7}{3}$	0	0	$-\dfrac{1}{3}$	$\dfrac{2}{3}$
$-\left(-\dfrac{1}{3}\right)$ NPE	$\dfrac{1}{3}$	0	–1	$\dfrac{1}{3}$	$\dfrac{1}{3}$
New Z	–2	0	–1	0	1
Old X_2	$\dfrac{2}{3}$	1	0	$-\dfrac{1}{3}$	$\dfrac{2}{3}$
$-\left(-\dfrac{1}{3}\right)$ NPE	$\dfrac{1}{3}$	0	–1	$\dfrac{1}{3}$	$\dfrac{1}{3}$
New X_2	1	1	–1	0	1

Formulate the basic table.

Basic	X_1	X_2	S_1	S_2	Solution
Z	–2	0	–1	0	1
S_2	1	0	–3	1	1
X_2	1	1	–1	0	1

All the values in the solution columns are positive. Therefore, the solution is optimum basic feasible solution.

$$\text{Minimise } Z = 1$$

subject to

$$X_1 = 0$$
$$X_2 = 1$$

EXERCISES

1. Obtain the dual problem of the following LPP.

$$\text{Maximise } Z = 2X_1 + 5X_2 + 6X_3$$

subject to

$$5X_1 + 6X_2 - X_3 \leq 3$$
$$-2X_1 + X_2 + 4X_3 \leq 4$$
$$X_1 - 5X_2 + 3X_3 \leq 1$$
$$-3X_1 - 3X_2 + 7X_3 \leq 6$$
$$X_1, X_2, X_3 \geq 0$$

2. Find the dual of this simplex primal problem.

$$\text{Minimize } Z = X_1 - 3X_2 - 2X_3$$

subject to

$$3X_1 - X_2 + 2X_3 \le 7$$
$$2X_1 - 4X_2 \ge 12$$
$$-4X_1 + 3X_2 + 8X_3 = 10$$
$$X_1, X_2 \ge 0 \text{ and } X_3 \text{ is unrestricted.}$$

3. Find the dual of the following LPP.

$$\text{Maximize } Z = 2X_1 + X_2$$

subject to

$$X_1 + 5X_2 \le 10$$
$$X_1 + 3X_2 \ge 6$$
$$2X_1 + 2X_2 \le 8$$
$$X_2 \ge 0 \text{ and } X_1 \text{ is unrestricted.}$$

4. Write the dual of the following LPP.

$$\text{Maximize } Z = 3X_1 - 2X_2 + 4X_3$$

subject to

$$3X_1 + 5X_2 + 4X_3 \ge 7$$
$$6X_1 + X_2 + 5X_3 \ge 3$$
$$4X_1 + 7X_2 - 2X_3 \ge 2$$

5. Write the dual of the following LPP.

$$\text{Minimise } Z = 10X_1 + 20X_2$$

subject to

$$3X_1 + 2X_2 \ge 18$$
$$X_1 + 3X_2 \ge 8$$
$$2X_1 - X_2 \le 6$$
$$X_1, X_2, X_3 \ge 0$$

6. Obtain the dual of the LPP given here.

$$\text{Maximise } Z = 8X_1 + 10X_2 + 5X_3$$

subject to

$$X_1 - X_3 \le 4$$
$$2X_1 + 4X_2 \le 12$$
$$X_1 + X_2 + X_3 \ge 2$$
$$3X_1 + 2X_2 - X_3 = 8$$
$$X_1, X_2, X_3 \ge 0$$

7. Obtain the dual of the following LPP.

$$\text{Maximise } Z = 3X_1 + 5X_2 + 7X_3$$

subject to

$$\begin{aligned} X_1 + X_2 + 3X_3 &\le 10 \\ 4X_1 - X_2 + 2X_3 &\ge 15 \\ X_1, X_2 &\ge 0 \text{ and } X_3 \text{ is unrestricted.} \end{aligned}$$

8. Obtain the dual of the following LPP.

$$\text{Maximise } Z = 3X_1 + 5X_2 + 7X_4 - 7X_5$$

subject to

$$\begin{aligned} X_1 + X_2 + 3X_4 - 3X_5 &\le 10 \\ -4X_1 + X_2 - 2X_4 + 2X_5 &\le -15 \\ X_1, X_2, X_4, X_5 &\ge 0 \end{aligned}$$

9. Obtain the dual of the following LPP.

$$\text{Maximise } Z = X_1 + X_2 + X_3$$

subject to

$$\begin{aligned} X_1 - 3X_2 + 4X_3 &= 5 \\ X_1 - 2X_2 &\le 3 \\ 2X_2 - X_3 &\ge 4 \\ X_1, X_2 &\ge 0 \text{ and } X_3 \text{ is unrestricted.} \end{aligned}$$

10. Find the dual of the following LPP.

$$\text{Maximise } Z = 6X_1 + 6X_2 + X_3 + 7X_4 + 5X_5$$

subject to

$$\begin{aligned} 3X_1 + 7X_2 + 8X_3 + 5X_4 + X_5 &= 2 \\ 2X_1 + X_2 + 3X_3 + 2X_4 + 9X_5 &= 6 \\ X_1, X_2, X_3, X_4 &\ge 0 \text{ and } X_5 \text{ is unrestricted.} \end{aligned}$$

11. Use dual simplex method to solve the following LPP.

$$\text{Minimize } Z = 3X_1 + 2X_2$$

subject to

$$\begin{aligned} 3X_1 + X_2 &\ge 3 \\ 4X_1 + 3X_2 &\ge 6 \\ X_1 + X_2 &\le 3 \\ X_1, X_2 &\ge 0 \end{aligned}$$

12. Use dual simplex method to solve this LPP.

$$\text{Minimize } Z = X_1 + X_2$$

subject to

$$2X_1 + X_2 \geq 4$$
$$X_1 + 7X_2 \geq 7$$
$$X_1, X_2 \geq 0$$

13. Solve by dual simplex method.

$$\text{Minimize } Z = 10X_1 + 6X_2 + 2X_3$$

subject to

$$-X_1 + X_2 + X_3 \geq 1$$
$$3X_1 + X_2 - X_3 \geq 2$$
$$X_1, X_2, X_3 \geq 0$$

14. Use dual simplex method to solve the following LPP.

$$\text{Minimize } Z = X_1 + 2X_2 + 3X_3$$

subject to

$$X_1 - X_2 + X_3 \geq 4$$
$$X_1 + X_2 + 2X_3 \leq 8$$
$$X_2 - X_3 \geq 2$$
$$X_1, X_2, X_3 \geq 0$$

15. Solve the following LPP by dual simplex method.

$$\text{Minimize } Z = 2X_1 + 9X_2 + 24X_3 + 8X_4 + 5X_5$$

subject to

$$2X_1 + 3X_2 - 5X_3 = 4$$
$$-X_1 + 9X_2 - X_3 \geq 3$$
$$4X_1 + 6X_2 + 3X_3 \leq 8$$
$$X_1, X_2, X_3 \geq 0$$

16. Use dual simplex method to solve the LPP.

$$\text{Minimize } Z = 3X_1 + 2X_2 + X_3 + X_4$$

subject to

$$2X_1 + 4X_2 + 5X_3 + X_4 \geq 10$$
$$3X_1 - X_2 + 7X_3 - 2X_4 \geq 2$$
$$5X_1 + 2X_2 + X_3 + 6X_4 \geq 15$$
$$X_1, X_2, X_3, X_4 \geq 0$$

17. Use dual simplex method to solve the LPP.

$$\text{Maximise } Z = -2X_1 - X_2$$

subject to

$$\begin{aligned} 3X_1 + X_2 &\geq 0 \\ 4X_1 + 3X_2 &\geq 6 \\ X_1 + 2X_2 &\geq 3 \\ X_1, X_2 &\geq 0 \end{aligned}$$

Chapter 5

TRANSPORTATION PROBLEMS

5.1 INTRODUCTION

The transportation problem is a class of allocation methods. Transportation problem arises when there are a number of sources each with a given quantity of products and there are a number of distribution centres to which the products are to be transported. The job of a distribution manager is to evolve an optimum plan which results in minimum cost of transportation.

The information given can be represented in the form of a table. There is one row for each factory and one column for each distribution centre. The number of units available at each factory is written to the right of each row and the number of units required at each distribution centre is written at the bottom of each column. The costs are written in the upper right small boxes.

5.2 FORMULATION OF LPP

	M_1	M_2	M_3	M_4	Supply
F_1	X_{11} [6]	X_{12} [8]	X_{13} [8]	X_{14} [5]	30
F_2	X_{21} [5]	X_{22} [11]	X_{23} [9]	X_{24} [7]	40
F_3	X_{31} [8]	X_{32} [9]	X_{33} [7]	X_{34} [13]	50
Demand	35	28	32	25	

The objective function is,

$$\text{Minimise } Z = 6X_{11} + 8X_{12} + 8X_{13} + 5X_{14} + 5X_{21} + 11X_{22} + 9X_{23} + 7X_{24} + 8X_{31} + 9X_{32} + 7X_{33} + 13X_{34}$$

subject to

$$\begin{aligned} X_{11} + X_{12} + X_{13} + X_{14} &= 30 \\ X_{21} + X_{22} + X_{23} + X_{24} &= 40 \\ X_{31} + X_{32} + X_{33} + X_{34} &= 50 \\ X_{11} + X_{21} + X_{31} &= 35 \\ X_{12} + X_{22} + X_{32} &= 28 \\ X_{13} + X_{23} + X_{33} &= 32 \\ X_{14} + X_{24} + X_{34} &= 25 \end{aligned}$$

In general,

$$\text{Minimise } Z = \sum_{i-1}^{m} \sum_{i-1}^{m} Cij \times Xij$$

we denote rows by '*i*' and columns by '*j*', subject to,

$$\begin{aligned} X_{i1} + X_{i2} + X_{i3} + \cdots + X_{im} &= a_i \\ X_{ij} + X_{2j} + X_{3j} + \cdots + X_{nj} &= b_j \end{aligned}$$

where $i = 1, 2, 3, \ldots, m$ and $j = 1, 2, 3, \ldots, n$.

Here a_i denotes the availability at the *i*th factory and b_j is the demand at the *j*th distribution centre.

In the matrix shown in section 5.2, there are $3 \times 4 = 12$ variables and $3 + 4 = 7$ restrictions. In general, there are $m \times n$ variables and $m + n$ constraints. A transportation problem is therefore a linear programming problem.

5.3 METHODS OF OBTAINING INITIAL BASIC FEASIBLE SOLUTION

A feasible solution to a transportation problem with m sources and n destination centres is said to be basic feasible solution if the number of allocations are $m + n - 1$. There are three methods to obtain initial basic feasible solution. They are

1. North-West Corner Rule (NWCR)
2. Least Cost Method (LCM)
3. Vogel's Approximation Method (VAM)

Example 5.1

Find the initial basic feasible solution for the following problem.

	I	II	III	Supply
A	2	7	4	5
B	3	3	1	8
C	5	4	7	7
D	1	6	2	14
Demand	7	9	18	

Which of the three methods known to you gives better solution?

Solution

North-West Corner Rule (NWCR)

Step 1 Allocate as much as you can to the cell situated at the North-West Corner. Here it is *A*-I. For this cell, *A* is the source and I is the distribution centre. *A* can supply 5 units and demand at distribution centre I is 7 units. Therefore, allocate 5 units to cell *A*-I and delete row *A*.

	I	II	III	Supply
A	2 (5)	7	4	~~5~~
B	3	3	1	8
C	5	4	7	7
D	1	6	2	4
Demand	~~7~~ 2	9	18	

Step 2 Distribution centre-I has already received 5 units from source *A* in the previous step. Its remaining capacity is only 2 units. Therefore, allocate to the cell *B*-I which is situated at North-West corner. Distribution centre-I receives the remaining 2 units from Source *B*. Therefore, delete column-I.

	I	II	III	Supply
B	3	3	1	~~8~~ 6
C	5	4	7	7
D	1	6	2	4
Demand	2	9	18	

Step 3 Allocate to the North-West Corner cell *B*-II. Source *B* has already supplied 2 units to the distribution centre-I. Its remaining supply capacity is 6 units only. Therefore, *B* supplies 6 units to the distribution centre-II. Delete row *B*.

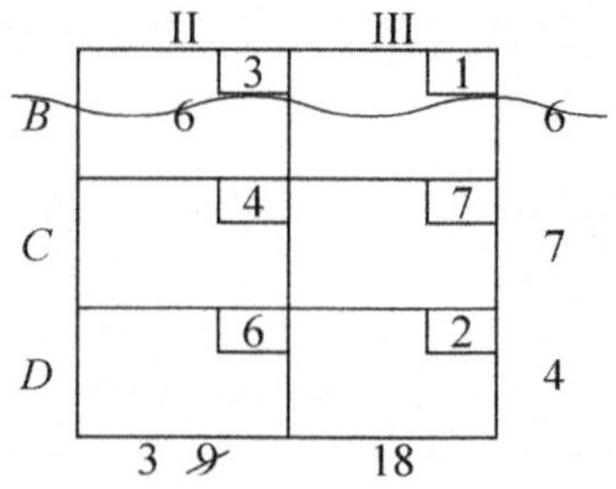

Step 4 Allocate to the North-West Corner cell C-II. Distribution centre II has already received 6 units from B. The capacity remaining is 3 units only. Therefore, transport 3 units from source 'C' to Destination centre-II. Delete Column II.

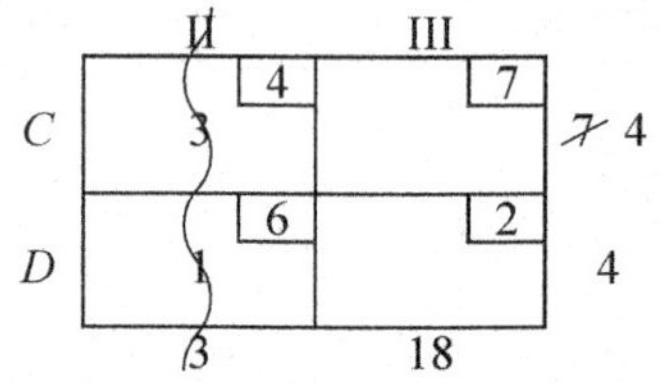

Step 5 Allocate the cell C-III. C already supplied 3 units to the distribution centre-II. Its supply capacity is only 4 units. Therefore, allocate 4 units to the distribution centre-III. Delete row C and allocate the remaining 14 units to the cell D-III.

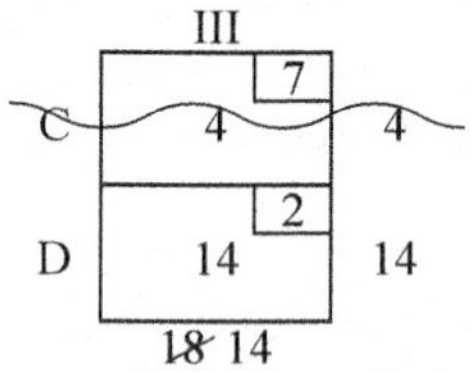

Allocation	Transportation cost
A-I (5 units)	$5 \times 2 =$ Rs. 10
B-I (2 units)	$2 \times 3 =$ Rs. 06
B-II (6 units)	$6 \times 9 =$ Rs. 18
C-II (3 units)	$3 \times 4 =$ Rs. 12
C-III (4 units)	$4 \times 7 =$ Rs. 28
D-III (14 units)	$14 \times 2 =$ Rs. 28
Total transportation cost = Rs. 102	

Least Cost Method (LCM)

Step 1 Look for the lowest transportation cost in the above table.

Allocate as much as you can to that cell. Here, cell B-III and cell D-I have the, least cost of 1. We can allocate 8 units to cell B-III and 7 units to cell D-I. Between these two, cell B-III is better. Therefore, allocate 8 units to cell B-III. Delete row B.

	I	II	III	Supply
A	[2]	[7]	[4]	5
~~B~~	~~[3]~~	~~[3]~~	~~[1] 8~~	~~8~~
C	[5]	[4]	[7]	7
D	[1]	[6]	[2]	14
Demand	7	9	~~18~~ 10	

Step 2 Allocate 7 units to cell *D*-I as it is having minimum transportation cost. Delete column I.

	~~I~~	II	III	Supply
A	~~[2]~~	[7]	[4]	5
C	~~[5]~~	[4]	[7]	7
D	~~[1] 7~~	[6]	[2]	~~14~~ 7
Demand	~~7~~	9	10	

Step 3 Cell *D*-III has minimum transportation cost. Therefore, allocate 7 units to this cell. Delete row *D*.

	II	III	Supply
A	[7]	[4]	5
C	[4]	[7]	7
~~D~~	~~[6]~~	~~[2] 7~~	~~7~~
Demand	9	~~10~~ 3	

Step 4 Allocate 7 units to cell *C*-II and delete row *C*. Allocate 2 units to cell *A*-II and 3 units to cell *A*-III.

	II	III	Supply
A	[7] 2	[4] 3	5
~~C~~	~~[4] 7~~	~~[7]~~	~~7~~
Demand	9	~~10~~ 3	

Allocation	Transportation cost
B-II	8 × 1 = Rs. 8
D-I	7 × 1 = Rs. 7
D-III	7 × 2 = Rs. 14
C-II	7 × 4 = Rs. 28
A-II	2 × 7 = Rs. 14
A-III	3 × 4 = Rs. 12
Total transportation cost = Rs. 83	

Vogel's Approximation Method (VAM)

Step 1 Consider each row. Find the difference between the least and the next least values. Consider each column and find the difference between the least and the next least cost.

Allocate to the row or column for which this difference is the maximum. Here, row A and row B have the cost difference of 2. Therefore, allocation is to be done either for row A or row B. The least cost in row A is 2 and that in row B is 1. Therefore, allocate to the cell B-III. Delete row B.

	I	II	III	Supply	
A	[2]	[7]	[4]	5	2
~~*B*~~	[3]	[3]	[1] 8	~~8~~	2
C	[5]	[4]	[7]	7	1
D	[1]	[6]	[2]	14	1
Demand	7	9	~~18~~ 10		
	1	1	1		

Step 2 Find the difference between the least and next least transportation cost in each row and column.

Here row *A* and column II and III have a value of 2. Minimum in row *A* is 2 column II is 4 and column III is 2. Therefore, allocation is to be done to either *A*-I or *D*-III. For *A*-I, we can allocate 5 units and for *D*-III we can allocate 10 units. Therefore, allocate 10 units to *D*-III. Delete column-III.

	I	II	III	Supply	
A	[2]	[7]	[4]	5	2
C	[5]	[4]	[7]	7	1
D	[1]	[6]	[2] (10)	~~14~~ 4	1
Demand	7	9	~~10~~		
	1	2	2		

Step 3 Row A and row D have a value of difference 5. Minimum cost in row A is 2 and row D is 1. Therefore, allocate to row D. Delete row D.

	I	II	Supply	
A	[2]	[7]	5	5
C	[5]	[4]	7	1
~~D~~	[1] 4	[6]	~~4~~	5
Demand	~~7~~ 3	9		
	1	2		

Step 4 Row A has the highest difference value 5. Minimum transportation cost in row A is 2. Therefore, allocate 3 units to cell A-I, allocate 2 units to cell A-II and 7 units to cell C-II.

	~~I~~	II	Supply	
A	[2] 3	[7] 2	~~5~~ 2	5
C	[5]	[4] 7	7	1
Demand	3	~~9~~ 7		
	~~3~~	3		

Allocation	Transportation cost
B-III	8 × 1 = Rs. 8
D-III	10 × 2 = Rs. 20
D-I	4 × 1 = Rs. 4
A-I	3 × 2 = Rs. 6
A-II	2 × 7 = Rs. 14
C-II	7 × 4 = Rs. 28
Total transportation cost = Rs. 80	

Of the three methods, VAM gives a better solution.

Example 5.2

Obtain an initial basic feasible solution using LCM and VAM.

	I	II	III	IV	
A	5	1	3	3	34
B	3	3	5	4	15
C	6	4	4	3	12
D	4	–1	4	2	19
	21	25	17	17	

Solution

Least Cost Method

	I	II	III	IV	Supply
A	[5]	[1] 6	[3] 11	[3] 17	34 28 11
B	[3] 15	[3]	[5]	[5]	15
C	[6] 6	[4]	[4] 6	[3]	12 6
D	[4]	[−1] 19	[4]	[2]	19
Demand	21 6	25 6	17 6	17	

Step 1 Allocate 19 units to the least cost cell *D*-II. Delete row *D*.

Step 2 Allocate 6 units to the next least cost *A*-II. Delete column II.

Step 3 Allocate 17 units to *A*-IV. Delete column IV.

Step 4 Allocate 15 units to cell *B*-I. Delete row *B*.

Step 5 Allocate 11 units to cell *A*-III. Delete row *A*.

Step 6 Allocate 6 units to cell *C*-III. Delete column-III.

Step 7 Allocate 6 units to cell *C*-I.

Allocation	Transportation cost
D-II	$-1 \times 19 =$ Rs. -19
A-I	$1 \times 6 =$ Rs. 6
A-III	$3 \times 17 =$ Rs. 51
B-II	$3 \times 15 =$ Rs. 45
A-II	$3 \times 11 =$ Rs. 33
C-III	$4 \times 6 =$ Rs. 24
C-I	$6 \times 6 =$ Rs. 36
Total transportation cost = Rs. 176	

Vogel's Approximation Method (VAM)

	I	II	III	IV	Supply
A	[5] 6	[1] 6	[3] 17	[3] 5	1 ~~34~~ ~~28~~ 2 0
B	[3] 15	[3]	[5]	[5]	~~15~~ 0 1
C	[6]	[4]	[4]	[3] 12	12 1 1
D	[4]	[−1] 19	[4]	[2]	~~19~~ 3
Demand	6 ~~21~~	~~25~~	~~17~~	17	
	6	6	6	5	
	1	2	1	1	
	2	2	1	0	
	1		1	0	

Step 1 Allocate 19 units to the cell *D*-II.

Step 2 Allocate 15 units to the cell *B*-I.

Step 3 Allocate 12 units to the cell *C*-IV.

Step 4 Allocate 17 units to the cell *A*-III.

Step 5 Allocate 6 units to the cell *A*-II.

Step 6 Allocate 6 units to the cell *A*-I.

Allocation	Transportation cost
D-II	−1 × 19 = Rs. −19
A-I	3 × 15 = Rs. 45
C-IV	3 × 12 = Rs. 36
A-III	3 × 17 = Rs. 51
A-IV	3 × 5 = Rs. 15
A-II	1 × 6 = Rs. 06
A-I	5 × 6 = Rs. 30
Total transportation cost = Rs. 164	

5.4 OPTIMALITY CHECK

The initial basic feasible solution is obtained by North-West Corner Rule, Least Cost Method or Vogel's Approximation Method. Cells to which allocation is made are called occupied cells or stone cells. Cells which do not have allocation are called empty cells or water cells. The distribution in the initial basic feasible solution is modified to check the optimality of the allocation. Thus the procedure used for solving the problem and to check the optimality of allocation is called Modified Distribution Method or MODI method. This method is also called as stepping stone method as the distribution is modified by putting steps from one cell to the other.

5.5 BALANCED TRANSPORTATION PROBLEM

In a transportation problem, if the total supply is equal to the total demand, such a problem is called a balanced transportation problem.

Example 5.3

There are three parties who supply the following quantities of coal.

Party 1: 14 *T*

Party 2: 12 *T*

Party 3: 5 *T*

There are three consumers who require coal as follows

Consumer *A*: 6 *T*

Consumer *B*: 10 *T*

Consumer *C*: 15 *T*

The cost matrix is as follows

	A	*B*	*C*
1	6	8	4
2	4	9	3
3	1	2	6

Find the schedule of a transportation policy which reduces the cost.

Solution

Initial basic feasible solution by North-West Corner Rule

	A	*B*	*C*	Supply
1	[6] 6	[8] 8	[4]	~~14~~ ~~8~~
2	[4]	[9] 2	[3] 10	~~12~~ 10
3	[1]	[6]	[2] 5	5
Demand	~~6~~	~~10~~ 2	15	31 / 31

The total supply is equal to the total demand. Therefore, the problem is a balanced transportation problem.

Allocation	Transportation cost
1-*A*	6 × 6 = Rs. 36
1-*B*	8 × 8 = Rs. 64
2-*B*	9 × 2 = Rs. 18
2-*C*	3 × 10 = Rs. 30
3-*C*	6 × 5 = Rs. 30
Total transportation cost = Rs. 178	

Check for optimality by MODI method

Step 1 Find U_i, V_j and $C_{ij} = U_i + V_j$ from the initial basic feasible solution.

	A	*B*	*C*	
1	cost 6, (6)	cost 8, (8)	cost 4	$U_1 = 0$
2	cost 4, 7	cost 9, (2)	cost 3, (10)	$U_2 = 1$
3	cost 1, 10	cost 2, 12	cost 6, (5)	$U_3 = 4$
	$V_A = 6$	$V_B = 8$	$V_C = 2$	

Initially U_i is assumed to be zero. Considering the allocated cells, find U_i and V_j.

Considering cell 1-*A*,

$$C_{1A} = U_1 + V_A$$
$$6 = 0 + V_A$$
$$\therefore V_A = 6$$

For cell 1-*B*,

$$C_{1B} = U_1 + V_B$$
$$8 = + V_B$$
$$\therefore V_B = 8$$

For cell 2-*B*,

$$C_{2B} = U_2 + V_B$$
$$9 = U_2 + 8$$
$$\therefore U_2 = 1$$

For cell 2-*C*,

$$C_{2C} = U_2 + V_C$$
$$3 = 1 + V_C$$
$$\therefore V_C = 2$$

For cell 3-*C*,

$$C_{3C} = U_3 + V_C$$
$$6 = U_3 + 2$$
$$U_3 = 6 - 2$$
$$\therefore U_3 = 4$$

Considering the empty cell, determine $(U_i + V_j)$ and put in the lower left box.

For empty cell 1-C, $U_1 + V_C = 0 + 2 = 2.$

For cell 2-A, $U_2 + V_A = 1 + 6 = 7.$

For cell 3-A, $U_3 + V_A = 4 + 6 = 10.$

For cell 3-B, $U_3 + V_B = 4 + 8 = 12.$

Step 2 Evaluate each of the empty cell for $C_{ij} - (U_i + V_j)$.

For cell 1-C, $C_{1C} - (U_1 + V_C) = 4 - 2 = 2.$

For cell 2-A, $C_{2A} - (U_2 + V_A) = 4 - 7 = -3.$

For cell 3-A, $C_{3A} - (U_3 + V_A) = 1 - 10 = -9.$

For cell 3-B, $C_{3B} - (U_3 + V_B) = 2 - 12 = -10.$

If all the values of $C_{ij} - (U_i + V_j)$ are positive or zero the solution is optimum. Otherwise proceed further.

Step 3 The cell with the most negative value is the one which enters the basic feasible solution. Determine the variable which should leave.

Find the path for modifying the distribution with cell 3-A as the starting point. Right angle turns are allowed at only stone cells and not at water cells. Put + sign at cell 3-B, alternatively, put – and + for subsequent cells in order.

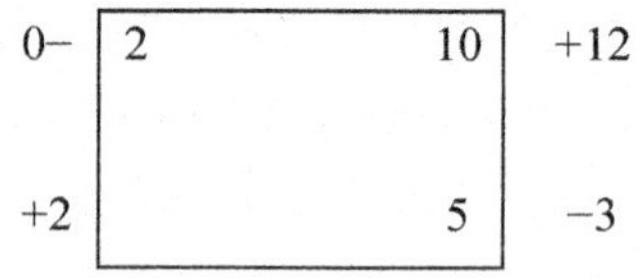

Path for modified distribution

Consider the minimum value at the negative sign. Here it is 2 therefore, move 2 units. If it is positive add 2 units and if it is negative deduct 2 units.

Write the modified distribution

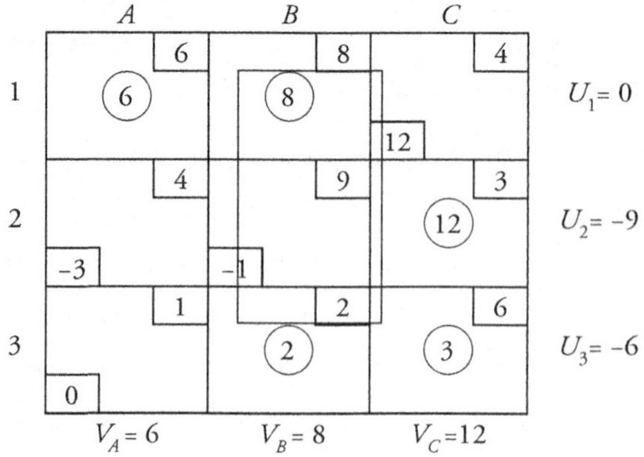

Find the values of U_i and V_j and evaluate empty cells for $C_{ij} - (U_i + V_j)$.

For cell 1-*C*, $C_{1C} - (U_1 + V_C) = 4 - 12 = -8.$

For cell 2-*A*, $C_{2A} - (U_2 + V_A) = 4 + 3 = 7.$

For cell 2-*B*, $C_{2B} - (U_2 + V_B) = 9 + 1 = 10.$

For cell 3-*A*, $C_{3A} - (U_3 + V_A) = 1 - 0 = 1.$

Therefore, solution is not optimum and hence cell 1-*C* enters the basic.

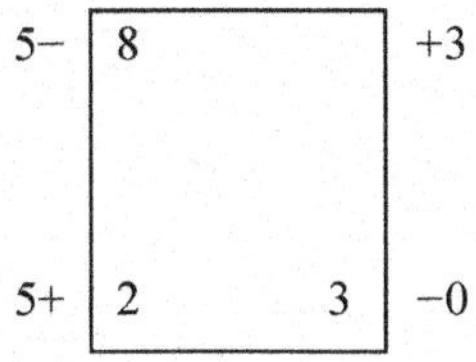

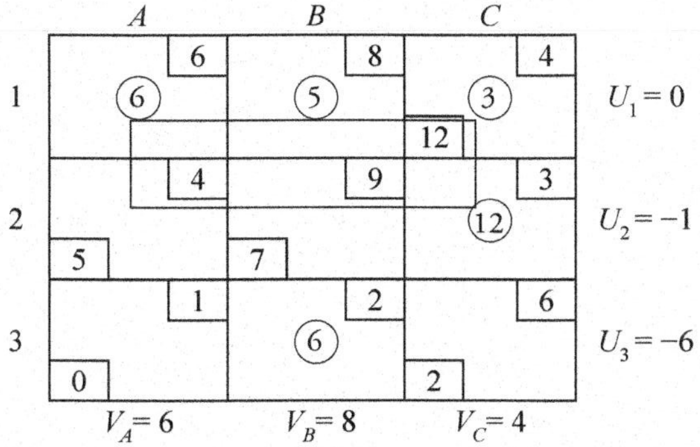

Find U_i and V_j and evaluate empty cells for $C_j - (U_i + V_j)$.

For cell 2-*A*, $C_{2A} - (U_2 + V_A) = 4 - 5 = -1.$

For cell 2-*B*, $C_{2B} - (U_2 + V_B) = 9 - 7 = 2.$

For cell 3-*A*, $C_{3A} - (U_3 + V_A) = 1 - 0 = 1.$

For cell 3-*C*, $C_{3C} - (U_3 + V_C) = 6 - 2 = 4.$

Hence the solution is not optimum. Cell 2-*A* enters the basic.

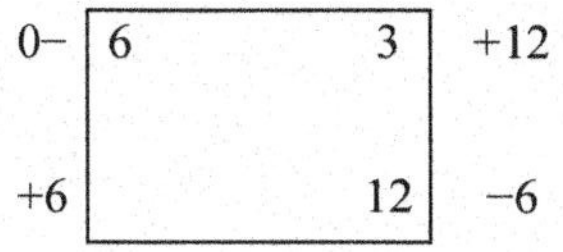

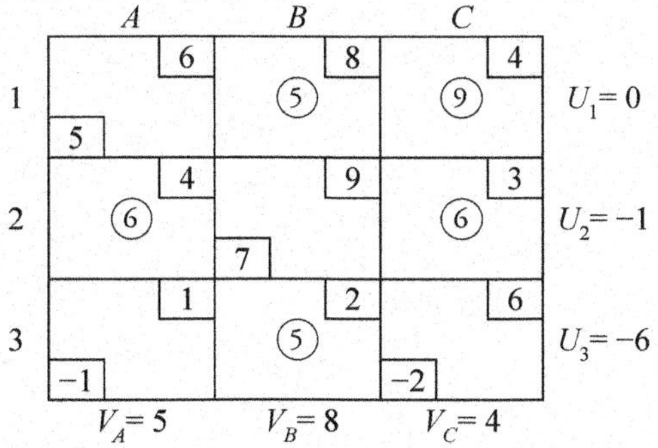

For cell 2-A, $C_{2A} - (U_2 + V_A) = 6 - 5 = -1.$

For cell 2-B, $C_{2B} - (U_2 + V_B) = 9 - 7 = 2.$

For cell 3-A, $C_{3A} - (U_3 + V_A) = 1 + 1 = 2.$

For cell 3-C, $C_{3C} - (U_3 + V_C) = 6 + 2 = 8.$

All the values are positive. Therefore, solution is optimum.

Optimum allocation	Transportation cost
1-*B*	5 units × 8 Rs./unit = Rs. 40
1-*C*	9 units × 4 Rs./unit = Rs. 36
2-*A*	6 units × 4 Rs./unit = Rs. 24
2-*C*	6 units × 3 Rs./unit = Rs. 18
3-*B*	5 units × 2 Rs./unit = Rs. 10
	Total transportation cost = Rs. 128

Example 5.4

Solve the following transportation problem.

	A	*B*	*C*	Available
F_1	4	8	8	56
F_2	16	24	16	82
F_3	8	16	24	77
Required	72	16	24	

Solution

Initial basic feasible solution by VAM

	A	*B*	*C*	Available
F_1	[4]	[8] 56	[8]	56 40
F_2	[16]	[24] 41	[16] 41	41 82 08
F_3	[8] 72	[24] 5	[16] 6	77 88 5
Required	72	102	41	
		46		
	4	8	8	

Allocation	Transportation cost
F_1-B	$8 \times 56 =$ Rs. 448
F_2-B	$24 \times 41 =$ Rs. 984
F_2-C	$16 \times 41 =$ Rs. 656
F_3-A	$8 \times 72 =$ Rs. 576
F_3-B	$16 \times 5 =$ Rs. 80
Total transportation cost = Rs. 2744	

Check for optimality by MODI method

	A	B	C	
F_1	4 / 0	8 / (56)	8 / 0	$U_1 = 0$
F_2	16 / 16	24 / (41)	16 / (41)	$U_2 = 16$
F_3	1 / (72)	6 / (5)	2 / 8	$U_3 = 8$
	$V_A = 6$	$V_B = 8$	$V_C = 4$	

Find U_i and V_j. Evaluate $C_{ij} - (U_i + V_j)$ for empty cells.

For cell F_1-A, $C_{F1A} - (U_{F1} + V_C) = 4 - 0 = 4.$

For cell F_1-C, $C_{F1C} - (U_{F1} + V_C) = 8 - 0 = 8.$

For cell F_2-A, $C_{F2A} - (U_{F2} + V_A) = 16 - 16 = 0.$

For cell F_3-C, $C_{F3C} - (U_{F3} + V_C) = 24 - 8 = 16.$

All the values are positive, therefore, solution is optimum.

Optimum allocation	Transportation cost
F_1-B	$8 \times 56 =$ Rs. 448
F_2-B	$24 \times 41 =$ Rs. 984
F_2-C	$16 \times 41 =$ Rs. 656
F_3-A	$8 \times 72 =$ Rs. 576
F_3-B	$16 \times 5 =$ Rs. 80
Total transportation cost = Rs. 2744	

Example 5.5

A company has factories F_1, F_2 and F_3 which supply warehouses at W_1, W_2 and W_3. Weekly factory capacities are 200, 160 and 90 units respectively. Weekly warehouse requirements are 180, 120 and 150 respectively. Unit shipping cost in rupees is as follows

	W_1	W_2	W_3
F_1	16	20	12
F_2	14	8	16
F_3	26	24	16

Determine the optimal distribution for this company to minimise the shipping cost.

Solution

Initial basic feasible solution by LCM

	W_1	W_2	W_3	
F_1	[16] 50	[20]	[12] 150	50 ~~200~~
F_2	[14] 40	[8] 120	[16]	~~160~~ 40
F_3	[26] 90	[24]	[16]	90
	180	120	150	450 / 450

Allocate to the cell having minimum transportation cost per unit.

Allocation	Transportation cost
F_1-W_1	16 × 50 = Rs. 800
F_1-W_3	12 × 150 = Rs. 1800
F_2-W_1	14 × 40 = Rs. 560
F_2-W_2	8 × 120 = Rs. 960
F_3-W_1	26 × 90 = Rs. 2340
Total transportation cost = Rs. 6460	

Check for optimality by MOD1 Method

	W_1	W_2	W_3	
F_1	16 (50)	20 [10]	12 (150)	$UF_1 = 0$
F_2	14 (40)	8 (120)	16 [10]	$UF_2 = -2$
F_3	26 (90)	24 [20]	16 [22]	$UF_3 = 10$
	$V_{W_1} = 16$	$V_{W_2} = 10$	$V_{W_3} = 12$	

Find U_i and V_j. Evaluate empty cells for $C_{ij} - (U_i + V_j)$.

For cell F_1-W_2, $C_{12} - (U_1 + V_2) = 20 - 10 = 10.$

For cell F_2-W_3, $C_{23} - (U_2 + V_3) = 16 - 10 = 6.$

For cell F_3-W_2, $C_{32} - (U_3 + V_2) = 24 - 20 = 4.$

For cell F_3-W_3, $C_{33} - (U_3 + V_3) = 16 - 22 = -6.$

Therefore, the solution is not optimum.

F_3-W_3 having $C_{ij} - (U_i + V_j)$, -6 enters the basic.

140+	50	150	−60
0−	90		+90

	W_1	W_2	W_3	
F_1	16 (140)	20 [10]	12 (60)	$U_1 = 0$
F_2	14 (40)	8 (120)	16 [10]	$U_2 = -2$
F_3	26 [20]	24 [14]	16 (90)	$U_3 = 4$
	$V_1 = 16$	$V_2 = 10$	$V_3 = 12$	

Find U_i and V_j. Evaluate empty cells for $C_{ij} - (U_i + V_j)$.

For cell F_1-W_2, $C_{12} = (U_1 + V_2) = 20 - 10 = 10.$

For cell F_2-W_3, $C_{23} = (U_2 + V_3) = 16 - 10 = 6.$

For cell F_3-W_1, $C_{31} = (U_3 + V_1) = 26 - 20 = 6.$

For cell F_3-W_2, $C_{32} = (U_3 + V_2) = 24 - 14 = 10.$

Hence the solution is optimum.

Optimum allocation	Transportation cost
F_1-W_1	16 × 140 = Rs. 2240
F_1-W_3	12 × 60 = Rs. 720
F_2-W_1	14 × 40 = Rs. 560
F_2-W_2	8 × 120 = Rs. 960
F_3-W_3	16 × 90 = Rs. 1440
Total transportation cost = Rs. 5920	

Example 5.6

A commodity has to be supplied to 3 warehouses *A*, *B*, and *C* whose requirements are 70, 100 and 40 tons respectively. It is available at 3 places *X*, *Y*, and *Z* in quantities of 55, 80 and 75 tons respectively. Transportation cost between different places are shown below.

	A	*B*	*C*
X	5	10	10
Y	20	30	20
Z	10	20	30

Find the least-cost transportation schedule.

Solution

Initial basic feasible solution by LCM

	A	*B*	*C*	Supply
X	[5] 55	[10]	[10]	55
Y	[20] 40	[30] 120	[20]	80
Z	[10] 15	[20] 60	[30]	75 → 60
Demand	70 → 15	100 → 40	40	210 / 210

It is a balanced transportation problem.

Give allocation to the cells with least cost.

Allocation	Transportation cost
X-A	5 × 55 = Rs. 275
Y-B	30 × 40 = Rs. 1200
Y-C	20 × 40 = Rs. 800
Z-A	10 × 15 = Rs. 150
Z-B	20 × 60 = Rs. 1200
Total Transport cost = Rs. 3625	

Check for optimality by MODI method

	A	B	C	
X	5 (55)	10 / 15	10 / 5	$U_x = 0$
Y	20 / 20	30 (40)	20 (40)	$U_y = 15$
Z	20 (15)	20 (60)	30 / 10	$U_z = 5$
	$V_A = 5$	$V_B = 15$	$V_C = 5$	

Find U_i and V_j. Evaluate empty cells for $C_{ij} - (U_i + V_j)$.

For cell *X-B*, $C_{XB} - (U_X + V_B) = 10 - 15 = -5.$

For cell *X-C*, $C_{XC} - (U_X + V_C) = 10 - 5 = 5.$

For cell *Y-A*, $C_{YA} - (U_Y + V_A) = 20 - 20 = 0.$

For cell *Z-C*, $C_{ZC} - (U_Z + V_C) = 30 - 10 = 20.$

Therefore, solution is not optimum. Cell *X-B*, with negative value enters the basic.

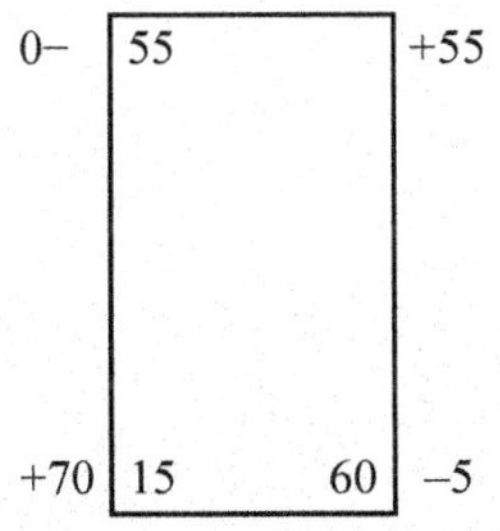

	A	B	C	
X	5 / 0	10 (55)	10 / 0	$U_X = 0$
Y	20 / 20	30 (40)	20 (40)	$U_y = 20$
Z	10 (70)	20 (5)	30 / 10	$U_z = 10$
	$V_A = 0$	$V_B = 10$	$V_C = 0$	

Find U_i and V_j. Evaluate empty cells for $C_{ij} - (U_i + V_j)$.

For cell X_A, $C_{XA} - (U_X + V_A) = 5 - 0 = 5.$

For cell X_C, $C_{XC} - (U_X + V_C) = 10 - 0 = 10.$

For cell Y_A, $C_{YA} - (U_Y + V_A) = 20 - 20 = 0.$

For cell Z_C, $C_{ZC} - (U_Z + V_C) = 30 - 10 = 20.$

Hence the solution is optimum.

Optimum allocation	Transportation cost
X-B	10 × 55 = Rs. 550
Y-B	30 × 40 = Rs. 1200
Y-C	20 × 40 = Rs. 800
Z-A	10 × 70 = Rs. 700
Z-B	20 × 5 = Rs. 100
Total transportation cost = Rs. 3350	

Example 5.7

Solve the following transportation problem.

	Store		a_i	
Factories	*A*	*B*	*C*	
F_1	10	8	8	8
F_2	10	7	10	7
F_3	11	9	7	9
F_4	12	14	10	4
	10	10	8	

Solution

Initial basic feasible solution by VAM

	A	*B*	*C*	Supply
F_1	[10] 5	[18] 3	[8]	5 ~~8~~ 0 2
~~F_2~~	[10]	[7] ~~7~~	[10]	~~7 3~~
F_3	[20] 1	[30]	[20] 8	1 ~~9~~ 2 2
F_4	[12] 4	[14]	[10]	4 2 2
Demand	10 0 1	~~10~~ 1 ~~3~~ 1	8 1 1	28 / 28

In this transportation problem, the total supply is equal to the total demand. Therefore, it is a balanced transportation problem.

Find the difference between the least and next least transportation cost in each row and each column. Consider the row or column having the highest difference value. Allocate to the cell with minimum transportation cost in that row or column.

Optimum allocation	Transportation cost
F_1-A	10 × 5 = Rs. 50
F_1-B	8 × 3 = Rs. 24
F_2-B	7 × 7 = Rs. 49
F_3-A	11 × 1 = Rs. 11
F_3-C	7 × 8 = Rs. 56
F_4-A	12 × 4 = Rs. 48
Total transportation cost = Rs. 238	

Check for optimality by MODI method

	A	B	C	
F_1	10 (5)	18 (3)	8, 6	$U_1 = 0$
F_2	10, 9	7 (7)	10, 5	$U_2 = -1$
F_3	11 (1)	9, 9	7 (8)	$U_3 = 1$
F_4	12 (4)	14, 9	10, 8	$U_4 = 2$
	$V_A = 10$	$V_B = 8$	$V_C = 6$	

Find U_i and V_j. Evaluate empty cells for $C_{ij} - (U_i + V_j)$.

For cell F_1-C, $C_{1C} - (U_i + V_C) = 8 - 6 = 2.$

For cell F_2-A, $C_{2A} - (U_2 + V_A) = 10 - 9 = 1.$

For cell F_2-C, $C_{2C} - (U_2 + V_C) = 10 - 5 = 5.$

For cell F_3-B, $C_{3B} - (U_3 + V_B) = 9 - 9 = 0.$

For cell F_4-B, $C_{4B} - (U_4 + V_B) = 14 - 9 = 5.$

For cell F_4-C, $C_{4C} - (U_4 + V_C) = 10 - 8 = 2.$

Hence the solution is optimum.

Optimum allocation	Transportation cost
F_1-A	10 × 5 = Rs. 50
F_1-B	8 × 3 = Rs. 24
F_2-B	7 × 7 = Rs. 49
F_3-A	11 × 1 = Rs. 11
F_3-C	7 × 8 = Rs. 56
F_4-A	12 × 4 = Rs. 48
Total transportation cost = Rs. 238	

Example 5.8

Power stations P_j, $j = 1, 2, 3, 4$ run on coal found in coal mines Q_i, $i = 1, 2, 3$. Quantity of coal produced at Q_i is a_i. Maximum possible production of power at P_j is b_j. The cost of unit quantity of coal at Q_i is C_i and its transportation cost to P_i is C_{ij}. A unit quantity of coal produces h_j units of power at P_j. How should coal be distributed so that the total cost of coal at power station is minimum. How much of power is produced at each station? Solve using the following data.

i	C_{i1}	C_{i2}	C_{i3}	C_{i4}	C_i	a_i
1	4	3	2	1	10	750
2	3	5	6	2	15	350
3	0	4	3	3	20	400
h_j	$\frac{1}{2}$	$\frac{1}{2}$	$\frac{1}{3}$	$\frac{1}{4}$	–	–
b_j	100	150	300	200	–	–
j	1	2	3	4	–	–

Solution

Formulation

Coal mines \ Power stations	1	2	3	4	Coal supply
1	2 [14]	[13]	[12]	[11]	750
2	[18]	[20]	[21]	[17]	350
3	[20]	[24]	[23]	[23]	400
Coal demand	200	300	900	800	1500 / 2200

C_{ij} value in the upper right boxes represent the total cost of coal at the power stations 1, 2, 3 and 4. In this problem, the total supply is not equal to the total demand. Therefore, it is not a balanced transportation problem. There is a shortage in supply. Therefore, add one row with zero transportation cost.

Initial basic feasible solution by VAM

Coal mines	Power supply 1	2	3	4	Coal supply			
F_1	[14] 2	[13]	[12] 700	[11] 50	750	1	1	1
F_2	[18]	[20]	[21]	[17] 350	350	1	3	4
F_3	[20]	[24]	[23]	[23] 400	400	3	0	0
F_4	[0] 200	[0] 300	[0] 200	[0]	200 ~~500~~ ~~400~~	0	0	0
Coal demand	200	300	~~900~~ 700	800	2200 / 2200			
	14	13	12 9	11 6				

Find the difference between the least and next least transportation cost. Consider the row or column having the highest difference. Allocate as much as you can to the cell with least transportation cost.

Allocation	Transportation cost
1-3	12 × 700 = Rs. 8400
2-4	17 × 350 = Rs. 5950
3-4	23 × 400 = Rs. 9200
4-1	0 × 200 = –
4-2	0 × 300 = –
4-3	0 × 200 = –
	Total transportation cost = Rs. 23,550

Check for optimality

Find U_i and V_j using $C_{ij} = U_i + V_j$ w.r.t. allocated cells.

Coal mines	Power stations 1	2	3	4	
F_1	[14]	[13]	[12] (700)	[11] (50)	$U_1 = 0$
F_2	[18] 28	[20] 28	[21] 28	[17] (350)	$U_2 = 16$
F_3	[20] 34	[24] 34	[23] 34	[23] (400)	$U_3 = 22$
F_4	[0] (200)	[0] (300)	[0] (200)	[0] −11	$U_4 = -12$
	$V_1 = 12$	$V_2 = 12$	$V_3 = 12$	$V_4 = 1$	

Evaluate empty cells for $C_{ij} - (U_i + V_j)$.

For cell 1-1, $C_{ij} - (U_i + V_j) = 14 - 12 = 2.$

For cell 1-2, $C_{12} - (U_1 + V_2) = 13 - 12 = 1.$

For cell 2-1,	$C_{21}-(U_2+V_1)$	=	$18-28=-10.$
For cell 2-2,	$C_{22}-(U_2+V_2)$	=	$20-28=-8.$
For cell 2-3,	$C_{23}-(U_2+V_3)$	=	$21-28=-7.$
For cell 3-1,	$C_{31}-(U_3+V_1)$	=	$20-34=-14.$
For cell 3-2,	$C_{32}-(U_3+V_2)$	=	$24-34=-10.$
For cell 3-3,	$C_{33}-(U_3+V_3)$	=	$23-34=-11.$
For cell 4-4,	$C_{44}-(U_4+V_4)$	=	$0+11=11.$

If $C_{ij}-(U_i+V_j)$ is zero or positive, the solution is optimum otherwise proceed further. The cell with most negative value is the one which enters the basic feasible solution. Here it is cell 3-1. Determine the variable which should leave.

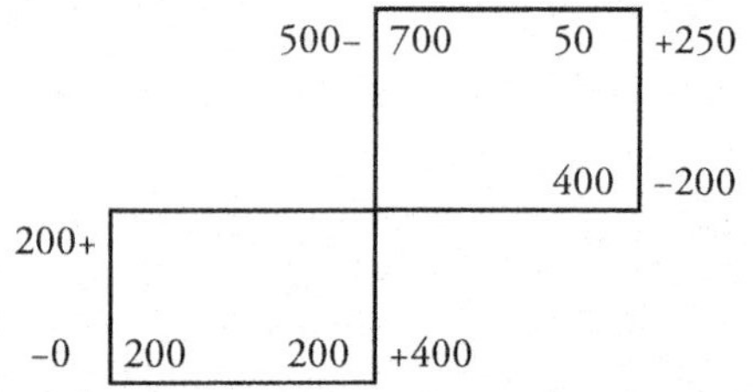

Starting with cell 3-1 draw the path-modified distribution by taking right-angle turns at stone cells. Consider the minimum value at the negative sign. Here, it is 200. Therefore, move 200 units. If it is positive, add 200 units. If it is negative, deduct 200 units.

	1	2	3	4	
1	14 / 8	13 / 12	12 / (500)	11 / (250)	$U_1 = 0$
2	18 / 14	20 / 18	21 / 18	17 / (350)	$U_2 = 16$
3	20 / (200) / 34	24 / 24	23 / 24	23 / (200)	$U_3 = 12$
4	0 / -4	0 / (300)	0 / (200)	0 / -1	$U_4 = -12$
	$V_1 = 8$	$V_2 = 12$	$V_3 = 12$	$V_4 = 11$	

Find out U_i and V_j using $C_{ij} = U_i + V_j$ w.r.t. allocated cells.

Evaluate empty cells for $C_{ij}-(U_i+V_j)$.

For cell 1-1,	$C_{11}-(U_1+V_1)$	=	$14-8=6.$
For cell 1-2,	$C_{12}-(U_1+V_2)$	=	$13-12=1.$
For cell 2-1,	$C_{21}-(U_2+V_1)$	=	$18-14=4.$
For cell 2-2,	$C_{22}-(U_2+V_2)$	=	$20-18=2.$

For cell 2-3,	$C_{23} - (U_2 + V_3)$	=	$21 - 18 = 3.$
For cell 3-2,	$C_{32} - (U_3 + V_2)$	=	$24 - 24 = 0.$
For cell 3-3,	$C_{33} - (U_3 + V_3)$	=	$23 - 24 = -1.$
For cell 4-1,	$C_{41} - (U_4 + V_3)$	=	$0 + 4 = 4.$
For cell 4-4,	$C_{44} - (U_4 + V_4)$	=	$0 + 1 = 1.$

The solution is not optimum. Cell 3-3 enters the basic solution. Starting with this cell 3-3, draw the path for modifying the distribution by taking right-angle turn at stone cells.

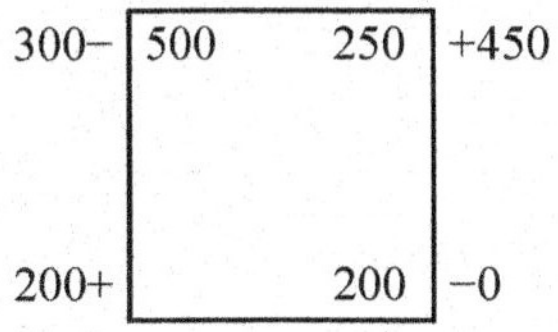

Consider the minimum value at the negative sign. Here it is 200. Therefore, move 200 units. If it is positive, add 200 units. If it is negative, deduct 200 units.

	1	2	3	4	
1	14; 9	13; 12	12; (300)	11; (450)	$U_1 = 0$
2	18; 15	20; 18	21; 18	17; (350)	$U_2 = 6$
3	20; (200)	24; 23	23; (200)	23; 22	$U_3 = 11$
4	0; -3	0; (300)	0; (400)	0; -1	$U_4 = -12$
	$V_1 = 9$	$V_2 = 12$	$V_3 = 12$	$V_4 = 11$	

Find out U_i and V_j using $C_{ij} = U_i + V_j$ w.r.t. allocated cells.
Evaluate empty cells for $C_{ij} - (U_i + V_j)$.

For cell 1-1,	$C_{11} - (U_1 + V_1)$	=	$14 - 9 = 5.$
For cell 1-2,	$C_{12} - (U_1 + V_2)$	=	$13 - 12 = 1.$
For cell 2-1,	$C_{21} - (U_2 + V_1)$	=	$18 - 15 = 3.$
For cell 2-2,	$C_{22} - (U_2 + V_2)$	=	$20 - 18 = 2.$
For cell 2-3,	$C_{23} - (U_2 + V_3)$	=	$21 - 18 = 3.$
For cell 3-2,	$C_{32} - (U_3 + V_2)$	=	$24 - 23 = 1.$
For cell 3-4,	$C_{34} - (U_3 + V_4)$	=	$23 - 22 = 1.$
For cell 4-1,	$C_{41} - (U_4 + V_1)$	=	$0 + 3 = 3.$
For cell 4-4,	$C_{44} - (U_4 + V_4)$	=	$0 + 1 = 1.$

The solution is optimum.

Allocation	Coal cost
*M*1 to *P*3	12 × 300 = Rs. 3600
*M*1 to *P*4	11 × 450 = Rs. 4950
*M*2 to *P*4	17 × 350 = Rs. 5950
*M*3 to *P*1	20 × 200 = Rs. 4000
*M*3 to *P*3	23 × 200 = Rs. 4600
*M*4 to *P*2	0 × 300 = 0
*M*4 to *P*3	0 × 400 = 0
	Total coal cost = Rs. 23,100

Power produced at each station

Allocation	Quantity of coal used units	Power generated per unit quantity of coal	Total power generated units
M_3-P_1	200	$\frac{1}{2}$ units	100
M_4-P_2	300	–	–
M_4-P_3	400	–	–
M_1-P_3	300	$\frac{1}{3}$ units	100
M_3-P_3	200	$\frac{1}{3}$ units	66.66
M_1-P_4	450	$\frac{1}{4}$ units	112.5
M_2-P_4	350	$\frac{1}{4}$ units	87.5

Power station 1 generates 100 units of power.

Power station 2 generates no power.

Power station 3 generates 166.66 units of power.

Power station 4 generates 200 units of power.

Example 5.9

A company has factories *A, B* and *C*. Monthly factory capacities are 250, 300 and 400 units respectively for regular production. If overtime production is permitted, the factories *A* and *B* can produce 50 and 75 additional units repectively. The overtime incremental costs are Rs. 4 and Rs. 5 at *A* and *B* respectively. The current warehouse requirements are 200, 225, 275 and 300 units respectively. Unit transportation costs in rupees from factories to warehouses are as follows. Determine the optimal distribution to minimise the cost.

From \ To		D	E	F	G
Factories	A	11	13	17	14
	B	16	18	14	10
	C	21	24	13	10

Solution

Formulation of transportation problem.

		Warehouses D	E	F	G	Supply
Factories	A	[11]	[13]	[17]	[14]	250
	B	[16]	[18]	[14]	[10]	300
	C	1 [21]	[24]	8 [13]	8 [10]	400
	A_{OT}	4 [15]	[17]	[19]	[15]	50
	B_{OT}	4 [21]	[23]	[19]	[15]	75
Demand		200	225	275	300	1075 / 1000

This is an unbalanced transportation problem. Supply is more than demand. Therefore, add a column with zero transportation cost.

	D	E	F	G	H	Supply	
A	200 [11]	50 [13]	[17]	[14]	[0]	~~250~~ 50	11 2 2 2
~~B~~	[16]	[18]	[14]	300 [10]	[0]	~~300~~	10 4
C	[21]	125 [24]	275 [13]	[10]	[0]	~~400~~ 125	10 3 8 3
A_{OT}	[15]	50 [17]	[19]	[15]	[0]	50	15 2 2 2
~~B_{OT}~~	[21]	[23]	[19]	[15]	75 [0]	~~75~~	15
Demand	200	225	275	300	75	1075 / 1075	
	4	4	1	0	0		

Initial basic feasible solution by VAM

Find the difference between the least and the next least in each row and column. Allocate to the cell with least value in the row or column that has maximum difference value.

Allocation	Cost
A to D	11 × 200 = Rs. 2200
A to E	13 × 30 = Rs. 650
A to E (Overtime)	17 × 50 = Rs. 850
B to G	10 × 300 = Rs. 3000
C to E	24 × 125 = Rs. 3000
C to F	13 × 275 = Rs. 3575
B to H (Overtime)	–
	Total cost = Rs. 13,275

Check for optimality

Find U_i and V_j using $C_{ij} = U_i + V_j$ w.r.t. allocated cells.

	D	E	F	G	H	
A	11 (200)	13 (50)	17; 2	14; 6	0; –4	$U_A = 4$
B	16; 15	18; 17	14; 4	10 (300)	0 (ε)	$U_B = 4\ -1$
C	21; 22	24 (125)	13 (275)	10; 7	0; 7	$U_C = 11$
A_{OT}	15; 15	17 (50)	21; 6	18; 10	0 (ε)	$U_{AOT} = 4$
B_{OT}	21; 15	23; 17	19; 6	15; 10	0 (75)	$U_{BOT} = 4\ 0$
	$V_D = 11$	$V_E = 13$	$V_F = 2$	$V_G = 6\ 11$	$V_H = -4\ 0$	

Evaluate empty cells for $C_{ij} - (U_i + V_j)$.

For cell A-F, $C_{AF} - (U_A + V_F) = 17 - 2 = 15.$

For cell A-G, $C_{AG} - (U_A + V_G) = 14 - 6 = 8.$

For cell A-H, $C_{AH} - (U_A + V_A) = 0 + 4 = 4.$

For cell B-D, $C_{BD} - (U_B + V_D) = 16 - 15 = 1.$

For cell B-E, $C_{BE} - (U_B + V_E) = 18 - 17 = 1.$

For cell B-F, $C_{BF} - (U_B + V_F) = 14 - 6 = 8.$

For cell C-D, $C_{CD} - (U_C + V_D) = 21 - 22 = -1.$

For cell C-G, $C_{CG} - (U_C + V_D) = 10 - 17 = -7.$

For cell C-H, $C_{CH} - (U_C + V_H) = 0 - 7 = -7.$

For cell A_{OT}-D,	$C_{AOTD} - (U_{AOT} + V_D)$	=	$15 - 15 = 0$.
For cell A_{OT}-F,	$C_{AOT\,F} - (U_{AOT} + V_F)$	=	$21 - 6 = 15$.
For cell A_{OT}-G,	$C_{AOT\,G} - (U_{AOT} + V_G)$	=	$18 - 10 = 8$.
For cell B_{OT}-D,	$C_{BOT\,D} - (U_{BOT} + V_D)$	=	$21 - 15 = 6$.
For cell B_{OT}-E,	$C_{BOT\,E} - (U_{BOT} + V_E)$	=	$23 - 17 = 6$.
For cell B_{OT}-F,	$C_{BOT\,F} - (U_{BOT} + V_F)$	=	$19 - 6 = 13$.
For cell B_{OT}-G,	$C_{BOT\,G} - (U_{BOT} + V_G)$	=	$15 - 10 = 5$.

The solution is not optimum. The cell with highest negative value enters the basic. Write the path for modifying the distribution.

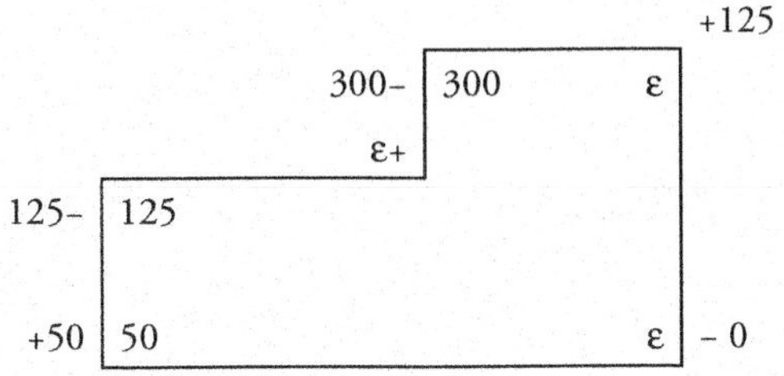

Write modified distribution and find U_i and V_j using $C_{ij} = (U_i + V_j)$ w.r.t. allocated cells.

	D	E	F	G	H	
A	11, (200)	13, (50)	17, 2	14, −1	0, −11	$U_A = 0$
B	16, 22	18	14, 13	10, (300)	0, (ε)	$U_B = 5$
C	21, 22	24, (125)	13, (275)	10, (ε)	0, 0	$U_C = 5$
A_{OT}	15, 15	17, (50)	21, 6	18, 3	0, −1	$U_{A_{OT}} = 4$
B_{OT}	21, 22	23, 24	19, 13	15, 10	0, (75)	$U_{B_{OT}} = 5$
	$V_D = 11$	$V_E = 13$	$V_F = 2$	$V_G = -1$	$V_H = -11$	

Evaluate empty cells for $C_{ij} - U_i + V_j$.

For cell A-F,	$C_{AF} - (U_A + V_F)$	=	$17 - 2 = 15$.
For cell A-G,	$C_{AG} - (U_A + V_F)$	=	$14 + 1 = 15$.
For cell A-H,	$C_{AH} - (U_A + V_H)$	=	$0 + 11 = 11$.

For cell *B-D*, $C_{BD} - (U_B + V_D) = 16 - 22 = -6$

For cell *B-E*, $C_{BE} - (U_B + V_E) = 18 - 24 = -6$

For cell *B-F*, $C_{BF} - (U_B + V_F) = 14 - 13 = 1$

For cell *C-D*, $C_{CD} - (U_C + V_D) = 21 - 22 = -1$

For cell *C-H*, $C_{CH} - (U_C + V_H) = 0 - 0 = 0$

For cell A_{OT}-*D*, $C_{AOTD} - (U_{AOT} + V_D) = 15 - 15 = 0$

For cell A_{OT}-*F*, $C_{AOTF} - (U_{AOT} + V_F) = 21 - 6 = 15$

For cell A_{OT}-*G*, $C_{AOTG} - (U_{AOT} + V_G) = 18 - 3 = 15$

For cell A_{OT}-*H*, $C_{AOTH} - (U_{AOT} + V_H) = 0 + 7 = 7$

For cell B_{OT}-*D*, $C_{BOTD} - (U_{BOT} + V_D) = 21 - 22 = -1$

For cell B_{OT}-*E*, $C_{BOTE} - (U_{BOT} + V_E) = 23 - 24 = -1$

For cell B_{OT}-*F*, $C_{BOTF} - (U_{BOT} + V_F) = 19 - 13 = 6$

For cell B_{OT}-*G*, $C_{BOTG} - (U_{BOT} + V_G) = 15 - 10 = 5$

The solution is not optimum. Cell *B-E* with highest negative value enters the basic. Therefore draw the path for modified distribution.

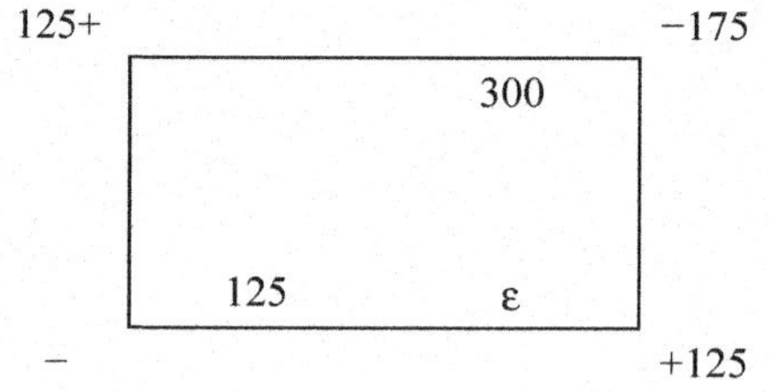

	D	E	F	G	H	
A	11 (200)	13 (50)	17 / 8	14 / 5	0 / −5	$U_A = 0$
B	16 / 16	18 (125)	14 / 13	10 (175)	0 (ε)	$U_B = 5$
C	21 / 16	24 / 18	13 (275)	10 (125)	0 / 0	$U_C = 5$
A_{OT}	15 / 15	17 (50)	21 / 12	18 / 9	0 / −1	$U_{AOT} = 4$
B_{OT}	21 / 16	23 / 18	19 / 13	15 / 10	0 (75)	$U_{BOT} = 5$
	$V_D = 11$	$V_E = 13$	$V_F = 8$	$V_G = 5$	$V_H = -5$	

Find U_i and V_j using $C_{ij} = U_i + V_j$ for stone cells.

Evaluate empty cells for $C_{ij} - (U_i + V_j)$.

For cell *A-F*, $C_{AF} - (U_A + V_F) = 17 - 8 = 9.$

For cell *A-G*, $C_{AG} - (U_A + V_G) = 14 - 5 = 9.$

For cell *A-H*, $C_{AH} - (U_A + V_H) = 0 + 5 = 5.$

For cell *B-D*, $C_{BD} - (U_B + V_D) = 16 - 16 = 0.$

For cell *B-F*, $C_{BF} - (U_B + V_F) = 14 - 13 = 1.$

For cell *C-D*, $C_{CD} - (U_C + V_D) = 21 - 16 = 5.$

For cell *C-E*, $C_{CE} - (U_C + V_E) = 24 - 18 = 6.$

For cell *C-H*, $C_{CH} - (U_C + V_H) = 0 - 0 = 0.$

For cell A_{OT}-*D*, $C_{AOTD} - (U_{AOT} + V_D) = 15 - 15 = 0.$

For cell A_{OT}-*F*, $C_{AOT\,F} - (U_{AOT} + V_F) = 21 - 12 = 9.$

For cell A_{OT}-*G*, $C_{AOT\,G} - (U_{AOT} + V_G) = 18 - 9 = 9.$

For cell A_{OT}-*H*, $C_{AOT\,H} - (U_{AOT} + V_H) = 0 + 1 = 1.$

For cell B_{OT}-*D*, $C_{BOT\,D} - (U_{BOT} + V_D) = 21 - 16 = 5.$

For cell B_{OT}-*E*, $C_{BOT\,E} - (U_{BOT} + V_E) = 23 - 18 = 5.$

For cell B_{OT}-*F*, $C_{BOT\,F} - (U_{BOT} + V_F) = 19 - 13 = 6.$

For cell B_{OT}-*G*, $C_{BOT\,G} - (U_{BOT} + V_G) = 15 - 10 = 5.$

The solution is optimum.

Allocation	Transportation cost
A-D	11 × 200 = Rs. 2200
A-E	13 × 50 = Rs. 650
B-E	18 × 125 = Rs. 2250
B-G	10 × 175 = Rs. 1750
C-F	13 × 275 = Rs. 3575
C-G	10 × 125 = Rs. 1250
A_{OT}-*E*	17 × 50 = Rs. 850
B_{OT}-*H*	–
	Total cost = Rs. 12,525

Example 5.10

Find the optimal solution for the following transportation problem. Use Vogel's Approximation Method to find the initial table. Verify for optimality.

		Destination				
		A	*B*	*C*	*D*	Availability
	I	21	16	25	13	11
Sources	II	17	18	14	23	13
	III	32	10	22	45	19
Requirement		6	10	12	15	

Solution

Initial basic feasible solution by VAM

	A	*B*	*C*	*D*	Availability
I	[21]	[16]	[25]	[13] 11	~~11~~ 3 8
II	[18] 6	[18]	[14] 3	[23] 4	~~9~~ ~~13~~ 3 3
III	[32]	[10] 10	[22] 9	[45]	~~19~~ 9 12 0
	6 4 15	10 6	12 8 8	~~15~~ 4 10 22	43 43

Find the difference between the least and next least in each row and column. Consider the row or column having highest difference. Give allocation to the cell with least cost in that row or column.

Allocation	Transportation cost
I-*D*	13 × 11 = Rs. 143
II-*A*	17 × 6 = Rs. 102
II-*C*	14 × 3 = Rs. 42
II-*D*	23 × 4 = Rs. 92
III-*B*	10 × 10 = Rs. 100
III-*C*	22 × 9 = Rs. 198
Total transportation cost = Rs. 677	

Check for optimality by D_1 MODI Method

	A	*B*	*C*	*D*	
I	[21] 7	[16] −8	[25] 4	[13] (11)	$U_1=0$
II	[18] (6)	[18] 7	[14] (3)	[23] (4)	$U_2=10$
III	[32] 25	[10] (10)	[22] (9)	[45] 31	$U_3=18$
	$V_A = 7$	$V_B = -8$	$V_C = 4$	$V_D = 13$	

Find U_i and V_j using $C_{ij} = U_i + V_j$ w.r.t. allocated cells.

Evaluate empty cells for $C_{ij} - (U_i + V_j)$.

For cell I-A,	$C_{\text{I-}A} - (U_{\text{I}} + V_A)$	$= 21 - 7 = 14.$
For cell I-B,	$C_{\text{I-}B} - (U_{\text{I}} + V_B)$	$= 16 + 8 = 24.$
For cell I-C,	$C_{\text{I-}C} - (U_{\text{I}} + V_C)$	$= 25 - 4 = 21.$
For cell II-B,	$C_{\text{II-}B} - (U_{\text{II}} + V_B)$	$= 18 - 2 = 16.$
For cell III-A,	$C_{\text{III-}A} - (U_{\text{III}} + V_A)$	$= 32 - 25 = 7.$
For cell III-D,	$C_{\text{III-}D} - (U_{\text{III}} + V_D)$	$= 45 - 31 = 14.$

The solution is optimum. Initial solution is the optimum solution. Total transportation cost = Rs. 677.

5.6 DEGENERACY IN TRANSPORTATION PROBLEM

The basic solution of the transportation problem with 'm' origins and 'n' destinations should have $(m + n - 1)$ positive basic variables (assigned boxes); otherwise the basic variables degenerate. It follows that whenever the number of assigned boxes is less than $(m + n - 1)$, the transportation problem is a degenerate one. The degeneracy can develop in two ways.

1. It develops while determining an initial assignment.
2. It develops at some iteration stage.

To resolve degeneracy, we use as many zero variables as necessary. These are selected in such a manner that $(m + n - 1)$ variables constitute a basic solution. For these zero variables, an extremely small positive value 'ε' is assigned. These boxes are treated as any other assigned boxes.

Example 5.11

A manufacturer wants to ship his products which have the following costs, availability and demand. Obtain an optimum product mix.

	A	B	C	Supply
1	7	3	4	2
2	2	1	3	3
3	3	4	6	5
Demand	4	1	4	9 \ 10

Solution

In this problem, the demand and supply are not equal, such problems are called as unbalanced transportation problem. The demand is less, therefore, add a dummy column. The cost associated with each of the cells in the dummy column is zero.

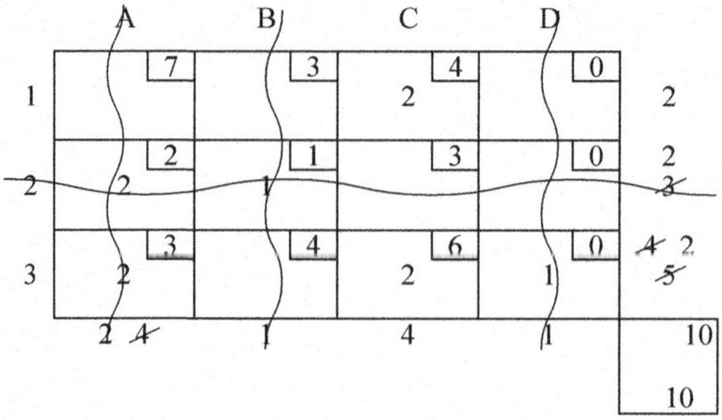

Initial basic feasible solution by LCM

Give allocation to the cell with minimum transportation cost.

Allocation	Transportation cost
1-*D*	4 × 2 = Rs. 8
2-*A*	2 × 2 = Rs. 4
2-*B*	1 × 1 = Rs. 1
3-*A*	2 × 3 = Rs. 6
3-*C*	2 × 6 = Rs. 12
3-*D*	0 × 1 = –
Total transportation cost = Rs. 31	

Check for optimality by MODI method

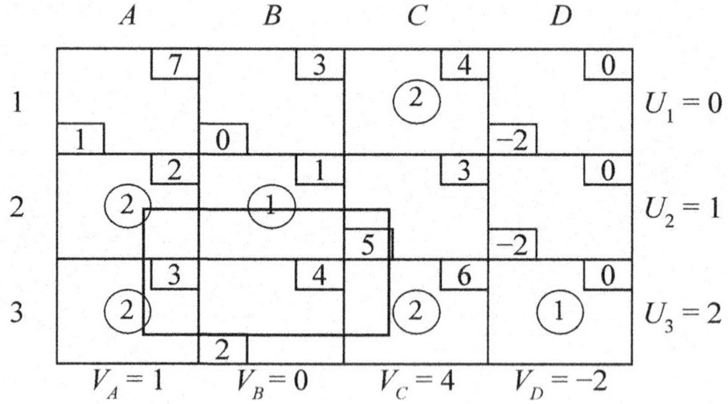

Find U_i and V_j, $C_{ij} = U_i + V_j$ w.r.t. allocated cells.

Evaluate empty cells for $C_{ij} - (U_i + V_j)$.

For cell 1-*A*, $C_{1A} - (U_1 + V_A) = 7 - 1 = 6.$

For cell 1-*B*, $C_{1B} - (U_1 + V_B) = 3 - 0 = 3.$

For cell 1-*D*, $C_{1D} - (U_1 + V_D) = 0 + 2 = 2.$

For cell 2-C, $C_{2C} - (U_2 + V_C) = 3 - 5 = -2.$

For cell 2-D, $C_{2D} - (U_2 + V_D) = 0 + 1 = 1.$

For cell 3-B, $C_{3B} - (U_3 + V_B) = 4 - 2 = 2.$

The solution is not optimum. Cell 2-C with negative value enters the basic. Starting with 2-C, draw the path for modified distribution.

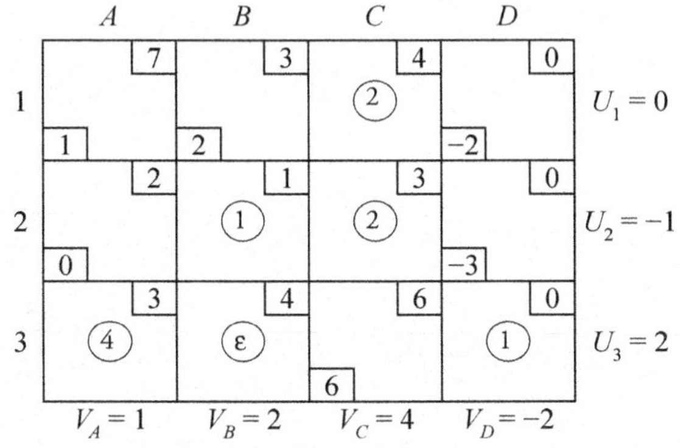

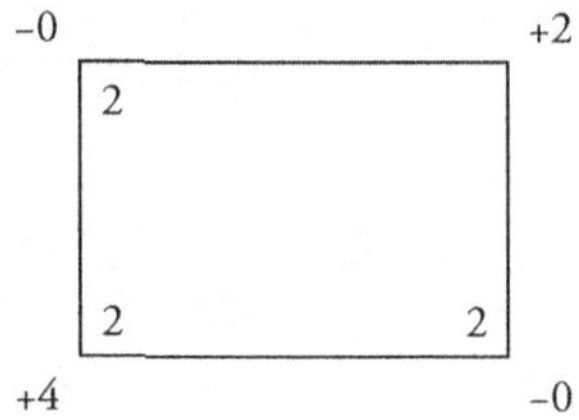

The number of assignments should have been $(m + n - 1) = 4 + 3 - 1 = 6$. But we have only 5 assignments. Therefore, the problem degenerates. Assign a small quanitity 'E' to an empty cell.

Find U_i and V_j using $C_{ij} = (U_i + V_j)$ w.r.t assigned cells.

Evaluate empty cells for $C_{ij} - (U_i + V_j)$.

For cell 1-A, $C_{1A} - (U_1 + V_A) = 7 - 1 = 6$

For cell 1-B, $C_{1B} - (U_1 + V_B) = 3 - 2 = 1$

For cell 1-D, $C_{1D} - (U_1 + V_D) = 0 + 2 = 2$

For cell 2-A, $C_{2A} - (U_2 + V_A) = 2 - 0 = 2$

For cell 2-D, $C_{2D} - (U_2 + V_D) = 0 + 3 = 3$

For cell 3-C, $C_{3C} - (U_3 + V_C) = 6 - 6 = 0$

The solution is optimum.

Allocation	Transportation cost
1-C	$4 \times 2 =$ Rs. 8
2-B	$1 \times 1 =$ Rs. 1
2-C	$3 \times 2 =$ Rs. 6
3-A	$3 \times 4 =$ Rs. 12
3-D	$0 \times 1 =$ Rs. 0
	Total transportation cost = Rs. 27

Example 5.12

Determine the optimum cost for the following transportation matrix.

	D_1	D_2	D_3	D_4	D_5	Capacities
O_1	4	7	3	8	2	4
O_2	1	4	7	3	8	7
O_3	7	2	4	7	7	2
O_4	4	8	2	4	7	2
Demand	3	4	2	2	4	15 / 15

Solution

Initial basic feasible solution by VAM

	D_1	D_2	D_3	D_4	D_5	Supply
O_1	[4]	[7]	[3]	[8]	4 [2]	4 1
O_2	3 [1]	2 [4]	[7]	2 [3]	[8]	4 7 2 1 1
O_3	[7]	2 [2]	[4]	[7]	[7]	2 2 2 5
O_4	[4]	[8]	2 [2]	[4]	[7]	2 2 6
Demand	3 2 3	2 4 2 2 2	2 1 2	2 1 1 4	4 5	

Find the difference between the least and the next least transportation cost. Consider row or column having highest difference. Allocate as much as you can to the cell with least transportation cost.

Allocation	Transportation cost
O_1-D_5	2 × 4 = Rs. 8
O_2-D_1	1 × 3 = Rs. 3
O_2-D_2	4 × 2 = Rs. 8
O_2-D_3	3 × 2 = Rs. 6
O_3-D_2	2 × 2 = Rs. 4
O_4-D_3	2 × 2 = Rs. 4
Total transportation cost = Rs. 33	

Check for optimality by MODI method

	D_1	D_2	D_3	D_4	D_5	
O_1	4; −2	7; 1	3; (ε)	8; 0	2; (4)	$U_1 = 0$
O_2	1; (3)	4; (2)	7; 6	3; (2)	8; 5	$U_2 = 3$
O_3	7; 7	2; (2)	4; (ε)	7; 1	7; 3	$U_3 = 1$
O_4	4; −3	8; 0	2; (2)	4; 1	7; 1	$U_4 = -1$
	$V_1 = -2$	$V_2 = 1$	$V_3 = 3$	$V_4 = 0$	$V_5 = 2$	

Find U_i and V_j using $C_{ij} = U_i + V_j$ w.r.t. allocated cells.

Evaluate empty cells for $C_{ij} - (U_i + V_j)$.

For cell O_1-D_1, $C_{11} - (U_1 + V_1) = 4 + 2 = 6.$

For cell O_1-D_2, $C_{12} - (U_1 + V_2) = 7 - 1 = 6.$

For cell O_1-D_4, $C_{14} - (U_1 + V_4) = 8 - 0 = 8.$

For cell O_2-D_3, $C_{23} - (U_2 + V_3) = 7 - 6 = 1.$

For cell O_2-D_5, $C_{25} - (U_2 + V_5) = 8 - 5 = 3.$

For cell O_3-D_1, $C_{31} - (U_3 + V_1) = 7 - 7 = 0.$

For cell O_3-D_4, $C_{34} - (U_3 + V_4) = 7 - 1 = 6.$

For cell O_3-D_5, $C_{35} - (U_3 + V_5) = 7 - 3 = 4.$

For cell O_4-D_1, $C_{41} - (U_4 + V_1) = 4 - 3 = 7.$

For cell O_4-D_2, $C_{42} - (U_4 + V_2) = 8 + 0 = 8.$

For cell O_4-D_4, $C_{44} - (U_4 + V_4) = 4 - 1 = 3.$

For cell O_4-D_5, $C_{45} - (U_4 + V_5) = 7 - 1 = 6.$

The solution is optimum. Cell O_3-D_3 with negative value enters the basic.

Allocation	Transportation cost
O_1-D_3	–
O_1-D_3	2 × 4 = Rs. 8
O_2-D_3	1 × 3 = Rs. 3
O_2-D_3	2 × 4 = Rs. 8
O_2-D_3	3 × 2 = Rs. 6
O_3-D_3	2 × 2 = Rs. 4
O_4-D_3	2 × 2 = Rs. 4
Total transportation cost = Rs. 33	

Example 5.13

Solve the following transportation problem for optimum allocated using MODI method. Cell enters represent unit cost of shipping.

12	4	9	5	9	55
8	1	6	6	7	45
1	12	4	7	7	30
10	12	6	9	1	20
40	20	50	30	40	150 / 180

Solution

This is an unbalanced transportation problem. Supply is less than demand. Therefore, take one extra row with zero transportation cost.

Initial solution by VAM

	1	2	3	4	5	Supply
1	12	4	9 (35)	5	9 (20)	55 1 5 0
2	8 (10)	1 (20)	6 (15)	6	7	15 ~~25~~ ~~45~~ 5 5 1
3	1 (30)	12	4	7	7	~~30~~ 3 3
4	10	12	6	9	1 (20)	~~20~~ 5 5
5	0	0	0	0 (30)	0	~~30~~ 0
Demand	~~40~~ 10	20	~~50~~ 35	30	40	180 / 180
	1	1	4	5	1	
	7	3	2		6	
	2	3	0		6	
	4	3	3		2	
	4		3		2	

Allocation	Transportation cost
1-3	35 × 9 = Rs. 315
1-5	9 × 20 = Rs. 180
2-1	8 × 10 = Rs. 80
2-2	1 × 20 = Rs. 20
2-3	6 × 15 = Rs. 90
3-1	1 × 30 = Rs. 30
4-1	1 × 20 = Rs. 20
5-4	0 × 30 = 0
Total transportation cost = Rs. 735	

Check for optimality by MODI method

	1	2	3	4	5	Supply
1	12, 11	4, 4	9, (35)	5, 4	9, (20)	$U_1 = 0$
2	8, (-10)	1, (20)	6, (15)	6, 1	7, 6	$U_2 = -3$
3	1, (30)	12, −6	4, −1	7, −6	7, −1	$U_3 = -10$
4	10, 3	12, −4	6, 2, 1	9, −4	1, (20)	$U_4 = -8$
5	0, (ε)	0, 7	0, −2	0, (30)	0, −2	$U_5 = -11$
Demand	$V_1 = 11$	$V_2 = 4$	$V_3 = 9$	$V_4 = 4$	$V_5 = 9$	

Find U_i and V_j, $C_{ij} = U_i + V_j$, for allocated cells.

Evaluate empty cells for $C_{ij} - (U_i + V_j)$.

For cell, 1-1, $C_{11} - (U_1 + V_1) = 12 - 11 = 1.$

For cell, 1-2, $C_{12} - (U_1 + V_2) = 4 - 4 = 0.$

For cell, 1-4, $C_{14} - (U_1 + V_4) = 5 - 4 = 1.$

For cell, 2-4, $C_{24} - (U_2 + V_4) = 6 - 1 = 5.$

For cell, 2-5, $C_{25} - (U_2 + V_5) = 7 - 6 = 1.$

For cell, 3-2, $C_{32} - (U_3 + V_2) = 12 + 6 = 18.$

For cell, 3-3, $C_{33} - (U_3 + V_3) = 4 + 1 = 5.$

For cell, 3-4, $C_{34} - (U_3 + V_4) = 7 + 6 = 13.$

For cell, 3-5, $C_{35} - (U_3 + V_5) = 7 + 1 = 8.$

For cell, 4-1, $C_{41} - (U_4 + V_1) = 10 - 3 = 7.$

For cell, 4-2, $C_{42} - (U_4 + V_2) = 12 + 4 = 16.$

For cell, 4-3, $C_{43} - (U_4 + V_3) = 6 - 1 = 5.$

For cell, 4-4, $C_{44} - (U_4 + V_4) = 9 + 4 = 13.$

For cell, 5-2, $C_{52} - (U_5 + V_2) = 0 + 7 = 7.$

For cell, 5-3, $C_{53} - (U_5 + V_3) = 0 + 2 = 2.$

For cell, 5-5, $C_{55} - (U_5 + V_5) = 0 + 2 = 2.$

The solution is optimum. Initial solution is the optimum solution. Therefore, optimum transportation cost is Rs. 735.

5.7 MAXIMISATION PROBLEMS

Initial solution by NWCR It is same as minimisation problem.

Initial solution by LCM Allocation is done to the cell having maximum profit.

Initial solution by VAM Find the difference between the highest and the next highest value in each row and column. Consider that row or column where the difference is highest. Allocation is to be done to this row or column. The cell is the one which is having the highest profit.

Moving towards optimality Evaluate each of the empty cells. If the value of $C_{ij} - (U_i + V_j)$ is either negative or zero, the solution is optimum. This is the reverse of minimisation.

Example 5.14

A canning company operates 2 canning plants; 3 growers are willing to supply fruits in the following amounts.

Grower 1: 200 quintals at Rs. 10 per quintal

Grower 2: 300 quintals at Rs. 9 per quintal

Grower 3: 400 quintals at Rs. 8 per quintal

Shipping costs per quintal are

From \ To	Plant	
	A	*B*
1	2	2.5
Grower 2	1	1.5
3	5	3

Plant capacity and labour cost are

	Plant *A*	*B*
Capacity in quintals	450	350
Labour cost/quintal	Rs. 25	Rs. 20

The canned fruits are sold at Rs. 50 per quintal to the distributors. How should the company plan its operations at the plant so as to maximise the profit?

Solution

It is to maximise the profit. Therefore,

Profit = Selling price – (Fruit cost + Transportation cost + Labour cost) of canned fruits.

Grower 1 – Plant *A*,	Profit	= 50 – (10 + 2 + 25) = Rs. 13.0.
Grower 2 – Plant *A*,	Profit	= 50 – (9 + 1 + 25) = Rs. 15.0.
Grower 3 – Plant *A*,	Profit	= 50 – (8 + 5 + 25) = Rs. 12.0.
Grower 1 – Plant *B*,	Profit	= 50 – (10 + 2.5 + 20) = Rs. 17.5.
Grower 2 – Plant *B*,	Profit	= 50 – (9 + 1.5 + 20) = Rs. 19.5.
Grower 3 – Plant *B*,	Profit	= 50 – (8 + 3 + 20) = Rs. 19.0.

Profit matrix is as follows

		Plant *A*	Plant *B*	Supply
Grower	1	13	17.5	200
	2	15	19.5	300
	3	12	19	400
Demand		450	350	900 / 800

It is an unbalanced transportation problem. Demand is less. Therefore, add a dummy column with zero profit.

Initial basic solution by LCM

Give allocation as much as you can to the cell with maximum profit.

	A	*B*	*C*	
~~1~~	[13] 200	[17.5]	[0]	~~200~~
~~2~~	[15]	[19.5]	[0]	~~300~~
3	[12] 250	[19] 50	[0] 100	~~400~~ 350
	~~450~~ 250	~~350~~ 50	100	900 / 900

Allocation	Profit
*G*1-*A*	13 × 200 = Rs. 2600
*G*2-*B*	19.5 × 300 = Rs. 5850
*G*3-*A*	12 × 250 = Rs. 3000
*G*3-*B*	19 × 50 = Rs. 950
*G*3-*C*	–
	Total profit = Rs. 12,400

Check for optimality by MODI method

	A	B	C	
1	13 (200)	17.5 [20]	0 [1]	$U_1 = 0$
2	15 [12.5]	19.5 (300)	0 [0.5]	$U_2 = -0.5$
3	12 (250)	19 (50)	0 (100)	$U_3 = -1$
	$V_A = 13$	$V_B = 20$	$V_C = 1$	

Find U_i and V_j using $C_{ij} = (U_i + V_j)$ w.r.t. stone cells.

Evaluate empty cells for $C_{ij} - (U_i + V_j)$.

For cell 1-*B*, $C_{1B} - (U_1 + V_B) = 17.5 - 20 = -2.5.$

For cell 1-*C*, $C_{1C} - (U_1 + V_C) = 0 - 1 = -1.$

For cell 2-*A*, $C_{2A} - (U_2 + V_A) = 15 - 12.5 = 2.5.$

For cell 2-*C*, $C_{2C} - (U_2 + V_C) = 0 - 0.5 = -0.5.$

This is a maximisation problem. Therefore, $C_{ij} - (U_i + V_j)$ should be either zero or negative. Here, the solution is not optimum. Look for the cell having the maximum positive value. It is cell 2-*A*.

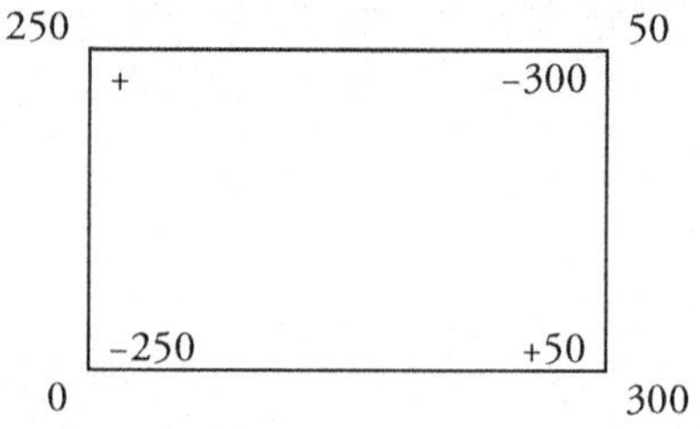

Draw the path for modified distribution.

	A	B	C	
1	13 (200)	17.5 [17.5]	0 [-1.5]	$U_1 = 0$
2	15 (250)	19.5 (50)	0 [0.5]	$U_2 = 2$
3	12 [14.5]	19 (300)	0 (100)	$U_3 = 1.5$
	$V_A = 13$	$V_B = 20$	$V_C = 1$	

Find U_i and V_j using $C_{ij} = (U_i + V_j)$ w.r.t. stone cells.

Evaluate empty cells for $C_{ij} - (U_i + V_j)$.

For cell 1-B, $C_{1B} - (U_1 + V_B) = 17.5 - 17.5 = 0$

For cell 1-C, $C_{1C} - (U_1 + V_C) = 0 + 1.5 = 1.5$

For cell 2-C, $C_{2C} - (U_2 + V_C) = 0 - 0.5 = - 0.5$

For cell 3-A, $C_{3A} - (U_3 + V_A) = 12 - 14.5 = -2.5$

The solution is not optimum, cell 1-C with positive value enters the basic.

Draw the path for modified distribution

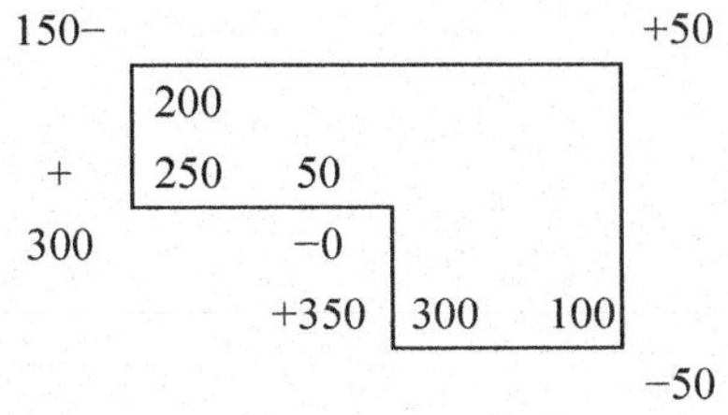

	A	B	C	
1	13 / (150)	17.5 / 19	0 / (50)	$U_1 = 0$
2	15 / (300)	19.5 / 21	0 / 2	$U_2 = 2$
3	12 / 13	19 / (350)	0 / (50)	$U_3 = 0$
	$V_A = 13$	$V_B = 19$	$V_C = 0$	

Find U_i and V_j using $C_{ij} = U_i + V_j$.

Evaluate empty cells for $C_{ij} - (U_i + V_j)$.

For cell 1-B, $C_{1B} - (U_1 + V_B) = 17.5 - 19 = -1.5$

For cell 2-B, $C_{2B} - (U_2 + V_B) = 19.5 - 21 = - 1.5$

For cell 2-C, $C_{2C} - (U_2 + V_C) = 0 - 2 = - 2$

For cell 3-A, $C_{3A} - (U_3 + V_A) = 12 - 12 = -1$

All the values are negative, therefore, the solution is optimum.

Allocation	Profit
1-A	13 × 150 = Rs. 1950
1-C	0 × 50 = –
2-A	15 × 300 = Rs. 4500
3-B	19 × 350 = Rs. 6650
3-C	0 × 50 = –
	Total profit = Rs. 13,100

Example 5.15

A job shop makes 3 products A, B and C. Demand for these products is 90, 210 and 120 units per week, respectively. The products can be manufactured by any one of the three methods 1, 2 and 3. The capacity of each method and the profit associated with each product and each manufacturing method are tabulated below. Determine the optimum method of manufacturing these products.

Method	Units per week	Product	Profit in method		
			1	2	3
1	160	*A*	139	140	137
2	120	*B*	209	207	210
3	140	*C*	254	255	255

Solution

Initial basic feasible solution by VAM

Products

Methods	*A*	*B*	*C*	Supply
1	[139] 90	[209] 70	[254]	160 45 70
2	[140]	[207]	[255] 120	~~120~~ 48
3	[137]	[210] 140	[255]	140 ~~45~~ 73
Demand	90 1	~~210~~ 70 1	120 0	420 / 420

It is a maximisation problem. Find the difference between the highest and next highest value in each row and column. Allocate to the cell with maximum profit in the row or column which has highest difference value.

Allocation	Profit
1 – *A*	139 × 90 = Rs. 12,510
1 – *B*	209 × 70 = Rs. 14,630
2 – *C*	255 × 120 = Rs. 30,600
3 – *B*	210 × 140 = Rs. 29,400
	Total profit = Rs. 87,140

Check for optimality by MODI method

	A	*B*	*C*	
1	[139] (90)	[209] (70)	[254] 254 (box)	$U_1 = 0$
2	[140] 140 (box)	[207] 210 (box)	[255] (120)	$U_2 = 1$
3	[137] 140 (box)	[210] (140)	[255] (ε)	$U_3 = 1$
	$V_A = 139$	$V_B = 209$	$V_C = 254$	

Find U_i and V_j using $C_{ij} = (U_i + V_j)$ for allocated cells. Degeneracy is there in the problem. Therefore, allocate ε to an empty cell.

Evaluate empty cells for $C_{ij} - (U_i + V_j)$.

For cell 1-C,	$C_{1C} - (U_1 + V_C)$	=	$254 - 254 = 0$.
For cell 2-A,	$C_{2A} - (U_2 + V_A)$	=	$140 - 140 = 0$.
For cell 2-B,	$C_{2B} - (U_2 + V_B)$	=	$207 - 210 = -3$.
For cell 3-A,	$C_{3A} - (U_3 + V_A)$	=	$137 - 140 = -3$.

The solution is optimum. Therefore, initial solution is the optimum solution.

Example 5.16

A departmental store wishes to stock the following quantities of a particular product in 3 types of containers.

Container type	1	2	3	4
Quantity	150	160	110	130

The store estimates the profit per unit that vary with the dealer as follows

		Container 1	2	3	4
Dealer	A	8	9	6	3
	B	6	11	5	10
	C	3	8	7	9

What should the distribution be to get the maximum profit?

Solution

Initial solution by VAM

Container \ Dealer	A	B	C	Capacity
1	[8] 150	[6]	[3]	~~150~~ 2 5
2	[9]	[11] 160	[8]	~~160~~ 2
3	[6] 20	[5] 140	[7] 90	110 1 1
4	[3]	[10] 40	[9] 90	90 ~~130~~ 1 6
	~~170~~ 90 1 2 2	~~210~~ 70 1 4	~~180~~ 90 1 2 4	550 / 550

It is a maximisation problem. Find the difference between the highest and the next highest value in each row and each column. Allocate to the cell with maximum profit in the row or column which has highest difference value.

Allocation	Profit
1 – *A*	8 × 150 = Rs. 1200
2 – *B*	11 × 160 = Rs. 1760
3 – *A*	6 × 20 = Rs. 120
3 – *C*	7 × 90 = Rs. 630
4 – *B*	10 × 40 = Rs. 400
4 – *C*	9 × 90 = Rs. 810
	Total profit = Rs. 4920

Check for optimality by MODI method

	A	*B*	*C*	
1	8 (150)	6 10	3 9	$U_1 = 0$
2	9 9	11 (160)	8 9	$U_2 = 1$
3	6 (20)	5 8	7 (90)	$U_3 = -2$
4	3 8	10 (40)	9 (90)	$U_3 = 0$
	$V_A = 8$	$V_B = 10$	$V_C = 9$	

Find U_i and V_j using $C_{ij} = (U_i + V_j)$ w.r.t. stone cells.

Evaluate empty cells for $C_{ij} - (U_i + V_j)$.

For cell, 1-*B*, $C_{1B} - (U_1 + V_B) = 6 - 10 = -4.$

For cell, 1-*C*, $C_{1C} - (U_1 + V_C) = 3 - 9 = -6.$

For cell, 2-*A*, $C_{2A} - (U_2 + V_A) = 9 - 9 = 0.$

For cell, 2-*C*, $C_{2C} - (U_2 + V_C) = 8 - 10 = -2.$

For cell, 3-*B*, $C_{3B} - (U_3 + V_B) = 5 - 8 = -3.$

For cell, 4-*A*, $C_{4A} - (U_4 + V_A) = 3 - 8 = -5.$

The solution is optimum. Therefore, initial solution is the optimum solution.

Example 5.17

A company has 3 plants located at different places, but produces an identical product. The cost of production, distribution cost from each plant to three different warehouses, the sales price at each warehouse and the individual capacity for both the plant and warehouse are as follows

Sales price at various warehouses		
W_1	W_2	W_3
Rs. 34	Rs. 32	Rs. 31

Plant	F_1	F_2	F_3
Raw material	Rs. 15	Rs. 18	Rs. 14
other exp	Rs. 10	Rs. 9	Rs.12

	Distribution cost			Warehouse capacity
	F_1	F_2	F_3	
W_1	3	9	5	80
W_2	1	7	4	110
W_3	3	8	3	150
Plant capacity	150	100	130	

Find the distribution pattern to maximise the profit.

Solution

It is a problem on profit maximisation. Profit = Sales price – (Raw material cost + other expenses + transportation cost

For F_1-W_1,	Profit = 34 – (15 + 10 + 3)	= Rs. 6
For F_1-W_2,	Profit = 32 – (15 + 10 + 1)	= Rs. 6
For F_1-W_3,	Profit = 31 – (15 + 10 + 3)	= Rs. 3
For F_2-W_1,	Profit = 34 – (18 + 9 + 9)	= Rs. –2
For F_2-W_2,	Profit = 32 – (18 + 9 + 7)	= Rs. –2
For F_2-W_3,	Profit = 31 – (18 + 9 + 8)	= Rs. –4
For F_3-W_1,	Profit = 34 – (14 + 12 + 5)	= Rs. 3
For F_3-W_2,	Profit = 32 – (14 + 12 + 4)	= Rs. 2
For F_3-W_3,	Profit = 31 – (14 + 12 + 3)	= Rs. 2

Profit matrix is as follows

	W_1	W_2	W_3	Supply
W_1	6	6	3	150
W_2	–2	–2	–4	100
W_3	3	2	2	130
Demand	80	110	150	380 / 380

It is an unbalanced transportation problem with less demand. Therefore, add a column with zero profit.

	W_1	W_2	W_3	W_4	
F_1	[6] 40	[6] 110	[3]	[0]	40 ~~150~~ 0 3
F_2	[–2] 40	[–2]	[–4] 20	[0] 40	100 2 2
F_3	[3]	[2]	[2] 130	[0]	130 1 1
	40 ~~80~~ 3 5	110 4	20 ~~150~~ 1 6	40 0 0	380 / 380

Initial basic feasible solution by VAM

Find the difference between the highest and the next highest value in each row and each column. Allocate to the cell, with maximum profit in that row or column which has maximum difference.

Allocation	Profit
F_1-W_1	6 × 40 = Rs. 240
F_1-W_2	6 × 110 = Rs. 660
F_2-W_1	–2 × 40 = Rs.–80
F_2-W_3	–4 × 20 = Rs.–80
F_2-W_4	0 × 40 = –
F_3-W_3	2 × 130 = Rs. 260
	Total profit = Rs. 1000

Check for optimality by MODI method

	W_1	W_2	W_3	W_4	
F_1	6 (40)	6 (110)	3 [4]	0 [8]	$U_1 = 0$
F_2	–2 (40)	–2 [–2]	–4 (20)	0 (40)	$U_2 = -8$
F_3	3 [4]	2 [4]	2 (130)	0 [6]	$U_3 = -2$
	$V_1 = 0$	$V_2 = 6$	$V_3 = 4$	$V_4 = 8$	

Find U_i and V_j, $C_{ij} = (U_i + V_j)$ w.r.t. stone cells.

Evaluate the empty cells for $C_{ij} - (U_i + V_j)$.

For cell F_1-W_3, $C_{13} - (U_1 + V_3) = 3 - 4 = -1$

For cell F_1-W_4, $C_{14} - (U_1 + V_4) = 0 - 8 = -8$

For cell F_2-W_2, $C_{22} - (U_2 + V_2) = -2 + 2 = 0$

For cell F_3-W_1, $C_{31} - (U_3 + V_1) = 3 - 4 = -1$

For cell F_3-W_2, $C_{32} - (U_3 + V_2) = 2 - 4 = -2$

For cell F_3-W_4, $C_{34} - (U_3 + V_4) = 0 - 6 = -6$

The solution is optimum. Optimum profit is Rs. 1000.

***Example* 5.18**

Obtain an optimum solution to the transportation problem shown using as initial solution the following feasible solution available.

							Available
	9	12	9	6	9	1	5
	7	3	7	7	5	5	6
	6	5	9	11	3	11	2
	6	8	11	2	2	10	9
Required	4	4	6	2	4	2	22 / 22

Solution

It is a balanced transportation problem. Table 1 shows C_{ij}, the unit transportation cost. Our aim is to minimise the total transportation cost.

3		2				5
	4	2				6
		2				2
1			2	4	2	9
4	4	6	2	4	2	

Given Initial basic feasible solution is as follows

	1	2	3	4	5	6	Availability
1	[9] (3)	[12]	[9] (2)	[6]	[9]	[1]	5
2	[7]	[3] (4)	[7] (2)	[7]	[5]	[5]	6
3	[6]	[5]	[9] (2)	[11]	[3]	[11]	2
4	[6] (1)	[8]	[11] 2	[2] (2)	[2] (4)	[10] (2)	9
Required	4	4	6	2	4	2	22 / 22

Allocation	Transportation cost
1-1	9 × 3 = Rs. 27
1-3	9 × 2 = Rs. 18
2 -2	3 × 4 = Rs. 12
2-3	7 × 2 = Rs. 14
3-3	9 × 2 = Rs. 18
4-1	6 × 1 = Rs. 6
4-4	2 × 2 = Rs. 4
4-5	2 × 4 = Rs. 8
4-6	10 × 2 = Rs. 20
Total transportation cost = Rs. 127	

Check for optimality by MODI method

Find U_i and V_j using $C_{ij} = (U_i + V_j)$ for allocated cells.

	1	2	3	4	5	6	
1	9 (3)	12, 5	9 (2)	6, 5	9, 5	1, 13	$U_1 = 0$
2	7, 7	3 (4)	7 (2)	7, 3	5, 3	5, 11	$U_2 = 0$
3	6, 9	5, 5	4 (2)	11, 5	3, 5	11, 13	$U_3 = 0$
4	6 (1)	8, 2	11, 6	2 (2)	2 (4)	10 (2)	$U_4 = -3$
	$V_1 = 9$	$V_2 = 5$	$V_3 = 9$	$V_4 = 5$	$V_5 = 5$	$V_6 = 13$	

Evaluate empty cells for $C_{ij} - U_i + V_j$.

For cell 1-2,	$C_{12} - (U_1 + V_2)$	=	12 – 5 = 7.
For cell 1-4,	$C_{14} - (U_1 + V_4)$	=	6 – 5 = 1.
For cell 1-5,	$C_{15} - (U_1 + V_5)$	=	9 – 5 = 4.
For cell 1-6,	$C_{16} - (U_1 + V_6)$	=	1 – 13 = –12.
For cell 2-1,	$C_{21} - (U_2 + V_1)$	=	7 – 7 = 0.
For cell 2-4,	$C_{24} - (U_2 + V_4)$	=	7 – 3 = 4.
For cell 2-5,	$C_{25} - (U_2 + V_5)$	=	5 – 3 = 2.
For cell 2-6,	$C_{26} - (U_2 + V_6)$	=	5 – 11 = –6.
For cell 3-1,	$C_{31} - (U_3 + V_1)$	=	6 – 9 = –3.
For cell 3-2,	$C_{35} - (U_3 + V_2)$	=	5 – 5 = 0.
For cell 3-4,	$C_{34} - (U_3 + V_4)$	=	11 – 5 = 6.
For cell 3-5,	$C_{35} - (U_3 + V_5)$	=	3 – 5 = –2.
For cell 3-6,	$C_{36} - (U_3 + V_6)$	=	11 – 13 = –2.
For cell 4-2,	$C_{42} - (U_4 + V_2)$	=	8 – 2 = 6.
For cell 4-3,	$C_{43} - (U_4 + V_3)$	=	11 – 6 = 5.

The solution is not optimum. Cell 1-6 with highest negative value enters the basic. Draw the path for modified distribution.

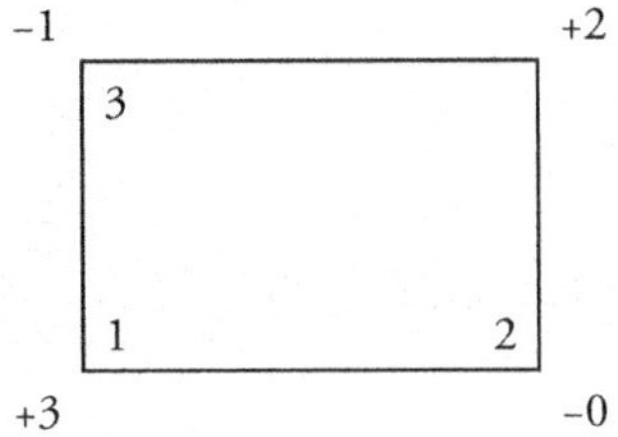

	1	2	3	4	5	6	
1	9 (1)	12, 5	9 (2)	6, 5	9, 5	1 (2)	$U_1 = 0$
2	7, 7	3 (4)	7 (2)	7, 3	5, 3	5, –1	$U_2 = -2$
3	6, 9	5, 5	9 (2)	11, 5	3, 5	11, 1	$U_3 = 0$
4	6 (3)	8, 2	12, 6	2 (2)	2 (4)	10, –2	$U_4 = -3$
	$V_1 = 9$	$V_2 = 5$	$V_3 = 9$	$V_4 = 5$	$V_5 = 5$	$V_6 = 1$	

Find U_i and V_j using $C_{ij} = (U_i + V_j)$ for allocated cells.

Evaluate empty cells for $C_{ij} - U_i + V_j$.

For cell 1-2,	$C_{12} - (U_1 + V_2)$	$= 12 - 5 = 7$
For cell 1-4,	$C_{14} - (U_1 + V_4)$	$= 6 - 5 = 1$
For cell 1-5,	$C_{15} - (U_1 + V_5)$	$= 9 - 5 = 4$
For cell 2-1,	$C_{21} - (U_2 + V_1)$	$= 7 - 7 = 0$
For cell 2-4,	$C_{24} - (U_2 + V_4)$	$= 7 - 3 = 4$
For cell 2-5,	$C_{25} - (U_2 + V_5)$	$= 5 - 3 = 2$
For cell 2-6,	$C_{26} - (U_2 + V_6)$	$= 5 + 1 = 6$
For cell 3-1,	$C_{31} - (U_3 + V_1)$	$= 6 - 9 = -3$
For cell 3-2,	$C_{32} - (U_3 + V_2)$	$= 5 - 5 = 0$
For cell 3-4,	$C_{34} - (U_3 + V_4)$	$= 11 - 5 = 6$
For cell 3-5,	$C_{35} - (U_3 + V_5)$	$= 3 - 5 = -2$
For cell 3-6,	$C_{36} - (U_3 + V_6)$	$= 11 - 1 = 10$
For cell 4-2,	$C_{42} - (U_4 + V_2)$	$= 8 - 2 = 6$
For cell 4-3,	$C_{43} - (U_4 + V_3)$	$= 12 - 6 = 6$
For cell 4-6,	$C_{46} - (U_4 + V_6)$	$= 10 + 2 = 12$

The solution is not optimum. The cell 3-1 with highest negative value enters the basic. Draw the path for modified distribution.

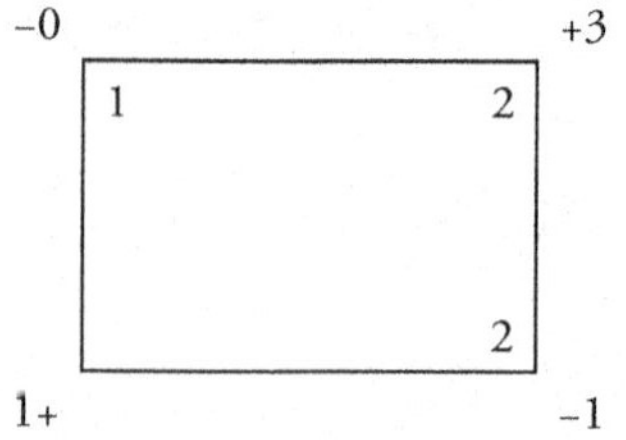

	1	2	3	4	5	6	
1	9 / 6	12 / 5	9 / (3)	6 / 2	9 / 2	1 / (2)	$U_1 = 0$
2	7 / 4	3 / (4)	7 / (2)	7 / 0	5 / 0	5 / −1	$U_2 = -2$
3	6 / (1)	5 / 5	9 / (1)	11 / 2	3 / 2	11 / 1	$U_3 = 0$
4	6 / (3)	8 / 5	12 / 9	2 / (2)	2 / (4)	10 / 1	$U_4 = 0$
	$V_1 = 6$	$V_2 = 5$	$V_3 = 9$	$V_4 = 2$	$V_5 = 2$	$V_6 = 1$	

Find U_i and V_j using $C_{ij} = (U_i + V_j)$ w.r.t. allocated cells.

Evaluate empty cells for $C_{ij} - (U_i + V_j)$.

For cell 1-1, $C_{11} - (U_1 + V_1) = 9 - 6 = 3$

For cell 1-2, $C_{12} - (U_1 + V_2) = 12 - 5 = 7$

For cell 1-4, $C_{14} - (U_1 + V_4) = 6 - 2 = 4$

For cell 1-5, $C_{15} - (U_1 + V_5) = 9 - 2 = 7$

For cell 2-1, $C_{21} - (U_2 + V_1) = 7 - 4 = 3$

For cell 2-4, $C_{24} - (U_2 + V_4) = 7 - 0 = 7$

For cell 2-5, $C_{25} - (U_2 + V_5) = 5 - 0 = 5$

For cell 2-6, $C_{26} - (U_2 + V_6) = 5 + 1 = 6$

For cell 3-2, $C_{32} - (U_3 + V_2) = 5 - 5 = 0$

For cell 3-4, $C_{34} - (U_3 + V_4) = 11 - 2 = 9$

For cell 3-5, $C_{35} - (U_3 + V_5) = 3 - 2 = 1$

For cell 3-6, $C_{36} - (U_3 + V_6) = 11 - 1 = 10$

For cell 4-2, $C_{42} - (U_4 + V_2) = 8 - 5 = 3$

For cell 4-3, $C_{43} - (U_4 + V_3) = 12 - 9 = 3$

For cell 4-6, $C_{46} - (U_4 + V_6) = 10 - 1 = 9$

The solution is optimum.

Optimum allocation	Transportation cost
1-3	9 × 3 = Rs. 27
1-6	1 × 2 = Rs. 2
2-2	3 × 4 = Rs. 12
2-3	7 × 2 = Rs. 14
3-1	6 × 1 = Rs. 6
3-3	9 × 1 = Rs. 9
4-1	6 × 3 = Rs. 18
4-4	2 × 2 = Rs. 4
4-5	2 × 4 = Rs. 8
Total transportation cost = Rs. 100	

***Example* 5.19**

The Kumar manufacturing company has several plants, three of which manufacture two principal products, a standard card table and a deluxe card table. A new delux card table will be introduced which must be combined in terms of selling price and cost.

The selling prices are

Standard	Deluxe	New deluxe
Rs. 14.95	Rs. 18.95	Rs. 21.95

Sales requirements of the three models are

Standard	Deluxe	New deluxe
450	1050	600

Variable cost in rupees, of models and plant capacities are given in the following table.

Variable cost

		Standard	Deluxe	New Deluxe	Available
	A	8.0	8.5	9.25	800
Plant	*B*	7.95	9.6	9.20	600
	C	8.1	8.45	9.30	700

Solve the problem by transportation method for the greatest contribution.

Contribution = Selling price – Variable cost

Solution

Following is the contribution matrix.

	S	D	N			
A	6.95 450	10.45 350	12.7	800	2.25	3.5
B	7	10.45	12.75 600	600	2.4	
C	6.85	10.5 700	12.65	700	2.15	3.65
	450 0.05	1058 350 0.05	600 0.05	2100 / 2100		

Initial feasible solution by VAM

Find the difference between the highest and the next highest value in each row and each column. Give allocation to the cell in the row or column having the highest difference.

Allocation	Profit
A-S	6.95 × 450 = Rs. 3127.5
A-D	10.45 = 350 = Rs. 3657.5
B-N	12.75 × 600 = Rs. 7650
C-D	10.5 × 700 = Rs. 7350
	Total profit = Rs. 21,785

Check for optimality by MODI method

	S	D	N	
A	6.95 (450)	10.45 (350)	12.7 (ε)	$U_A = 0$
B	7 7	10.35	12.75 (600)	$U_B = 0.5$
C	6.85 7	10.5 (700)	12.65 12.75	$U_C = 0.05$
	$V_S = 6.95$	$V_D = 10.45$	$V_N = 12.7$	

Find U_i and V_j using $C_{ij} = (U_i + V_j)$ for stone cells. This problem degenerates. Therefore, allocate 'ε' to cell *A-N*.

Evaluate empty cells for $C_{ij} - (U_i + V_j)$.

For cell *B-S*, $C_{BS} - (U_B + V_S) = 7 - 7 = 0$

For cell *B-D*, $C_{BD} - (U_B + V_D) = 10.35 - 10.5 = -0.15$

For cell *C-S*, $C_{CS} - (U_C + V_S) = 6.85 - 7 = -0.15$

For cell *C-N*, $C_{CN} - (U_C + V_N) = 12.65 - 12.75 = -0.1$

The solution is optimum.

EXERCISES

1. A manufacturer must produce a product in sufficient quantity to meet contractual sales in the next four months. The production capacity and the unit cost of production vary from month to month. The product produced in one month may be held for sale in the later month but at the estimated storage cost of Re.1 per unit per month. No storage cost is incurred for goods sold in the same month in which, they are produced. There is no opening inventory and no decision is taken at the end of the month. The necessary details are given in the following table.

Month	Sales	Maximum production	Unit cost of production
1	20	40	14
2	30	50	16
3	50	30	15
4	40	50	17

How should the manufacturer produce each month, to minimise the total cost?

2. A manufacturer wishes to ship 8 loads of the products as shown below. The matrix gives the mileage from origin O to the destination D.

	P_1	P_2	P_3	
O_1	50	30	20	1
O_2	90	45	170	1
O_3	50	200	50	1
	4	2	2	

Shipping cost is Rs. 10 per load per mile. Which shipping schedule must be used?

3. A company is developing a master plan for the production of an item over a four month period. The demand for the 4 months are 100, 200, 180 and 300 units respectively. The variable production cost per unit in any month is Rs. 4. The holiday cost is Rs. 0.5 per unit per month. The production capacity for the next four months are 50, 180, 280 and 270 units respectively. Back-ordered items incur a penalty cost of Rs. 2 per unit per month. What is the optimum production plan?

4. Solve the following transportation problem.

From	To			Available
	A	B	C	
I	50	30	220	1
II	90	45	170	3
III	250	200	50	4
Requirement	4	2	2	

5. A firm has three reservoirs with daily supplies of 15, 20 and 25 million litres of fresh water respectively. On each day we must supply to four cities A, B, C and D whose demands are 8, 10, 12 and 15, respectively. The cost of pumping per million litres is given below.

		Cities				
		A	B	C	D	
Reservoirs	1	2	3	4	5	
	2	3	2	5	2	
	3	4	1	2	3	

Use the transportation method to determine the cheapest pumping schedule if excess water can be disposed at no cost.

6. Faced with a court order to desegregate its schools, a country school board decides to redistribute its minority students through bussing. The plan calls for bussing 50 students from each of the three cities White, Black and Brown to the four schools East, West, North and South. For a perfect desegregation, the schools need 20, 40, 30 and 60 minority students respectively. The cost of bussing each student is as follows

		School			
		East	West	North	South
	White	7	6	5	4
City	Black	9	7	3	6
	Brown	8	8	7	3

The school board wishes to meet the court order with the least cost. Determine the optimal bussing plan using MODI method.

7. A manufacturer is interested in developing an advertising campaign that will reach four different age groups. Advertising campaigns can be conducted through TV, radio and magazines. The following table gives the estimated cost in paise per exposure for each age group according to the medium employed. In addition, maximum exposure levels possible in each of the media, namely, TV, radio and magazines are 40, 30 and 20 millions respectively. Also, the minimum desired exposures within each age group, namely, 13–18, 19–25, 26–35 and 36 and older are 30, 25, 15 and 10 millions respectively.

Media	Age groups			
	13–18	19–25	26–35	36 and over
TV	12	7	10	10
Radio	10	9	12	10
Magazines	14	12	9	12

Solve this problem to minimise the cost.

8. A cement company has three factories which manufacture cement, which is then transported to four distribution centres. The quantity of the monthly production of each factory, the demand of each distribution centre and the associated transportation cost per quintal are given below.

		Distribution centre				Monthly production in quintals
		W	*X*	*Y*	*Z*	
Factories	*A*	10	8	5	4	7000
	B	7	9	15	8	8000
	C	6	10	14	8	10,000
Monthly demand in quintols		6000	6000	8000	5000	

Solve this problem and suggest the optimal transportation schedule.

9. A company has four factories situated in different locations in the country and four sales agencies located in four other locations in the country. The cost of production (Rs. per unit), the sale price (Rs. per unit), shipping cost (Rs. per unit), are given in the cells of the matrix, monthly capacities and monthly requirements are given below.

		Sales agency				Monthly capacity (units)	Cost of production
		1	2	3	4		
Factory	*A*	7	5	6	4	10	10
	B	3	5	4	2	15	15
	C	4	6	4	5	20	16
	D	8	7	6	5	15	15

Monthly requirements				
Units	8	12	18	22
Sales price	20	22	25	18

Find the monthly production and distribution schedule which will maximise profits.

10. The *XYZ* tobacco company purchases tobacco and stores in warehouses located in the following four cities.

Warehouse location	Capacity (tonnes)
City *A*	90
City *B*	50
City *C*	80
City *D*	60

The warehouses supply tobacco to the cigarette companies in three cities that have the following demand.

Cigarette company	Demand (Tonnes)
Bharath	120
Janata	100
Red Lamp	110

The following rail road shipping costs per tonne (in hundred rupees) have been determined.

To / From	Bharat	Janata	Redlamp
A	7	10	5
B	12	9	4
C	7	3	11
D	9	5	7

Find the optimal distribution for XYZ tobacco company.

11. A company has three open earth furnaces and four rolling mills. The capacities of the furnaces are 20, 60 and 70 quintals respectively. The requirements at the different rolling mills are 30, 30, 40 and 50 quintals respectively. Transportation cost (Rs. per quintal) for shipping steel from furnaces to rolling mills are shown in the following table.

		Rolling mills			
Furnaces	F_1	3	1	5	10
	F_2	7	4	2	3
	F_3	8	8	5	2

What is the optimal shipping cost for the company?

12. Solve the following transportation problem.

From	To				Supply
	1	2	3	4	
1	8	8	5	12	7
2	6	7	11	9	7
3	10	15	6	13	10
4	6	8	7	8	6
5	11	10	11	13	5
6	8	14	5	12	6
Demand	9	10	8	14	

13. The oil company has three refineries and four depots. Transportation cost per ton, capacities and requirements are as follows

	D_1	D_2	D_3	D_4	Capacity (tons)
R_1	5	7	13	10	700
R_2	8	6	14	13	400
R_3	12	10	9	11	800
Requirement (tons)	300	600	700	400	

Determine the optimal allocation of output.

14. A manufacturer has distribution centres at *X, Y* and *Z*. These centres have availability of 40, 20 and 40 units respectively of his product. His retail outlets at *A, B, C, D* and *E* require 25, 10, 20, 30 and 15 units, respectively. The transport cost (in rupees) per unit between each centre outlet is given below.

Distribution centre	Retail outlet				
	A	*B*	*C*	*D*	*E*
X	55	30	40	50	50
Y	35	30	100	45	60
Z	40	60	95	35	30

Determine the optimal distribution to minimise the cost of transportation.

15. A wholesale company has three warehouses from which supplies are drawn for four retail customers. The company deals with a single product. The following gives the supply demand details.

Warehouse no.	Supply units	Customer no.	Demand (units)
1	20	1	15
2	28	2	19
3	17	3	13
		4	18

Conveniently, total supply at the warehouses is equal to the total demand from the customers. The following table gives the transportation costs per unit shipment from each warehouse to each customer.

		Customer			
		1	2	3	4
	1	3	6	8	5
Warehouse	2	6	1	2	5
	3	7	8	3	9

Determine what supplies to despatch from each of the warehouses to each customer so as to minimise the total transportation cost.

16. A particular product is manufactured in factories *A, B, C* and *D* and is sold at centres 1, 2 and 3. The cost (in rupees) of product per unit and capacity (in kg) per unit time of each plant is given below

Factory	Cost (Rs.) per unit	Capacity (kg) per unit
A	12	100
B	15	20
C	11	60
D	13	80

The sales price in rupees per unit and the demand in kg per unit time are as follows

Sales centre	Sales price (Rs.) per unit	Demand (kg) per unit time
1	15	120
2	14	140
3	16	60

Find the optimum sales distribution.

17. Solve the following transportation problem to maximise profit.

Origin	Profit (Rs.) percent destination				Supply
	1	2	3	4	
A	42	27	24	35	200
B	46	37	32	32	60
C	40	40	30	32	140
Demand	80	40	120	60	

Chapter 6

ASSIGNMENT PROBLEMS

6.1 INTRODUCTION

Assignment problem is a special type of transportation problem. In assignment model, the resources are allocated to the activities on a one-to-one basis.

Let n jobs be assigned to 'n' facilities. Let C_{ij} be the cost associated with assignment. The objective is to determine how all assignments should be made to minimise the total cost. Since there are 'n' assignees and 'n' assignments, the problem can be represented by $n \times n$ matrix.

	1	2	3 ···	n
1	$C_{11}\ a_1$	$C_{12}\ b_2$	$C_{13}\ b_3$ ···	$C_{in}\ b_n$
2	$C_{21}\ a_2$			
3	$C_{31}\ a_3$			
.	.			
.	.			
.	.			
n	$C_{n1}\ a_n$			C_{nn}

Here, $a_i = b_j = 1$, where $i = 1, 2, 3, ..., n$ and $j = 1, 2, 3, ..., n$

(a_i) assignees are similar to (b_j) sources, each having one unit of supply. C_{ij} is the cost associated with the cell corresponding to the ith row and jth column.

Mathematically,

$$\text{Minimise } Z = \sum_{i=1}^{n}\sum_{j=1}^{n} C_{ij}$$

subject to

$$\sum_{i=1}^{n} a_i = 1$$

where $i = 1, 2, 3, ..., n$

$j = 1, 2, 3, ..., n$

$$\sum_{j=1}^{n} b_j = 1$$

6.2 HUNGARIAN METHOD

The Hungarian method is the mathematical procedure used to solve the assignment problems and determine the one-to-one allocation of resources to the activities given in the problem. The step-by-step procedure of the Hungarian method is illustrated in Example 6.1.

Example 6.1

A project work consists of four major jobs for which four contractors have submitted tenders. The tender amounts quoted in lakhs of rupees are given in the matrix below. Each contractor has to be assigned at least one job. Determine which job is to be assigned to which contractor.

	A	*B*	*C*	*D*
1	10	24	30	15
2	16	22	28	12
3	12	20	32	10
4	9	26	34	16

Solution

Step 1 Consider each row. Deduct the smallest value in each row from all the values in that row.

	A	*B*	*C*	*D*
1	0	14	20	5
2	4	10	16	0
3	2	10	22	0
4	0	17	25	7

Step 2 Consider each column. Deduct the smallest value in each column from all the values in that column.

	A	*B*	*C*	*D*
1	[0]	4	4	5
2	4	0	[0]	0
3	2	0	6	[0]
4	0	7	9	7

Step 3 Consider the first row; if there is only one zero, enrectangle it. As no assignment can be done in that column, draw a line through that column. Repeat the procedure for other rows.

Step 4 Consider each column, if there is only one zero in that column, enrectangle it and draw a line through that row. If the assignment is made in all rows stop otherwise proceed.

Step 5 Determine the minimum value out of those through which the lines are not drawn. Subtract this minimum value from all the values through which lines are not drawn. Add this value to the value at the junction.

	A	*B*	*C*	*D*
1	0	0	[0]	1
2	8	[0]	0	0
3	6	0	6	[0]
4	[0]	3	5	3

Here the tie occurs. Look to the original matrix. The minimum cost is 10 pertaining to the cell 3-*D*. Therefore, enrectangle this zero. As no assignment can be done either to the row 3 or column *D*, draw lines through them.

Assign	Cost
1 to *C*	30
2 to *B*	22
3 to *D*	10
4 to *A*	9
Total project cost = Rs. 71 lakhs	

Example 6.2

A department head has 5 subordinates and 5 tasks are to be performed. The subordinates differ in efficiency and the tasks differ in their intrinsic difficulty. His estimate of the time each man would take to perform each task is given in the effective matrix below. How should the tasks to be allocated so as to minimise man-hours?

		Subordinate				
		A	*B*	*C*	*D*	*E*
	1	9	3	4	2	10
	2	12	10	8	1	9
Task	3	11	2	9	0	8
	4	8	0	10	3	7
	5	7	5	7	2	9

Solution

Step 1 Consider each row. Deduct the smallest value in each row from all the values in that row.

	A	*B*	*C*	*D*	*E*
1	7	1	2	0	8
2	11	9	7	0	8
3	11	2	9	0	8
4	8	0	10	3	7
5	5	3	5	0	7

Step 2 Consider each column. Deduct the smallest value in each column from all the values in that column.

	A	*B*	*C*	*D*	*E*
1	2	1	**[0]**	0	1
2	9	9	5	**[0]**	1
3	9	2	7	0	1
4	3	0	8	3	0
5	**[0]**	3	3	0	0

Step 3 Consider the first row *i* of there is only one zero enrectangle it. As no assignment can be done in that column, draw a line through that column. Repeat the procedure for other rows.

Step 4 Consider each column. If there is only one zero in that column, enrectangle it and draw a line through that row. If the assignment is made in all rows, stop; otherwise proceed.

Step 5 Determine the minimum value out of these, through which the lines are not drawn. Subtract this minimum value from all the values through which lines are not drawn. Add this value to the value at the junction.

	A	*B*	*C*	*D*	*E*
1	2	1	[0]	1	1
2	8	8	4	0	[0]
3	8	1	6	[0]	0
4	3	[0]	8	4	0
5	[0]	3	3	1	0

Assign	Time
1 to *C*	4
2 to *E*	9
3 to *D*	0
4 to *B*	0
5 to *A*	7
Total = 20 time units	

Example 6.3

In a textile sales emporium, salesmen *A*, *B*, *C* and *D* are available to handle 4 counters *W*, *X*, *Y* and *Z*. Each salesman can handle any counter. The service time in hours of each counter, when manned by each salesman is as follows

		Salesman			
		A	*B*	*C*	*D*
Counter	*W*	41	72	39	52
	X	22	29	49	65
	Y	27	39	60	51
	Z	45	50	48	52

How should the salesmen be assigned to counters so as to minimise the total service time. Each salesman should handle any one counter.

Solution

Step 1 Consider each row. Deduct the smallest value in each row by all the values in that row.

	A	*B*	*C*	*D*
W	2	33	0	13
X	0	7	27	43
Y	0	12	33	24
Z	0	5	3	7

Step 2 Consider each column. Deduct the smallest value in each column by all values in that column.

	A	*B*	*C*	*D*
W	2	28	[0]	6
X	[0]	2	27	36
Y	0	7	33	17
Z	0	[0]	0	0

Step 3 Consider the first row if there is only one zero, enrectangle it. As no assignment can be done in that column, draw a line through that column. Repeat the procedure for other rows.

Step 4 Consider each column if there is only one zero in that column enrectangle it and draw a line through that row. If the assignment is made in all rows, stop; otherwise proceed.

Step 5 Determine the minimum value out of those through which the lines are not drawn. Subtract this minimum value from all the values through which lines are not drawn. Add this value to the values at the junction.

	A	*B*	*C*	*D*
W	2	26	[0]	4
X	0	[0]	27	34
Y	[0]	5	33	15
Z	2	0	5	[0]

Assign	Time
W to *C*	39
X to *B*	29
Y to *A*	27
Z to *D*	52
Total service time = 147 hours	

6.3 TICKING METHOD

Ticking method is the modified procedure used to solve assignment problems. This method reduces the number of iteration required to get the required one to one allocates of resources to the activities. The step-by-step procedure is illustrated is Example 6.4.

Example 6.4

There are five jobs to be assigned one each to five machines and the associated cost matrix is as follows. Find the optimum assignment of jobs and the minimum total cost to produce all the jobs.

		Machines				
		P	*Q*	*R*	*S*	*T*
	1	11	17	8	16	20
	2	9	7	12	6	15
Jobs	3	13	16	15	12	16
	4	21	24	17	28	26
	5	14	10	12	11	15

Solution

Ticking method

Step 1 Consider each row. Deduct the smallest value in each row from all the values in that row.

		Machines				
		P	*Q*	*R*	*S*	*T*
	1	3	9	0	8	12
	2	3	1	6	0	9
Jobs	3	1	4	3	0	4
	4	4	7	0	11	9
	5	4	0	2	1	5

Step 2 Consider each column. Deduct the smallest value in each column from all the values in that column.

	P	*Q*	*R*	*S*	*T*	
1	2	9	[0]	8	8	√
2	2	1	6	[0]	5	
3	[0]	4	3	~~0~~	~~0~~	
4	3	7	~~0~~	11	5	√
5	3	[0]	2	1	1	
			√			

Step 3 Consider the first row; if there is only one zero, enrectangle it. As no assignments can be done in this column, cross all zeroes in that column.

Step 4 Consider each column; if there is only one zero, enrectangle it. As no assignments can be done in that row, cross all zeroes in that row.

Step 5 Consider the row to which an assignment is not made and write a tick mark against that row.

Step 6 Look for zeroes in that ticked row. Put a tick mark in the corresponding column.

Step 7 Observe if in that ticked column if there is an assigned cell; put a tick mark against that row.

Step 8 Draw lines through the unticked rows and the ticked column.

Step 9 Determine the minimum value out of those through which lines are not drawn. Subtract this value from all the values through which lines are not drawn. Add this value to the value at the junction.

	P	*Q*	*R*	*S*	*T*
1	[0]	7	~~0~~	6	6
2	2	1	8	[0]	5
3	~~0~~	4	5	~~0~~	[0]
4	1	5	[0]	9	3
5	3	[0]	2	1	1

Step 10 Repeat Steps 3 and 4. Here a one-to-one assignment is done.

Allocate	Cost
1 to *P*	11
2 to *S*	6
3 to *T*	16
4 to *R*	17
5 to *Q*	10
Total cost = Rs. 60	

Example 6.5

A national car rental service has a surplus of one car in each of the cities 1, 2, 3, 4, 5 and 6 and deficient of a car in each of the cities 7, 8, 9, 10, 11 and 12. The distance between the cities which have surplus and the cities which are deficient are displayed in the matrix below. How should the cars be assigned so as to minimise the total mileage travelled?

	7	8	9	10	11	12
1	41	72	39	52	25	51
2	22	29	49	65	81	50
3	27	39	60	51	32	32
4	45	50	48	52	37	43
5	29	40	39	26	30	33
6	82	40	40	60	51	30

Solution

Ticking method

Step 1 Consider each row. Deduct the smallest value in each row from all the values in that row.

	7	8	9	10	11	12
1	16	47	14	27	0	26
2	0	7	27	43	59	28
3	0	12	33	14	5	5
4	8	13	11	15	0	6
5	3	14	13	0	4	7
6	52	10	10	30	21	0

Step 2 Consider each column. Deduct the smallest value in each column from all the values in that column.

	7	8	9	10	11 √	12	
1	16	40	4	27	[0]	26	√
2	~~0~~	[0]	17	43	59	28	
3	[0]	5	23	14	5	5	
4	8	6	1	15	~~0~~	6	√
5	3	7	3	[0]	4	7	
6	52	3	[0]	3	21	~~0~~	

Step 3 Consider the first row; if there is only one zero, enrectangle it. As no assignments can be done in this column, cross all zeroes in that column.

Step 4 Consider each column; if there is only one zero enrectangle it. As no assignment can be done in that row, cross all zeroes in that row.

Step 5 Consider the row to which assignment is not made; put a tick mark against that row.

Step 6 Look for zeroes in that ticked row. Put a tick mark in the corresponding column.

Step 7 Observe if in the ticked column there is an assigned cell; put a tick mark against that row.

Step 8 Draw lines through unticked rows and ticked columns.

Step 9 Determine the minimum value out of these through which lines are not drawn. Subtract this value from all the values through which lines are not drawn. Add this value to the value at the junction.

	7	8	9	10	11	12
1	15	39	3	26	[0]	25
2	~~0~~	[0]	17	43	60	28
3	[0]	5	23	14	6	5
4	7	5	[0]	14	~~0~~	5
5	3	7	3	[0]	5	7
6	5	3	~~0~~	3	22	[0]

Step 10 Repeat Steps 3 and 4.

Assign	Distance in miles
Surplus at city 1 to city 11	25
Surplus at city 2 to city 8	29
Surplus at city 3 to city 7	27
Surplus at city 4 to city 9	48
Surplus at city 5 to city 10	26
Surplus at city 6 to city 12	30
Total distance = 185 miles	

6.4 MAXIMISATION PROBLEM

Hungarian method is meant for minimisation problems. If the given problem is maximisation type, it has to be converted to minimisation type. Consider the highest value in the matrix and subtract all the values from this highest value. The problem becomes a minimisation problem.

Example 6.6

Five jobs are to be processed and five machines are available. Any machine can process any job and the resulting profit in rupees is as follows

		Machines				
		A	*B*	*C*	*D*	*E*
	1	32	38	40	28	40
	2	40	24	28	21	36
Jobs	3	41	27	33	30	37
	4	22	38	41	36	36
	5	29	33	40	35	39

What is the maximum profit, if the optimum assignment is made?

Solution

This is a maximisation problem. Therefore, convert this into a minimisation problem. Maximum profit is 41. Deduct all the values from 41.

	A	*B*	*C*	*D*	*E*
1	9	3	1	13	1
2	1	17	13	20	5
3	0	14	8	11	4
4	9	3	0	5	5
5	12	8	1	6	2

Step 1 Consider each row. Deduct the smallest value in each row from all the values in that row.

	A	*B*	*C*	*D*	*E*
1	8	2	0	12	0
2	0	16	12	19	4
3	0	14	8	11	4
4	19	3	0	5	5
5	11	7	0	5	1

Step 2 Consider each column. Deduct the smallest value in each column from all the values in that column.

	A	*B*	*C*	*D*	*E*
1	8	[0]	0	7	0
2	[0]	14	12	14	4
3	0	12	8	6	4
4	19	1	[0]	0	5
5	11	5	0	[0]	1

Step 3 Consider the first row, if there is only one zero, enrectangle it. As no assignment can be done in that column, draw a line through that column. Repeat the procedure for other rows.

Step 4 Consider each column. If there is only one zero in that column, enrectangle it and draw a line through that row. If the assignment is made in all rows, stop; otherwise proceed.

Step 5 Determine the minimum value out of those through which the lines are not drawn. Subtract this minimum value from all the values through which lines are not drawn. Add this value at the junction.

	A	*B*	*C*	*D*	*E*
1	9	[0]	1	8	0
2	[0]	13	12	14	3
3	0	11	8	6	3
4	20	1	1	1	5
5	1	4	[0]	0	0

Step 6 Repeat Step 5.

	A	*B*	*C*	*D*	*E*
1	10	[0]	1	8	0
2	[0]	12	11	13	2
3	0	10	7	5	2
4	20	0	[0]	0	4
5	12	4	0	[0]	0

Step 7 Repeat Steps 3, 4 and 5.

	A	*B*	*C*	*D*	*E*
1	12	0	3	8	0
2	0	10	11	11	0
3	0	8	7	3	0
4	22	0	2	0	4
5	14	4	2	0	0

Step 8 Each row and each column must have at least one zero. Column '*C*' does not have zero. Therefore, deduct the least value from all the values.

	A	B	C	D	E
1	12	[0]	1	8	0
2	0	10	9	11	[0]
3	[0]	8	5	3	0
4	22	0	[0]	0	0
5	14	4	0	[0]	0

Assign	Profit
1 to *B*	38
2 to *E*	36
3 to *A*	41
4 to *C*	41
5 to *D*	39
Total profit = Rs. 195	

Example 6.7

A company has six jobs to be processed on six machines. The following table gives the return in rupees, when the *i*th job is assigned to the *j*th machine. How should the job be assigned to the machines so as to maximise the overall return?

		Machines					
		1	2	3	4	5	6
Jobs	1	7	20	56	9	17	25
	2	41	76	70	48	61	46
	3	39	26	89	35	43	31
	4	72	40	25	47	37	30
	5	34	9	55	20	23	16
	6	10	54	51	29	15	26

Solution

This is a maximisation problem. Therefore, convert this into a minimisation problem. Maximum return is Rs. 89. Deduct 89 from all the values.

	1	2	3	4	5	6
1	82	69	33	80	72	64
2	48	13	19	41	28	43
3	50	63	0	54	46	58
4	17	49	64	42	52	59
5	55	80	34	69	66	73
6	79	35	38	60	74	63

Step 1 Consider each row. Deduct the smallest value in each row from all the values in that row.

	1	2	3	4	5	6
1	49	36	0	47	39	31
2	35	0	6	28	15	30
3	50	63	0	54	46	58
4	0	32	47	25	35	42
5	21	46	0	35	32	39
6	39	0	3	25	39	28

Step 2 Consider each column. Deduct the smallest value in each column from all the values in that column.

	1	2	3	4	5	6
1	49	36	0	22	24	3
2	35	0	6	3	0	2
3	50	63	0	29	31	30
4	0	32	47	0	20	14
5	21	46	0	10	17	11
6	39	0	3	0	24	0

Step 3 Consider the first row; if there is only one zero, enrectangle it. As no assignment can be done in that column, draw a line through that column. Repeat the procedure for other rows.

Step 4 Consider each column. If there is only one zero in that column, enrectangle it and draw a line through that row. If the assignment is made in all rows, stop; otherwise proceed.

Step 5 Determine the minimum value out of those through which the lines are not drawn. Subtract this minimum value from all the values through which lines are not drawn. Add this value at the junction.

	1	2	3	4	5	6
1	46	33	0	19	21	0
2	35	0	9	3	0	2
3	47	60	0	26	28	27
4	0	32	50	0	20	14
5	18	43	0	7	14	8
6	39	0	6	0	24	0

Step 6 Repeat Step 5.

	1	2	3	4	5	6
1	46	33	7	19	21	0
2	35	0	16	3	0	2
3	40	53	0	19	21	20
4	0	32	57	0	20	14
5	11	36	0	0	7	1
6	39	0	13	0	24	0

Step 7 Repeat Steps 3, 4 and 5

Assignment	Return
Job 1 to machine 6	25
Job 2 to machine 5	61
Job 3 to machine 3	89
Job 4 to machine 1	72
Job 5 to machine 4	20
Job 6 to machine 2	54
Total revenue = Rs. 321	

Example 6.8

The marketing director of a multi-unit company faces problem of assigning 5 senior managers to 6 zones. From past experience, he knows that the efficiency percentage judged by sales, operating costs, etc., depends on the managers–zone combination. The efficiency of distinct managers is given below.

		Zones					
		1	2	3	4	5	6
	A	73	91	87	82	78	80
Managers	*B*	81	85	69	76	74	85
	C	75	72	83	86	78	91
	D	93	96	86	91	83	82
	E	90	91	79	89	69	76

Find out which zone will be managed by a junior manager due to non-availability of the senior manager.

Solution

As the assignees and resources are not equal, this is an unbalanced assignment problem.

Let '*F*' be the junior manager whose efficiency is zero. As the efficiency is given, this is a maximisation problem.

		Zones					
		1	2	3	4	5	6
	A	73	91	87	82	78	80
	B	81	85	69	76	74	85
Managers	*C*	75	72	83	84	78	91
	D	93	96	86	91	83	82
	E	90	91	79	89	69	76
	F	0	0	0	0	0	0

Convert this into a minimisation problem. Maximum efficiency is 96. Deduct all the values from 96.

Zones

		1	2	3	4	5	6
	A	23	5	9	14	18	16
	B	15	11	27	20	22	11
Managers	*C*	21	24	13	12	18	5
	D	3	0	10	5	13	14
	E	6	5	17	7	27	20
	F	96	96	96	96	96	96

Step 1 Consider each row. Deduct the smallest value in each row by all the values in that row.

	1	2	3	4	5	6
A	18	[0]	4	9	13	11
B	4	0	16	9	11	[0]
C	16	19	8	7	13	0
D	3	0	10	5	13	14
E	1	0	12	2	22	15
F	[0]	0	0	0	0	0

Step 2 Each column has a minimum value of zero. Therefore, same matrix results after the deduction of zero from all the column values.

Step 3 Consider each row; if there is only one zero in that column, enrectangle it and draw a line through that column. Similarly, consider each column; if there is only one zero in that column, enrectangle it and draw a line through that row.

Step 4 Determine the minimum value out of those through which lines are not drawn. Subtract the minimum value from all the values through which lines are not drawn. Add this value at the junction.

	1	2	3	4	5	6
1	17	[0]	3	8	12	11
2	3	0	15	8	10	[0]
3	15	19	7	6	12	0
4	2	0	9	4	12	14
5	[0]	0	11	1	21	15
6	0	1	[0]	0	0	1

Step 5 Repeat Step 4.

	1	2	3	4	5	6
1	17	[0]	2	7	11	11
2	3	0	14	7	9	[0]
3	15	19	6	5	11	0
4	2	0	8	3	11	14
5	[0]	0	10	0	20	15
6	4	2	[0]	0	0	0

Step 6 Repeat Steps 3 and 4.

	1	2	3	4	5	6
1	17	0	0	5	9	11
2	3	0	12	5	7	0
3	15	19	4	3	9	0
4	2	0	6	1	9	14
5	2	2	10	0	20	17
6	3	4	0	0	0	4

Step 7 First column does not have zero. Therefore, deduct the minimum value 2 from all the values in that column.

		Zones					
		1	2	3	4	5	6
	1	15	0	[0]	5	9	11
	2	1	[0]	12	5	7	0
Managers	3	13	19	4	3	9	[0]
	4	[0]	0	6	1	9	14
	5	0	2	10	[0]	20	17
	6	1	4	0	0	[0]	4

Step 8 Repeat Steps 3 and 4.

Assignment	Efficiency
Manager *A* to zone 3	87
Manager *B* to zone 2	85
Manager *C* to zone 6	91
Manager *D* to zone 1	93
Manager *E* to zone 4	69
Manager *F* to zone 5	0

Example 6.9

A manager has to assign four jobs to three production facilities. The assignment of job will increase profits by the amount shown below.

		Facilities		
		1	2	3
Jobs	*A*	1200	900	1350
	B	1000	1100	1250
	C	1150	1000	1000
	D	1300	1200	1050

If only one job can be assigned to a production facility, determine the optimal assignment.

Solution

As the jobs and facilities are not equal, this is an unbalanced assignment problem. Let 4 be the facility with profit zero. This is a maximisation problem; convert this into a minimisation problem.

		Facilities			
		1	2	3	4
Jobs	*A*	1200	900	1350	0
	B	1000	1100	1250	0
	C	1150	1000	1000	0
	D	1300	1200	1050	0

Maximum profit is Rs. 1350. Deduct all the values from Rs. 1350.

	1	2	3	4
A	150	450	0	1350
B	350	250	100	1350
C	200	350	350	1350
D	50	150	300	1350

By Ticking method

Step 1 Consider each row. Deduct the smallest value in each row from all the values in that row.

	1	2	3	4
A	150	450	0	1350
B	250	150	0	1250
C	0	150	150	1150
D	0	100	250	1300

Step 2 Consider each column. Deduct the smallest value in each column from all the values in that row.

	1	2	3	4	
A	150	350	[0]	200	√
B	250	50	X	100	√
C	X	50	150	[0]	
D	X	[0]	250	150	
			√		

Step 3 Consider each row, if there is only one zero, enrectangle it, and cross the zeroes in that column. Similarly, consider each column, if there is only one zero, enrectangle it and cross the zeroes in that row.

Step 4 Consider the row to which assignment is not made and put a tick mark against it.

Step 5 Look for zeroes in that ticked row. Put a tick mark in the corresponding column.

Step 6 Observe in that ticked column. If there is an assigned cell, put a tick mark in that row.

Step 7 Draw lines through unticked rows and ticked columns.

Step 8 Determine the minimum value out of these through which lines are not drawn. Subtract the minimum values from all the values through which lines are not drawn. Add this value at the junction.

	1	2	3	4
A	100	300	[0]	150
B	200	[0]	X	50
C	X	50	200	[0]
D	[0]	X	300	150

Step 9 Repeat Steps 3 and 4.

Assign	Profit
Job *A* to facility 3	1350
Job *B* to facility 2	1100
Job *C* to facility 4	0
Job *D* to facility 1	300
Total profit = Rs. 3750	

Example 6.10

Five workers are available to work with five machines and the respective costs in rupees associated with each worker and machine assignment is given below. A sixth machine is available to replace one of the existing machines and the associated costs are also given below. Determine whether the new machine can be accepted.

		Machine					
		m_1	m_2	m_3	m_4	m_5	m_6
	W_1	12	3	6	∞	5	9
	W_2	4	11	∞	5	∞	8
Worker	W_3	8	2	10	9	7	5
	W_4	∞	7	8	6	12	10
	W_5	5	8	9	4	6	∞

Solution

First, find out the cost with 5 machines and 5 workers. Then find out the cost with 5 workers and 6 machines. If the cost is less in the second case, replacement is necessary.

The given matrix represents the machining cost. Therefore, it is a minimisation problem.

	m_1	m_2	m_3	m_4	m_5
W_1	12	3	6	∞	5
W_2	4	11	∞	5	∞
W_3	8	2	10	9	7
W_4	∞	7	8	6	12
W_5	5	8	9	4	6

By ticking method

Step 1 Consider each row. Deduct the smallest value in each row from all the values in that row.

	m_1	m_2	m_3	m_4	m_5
W_1	9	0	3	∞	2
W_2	0	7	∞	1	∞
W_3	6	0	8	7	5
W_4	∞	1	2	0	6
W_5	1	4	5	0	2

Step 2 Consider each column. Deduct the smallest value in each column from all the values in that column.

	m_1	m_2	m_3	m_4	m_5
W_1	9	~~0~~	1	∞	[0]
W_2	[0]	7	∞	1	∞
W_3	6	[0]	6	7	3
W_4	∞	1	[0]	~~0~~	4
W_5	1	4	3	[0]	~~0~~

Step 3 Consider each row; if there is only one zero, enrectangle it and cross the zeroes in that column. Similarly, consider each column. If there is only one zero, enrectangle it and cross the zeroes in that row.

Assign	Cost
Worker 1 to machine m_5	5
Worker 2 to machine m_1	4
Worker 3 to machine m_2	2
Worker 4 to machine m_3	8
Worker 5 to machine m_4	4
Total cost = Rs. 23	

	m_1	m_2	m_3	m_4	m_5	m_6
W_1	12	3	6	∞	5	9
W_2	4	11	∞	5	∞	8
W_3	8	2	10	9	7	5
W_4	∞	7	8	6	12	10
W_5	5	8	9	4	6	∞
W_6	0	0	0	0	0	0

Repeat Steps 1, 2 and 3.

	m_1	m_2	m_3	m_4	m_5	m_6	
W_1	9	[0]	3	∞	2	6	√
W_2	[0]	7	∞	1	∞	4	
W_3	6	✗	8	7	5	3	√
W_4	∞	1	2	[0]	6	4	√
W_5	1	4	5	✗	2	∞	√
W_6	✗	✗	[0]	✗	✗	✗	
		√		√			

Step 4 Consider the row without assignment. Put a tick mark against it.

Step 5 Look for zeroes in that ticked row. Put a tick mark against the column containing zero.

Step 6 Observe in that ticked column. If there is an assigned cell, put a tick mark in that row.

Step 7 Draw lines through the ticked column and the unticked rows.

Step 8 Determine the minimum value out of those through which lines are not drawn. Subtract this value from all the values through which lines are not drawn. Add this value at the junction.

	m_1	m_2	m_3	m_4	m_5	m_6	
W_1	8	[0]	2	∞	1	5	√
W_2	[0]	8	∞	2	∞	4	
W_3	5	✗	7	7	4	2	√
W_4	∞	1	1	[0]	5	3	√
W_5	✗	4	4	✗	1	∞	√
W_6	✗	1	[0]	1	✗	✗	
		√		√			

Step 9 Repeat Steps 4, 5, 6 and 7.

	m_1	m_2	m_3	m_4	m_5	m_6
W_1	7	☒	1	∞	[0]	4
W_2	[0]	9	∞	3	∞	4
W_3	4	[0]	6	7	3	1
W_4	∞	1	[0]	☒	4	2
W_5	☒	4	3	[0]	☒	∞
W_6	☒	2	☒	2	☒	[0]

Assign	Cost
Worker 1 to machine 5	5
Worker 2 to machine 1	4
Worker 3 to machine 2	2
Worker 4 to machine 3	8
Worker 5 to machine 4	4
Worker 6 to machine 6	0
Total cost = Rs. 23	

Replacement of machine is not necessary.

Example 6.11

A company has purchased 4 new machines of different types. There are four locations available in the shop where the machines could be installed. Some of these locations are more desirable than others because of their proximities to work centres that would have a heavy work floor to and from these machine. The estimated cost per unit time of material handling involving each machine is given below.

		Locations			
		1	2	3	4
Machines	*A*	4	6	10	5
	B	7	4	–	5
	C	–	6	9	6
	D	9	3	7	2

Solution

Assign new machines to available location so as to minimise the total cost of material handling. Machine *B* cannot be assigned to location 3 and Machine *C* cannot be assigned to location 1. Therefore, assign ∞ to these cells.

		Locations			
		1	2	3	4
Machines	*A*	4	6	10	5
	B	7	4	∞	5
	C	∞	6	9	6
	D	9	3	7	2

Step 1 Consider each row. Deduct the smallest value in each row from all the values in that row.

Locations

Machines		1	2	3	4
	A	0	2	6	1
	B	3	0	∞	1
	C	∞	0	3	0
	D	7	1	5	0

Step 2 Consider each column. Deduct the smallest value in each column from all the values in that column.

	1	2	3	4
A	[0]	2	3	1
B	3	[0]	∞	1
C	∞	~~0~~	[0]	~~0~~
D	7	1	2	[0]

Step 3 Consider each row. If there is only one zero, enrectangle it and cross the zeroes in that column. Similarly, consider each column; if there is only one zero, enrectangle it and cross the zeroes in that row.

Assign	Handling cost
Machine *A* to location 1	4
Machine *B* to location 2	4
Machine *C* to location 3	9
Machine *D* to location 4	2
Total handling cost = Rs. 19	

Example 6.12

A company has four territories, and four salesmen are available for the assignment. The territories are not equally rich in their potential. It is estimated that a typical salesman operating in each territory would bring the following annual sales.

Territory 1 Rs. 60,000

Territory 2 Rs. 50,000

Territory 3 Rs. 40,000

Territory 4 Rs. 30,000

The four salesmen are also considered to differ in their ability. It is estimated that working under the same condition, their yearly sales would be proportionally as follows

$A = 7$

$B = 5$

$C = 5$

$D = 4$

If the criteria is to maximise the expected sales, the intuitive answer is to assign the best salesman to the richest territory, the next best to the second richest territory, etc. Verify this answer by the assignment method.

Solution

Sales when A is assigned to territory 1 $= \frac{7}{21} \times 60,000$

Sales when A is assigned to territory 2 $= \frac{7}{21} \times 50,000$

Sales when A is assigned to territory 3 $= \frac{7}{21} \times 40,000$

Sales when A is assigned to territory 4 $= \frac{7}{21} \times 30,000$

Similarly,

Sales when B/C is assigned to territory 1 $= \frac{5}{21} \times 60,000$

Sales when B/C is assigned to territory 2 $= \frac{5}{21} \times 50,000$

Sales when B/C is assigned to territory 3 $= \frac{5}{21} \times 40,000$

Sales when B/C is assigned to territory 4 $= \frac{5}{21} \times 30,000$

and

Sales when D is assigned to territory 1 $= \frac{4}{21} \times 60,000$

Sales when D is assigned to territory 2 $= \frac{4}{21} \times 50,000$

Sales when D is assigned to territory 3 $= \frac{4}{21} \times 40,000$

Sales when D is assigned to territory 4 $= \frac{4}{21} \times 30,000$

As the base 21 is common, we can ignore the base value and 1000s.

		Territory			
		1	2	3	4
Salesman	*A*	42	35	28	21
	B	30	25	20	15
	C	30	25	20	15
	D	24	20	16	12

It is a problem on maximisation of sales. Therefore, convert this into a minimisation problem. Deduct all values in the matrix from the highest value 42.

	1	2	3	4
A	0	7	14	21
B	12	17	22	27
C	12	17	22	27
D	18	22	26	30

Step 1 Consider each row. Deduct the smallest value in each row from all the values in that row.

	1	2	3	4
A	0	7	14	21
B	0	5	10	15
C	0	5	10	15
D	0	4	8	12

Step 2 Consider each column. Deduct the smallest value in each column from all the values in that column.

	1 √	2	3	4	
A	[0]	3	6	9	
B	✗	1	2	3	√
C	✗	1	2	3	√
D	✗	[0]	✗	✗	
	√				

Step 3 Consider each row; if there is only one zero, enrectangle it and cross the zeroes in that column. Similarly, consider each column, if there is only one zero, enrectangle it and cross the zeroes in that row.

Step 4 Consider each row to which assignment is not made and put a tick mark against that row.

Step 5 Look for zeroes in that ticked row and put a tick mark in the corresponding column.

Step 6 Observe in that ticked column, if there is an assigned cell and put a tick mark in that row.

Step 7 Draw lines through the ticked column and the unticked rows.

Step 8 Determine the minimum value out of those through which the lines are not drawn. Subtract this value from all the values through which lines are not drawn. Add this value at the junction.

	1	2	3	4
A	1	3	6	9
B	0	0	1	2
C	0	0	1	2
D	1	0	0	0

Step 9 Row A does not have zero. Therefore, deduct the least value 1 from all the values in that row.

	1	2	3	4	
A	[0]	2	5	8	√
B	✗	[0]	1	2	√
C	✗	✗	1	2	√
D	1	✗	[0]	✗	
	√	√			

Step 10 Repeat Steps 3 through 8.

	1	2	3	4
A	[0]	2	4	7
B	✗	[0]	✗	1
C	✗	✗	[0]	1
D	2	1	✗	[0]

Assign	Sales
A to territory 1	20,000
B to territory 2	11,905
C to territory 3	9524
D to territory 4	5714
Total sales = 47,143 units	

Example 6.13

An airline that operates 7 days a week has the time table shown below. The crew must have a minimum layover of 5 hours between flights. Obtain the pairing of flights that minimises layover time away from home for any given pairs. The crew will be based at the city that results the smaller layover time. For each pair, also mention the town where the crew should be based.

Flight No.	Delhi departure	Jaipur arrival	Flight No.	Jaipur departure	Delhi arrival
1	7.00 AM	8.00 AM	101	8.00 AM	9.15 AM
2	8.00 AM	9.00 AM	102	8.30 AM	9.43 AM
3	1.30 PM	2.30 PM	103	12 noon	1.15 PM
4	6.30 PM	7.30 PM	104	5.30 PM	6.45 PM

Solution

Flight based at Delhi	Minimum Layover time = 5 hours		
	Reaches Jaipur	Leaves Jaipur	Layover time
Reaches Jaipur as $F1$ and leaves Jaipur as $F101$. The flight cannot leave same day, as the minimum layover time is 5 hours. Therefore, it leaves the next day at 8 AM.	8.00 AM	8.00 AM	24.00 hours
Reaches Jaipur as $F1$ and leaves Jaipur as $F102$	8.00 AM	8.30 AM	24.30 hours
Reaches Jaipur as $F1$ and leaves Jaipur as $F103$	8.00 AM	12 noon	28.00 hours
Reaches Jaipur as $F1$ and leaves Jaipur as $F104$ Flight can leave as $F104$ on the same day	8.00 AM	5.30 PM	9.30 hours
Reaches Jaipur as $F2$ and leaves Jaipur as $F101$	9.00 AM	8.00 AM	23.00 hours
Reaches Jaipur as $F2$ and leaves Jaipur as $F102$	9.00 AM	8.30 AM	23.30 hours
Reaches Jaipur as $F2$ and leaves Jaipur as $F103$	9.00 AM	12 noon	27.00 hours
Reaches Jaipur as $F2$ and leaves Jaipur as $F104$	9.00 AM	5.30 PM	8.30 hours
Reaches Jaipur as $F3$ and leaves Jaipur as $F101$	2.30 PM	8.00 AM	17.30 hours
Reaches Jaipur as $F3$ and leaves Jaipur as $F102$	2.30 PM	8.30 AM	18.00 hours
Reaches Jaipur as $F3$ and leaves Jaipur as $F103$	2.30 PM	8.30 AM	21.30 hours
Reaches Jaipur as $F3$ and leaves Jaipur as $F104$	2.30 PM	5.30 PM	27.00 hours
Reaches Jaipur as $F4$ and leaves Jaipur as $F101$	7.30 PM	8.00 AM	12.30 hours
Reaches Jaipur as $F4$ and leaves Jaipur as $F102$	7.30 PM	8.30 AM	13.00 hours
Reaches Jaipur as $F4$ and leaves Jaipur as $F103$	7.30 PM	12 noon	16.30 hours
Reaches Jaipur as $F4$ and leaves Jaipur as $F104$	7.30 PM	5.30 PM	22.00 hours

Flight based at Jaipur	Reaches Delhi	Leaves Delhi	Layover time
Reaches Delhi as $F101$ and leaves Delhi as $F1$	9.15 AM	7.00 AM	21.45 hours
Reaches Delhi as $F101$ and leaves Delhi as $F2$	9.15 AM	8.00 AM	22.45 hours
Reaches Delhi as $F101$ and leaves Delhi as $F3$	9.15 AM	1.30 PM	28.15 hours
Reaches Delhi as $F101$ and leaves Delhi as $F4$	9.15 AM	6.30 PM	9.15 hours
Reaches Delhi as $F102$ and leaves Delhi as $F1$	9.45 AM	7.00 AM	21.15 hours
Reaches Delhi as $F102$ and leaves Delhi as $F2$	9.45 AM	8.00 AM	22.15 hours
Reaches Delhi as $F102$ and leaves Delhi as $F3$	9.45 AM	1.30 PM	27.45 hours
Reaches Delhi as $F102$ and leaves Delhi as $F4$	9.45 AM	6.30 PM	8.45 hours
Reaches Delhi as $F103$ and leaves Delhi as $F1$	1.15 PM	7.00 AM	17.45 hours
Reaches Delhi as $F103$ and leaves Delhi as $F2$	1.15 PM	8.00 AM	18.15 hours
Reaches Delhi as $F103$ and leaves Delhi as $F3$	1.15 PM	1.30 PM	24.15 hours
Reaches Delhi as $F103$ and leaves Delhi as $F4$	1.15 PM	6.30 PM	5.15 hours
Reaches Delhi as $F104$ and leaves Delhi as $F1$	6.45 PM	7.00 AM	12.45 hours
Reaches Delhi as $F104$ and leaves Delhi as $F2$	6.45 AM	8.00 AM	13.15 hours
Reaches Delhi as $F104$ and leaves Delhi as $F3$	6.45 AM	1.30 PM	18.45 hours
Reaches Delhi as $F104$ and leaves Delhi as $F4$	6.45 AM	6.30 PM	23.45 hours

Formulated assignment matrices are as follows

Delhi Based

	101	102	103	104
1	24	24.30	28	9.30
2	23	23.30	27	8.30
3	17.30	18	21.30	27
4	12.30	13	16.30	22

Jaipur Based

	101	102	103	104
1	21.45	21.15	17.45	12.45
2	22.45	22.15	18.15	13.15
3	28.15	27.45	24.15	18.45
4	9.15	8.45	5.15	23.45

Our objective is to minimise the layover time. Therefore, we take the minimum of the corresponding layover times.

	101	102	103	104
1	21.45	31.15	17.45	9.30*
2	22.45	22.15	18.15	9.30*
3	17.30*	18*	21.3*	18.45
4	9.15	8.45	5.15	22*

* indicates the crew based at Delhi

Convert the hour–minutes matrix to hours matrix by multiplying all the values by 4.

	101	102	103	104
1	87	85	71	38*
2	91	87	73	34*
3	70*	72*	86*	75
4	37	35	21	88*

Solution by Hungarians (ticking) method

Step 1 Consider each row. Deduct the smallest value in each row from all the values in that row.

	101	102	103	104
1	49	47	33	0*
2	57	55	39	0*
3	0*	2*	16*	5
4	16	14	0	67*

Step 2 Consider each column. Deduct the smallest value in each column from all the values in that column.

	101	102	103	104	
1	49	45	33	[0]*	√
2	57	53	39	~~0~~*	√
3	[0]*	~~0~~*	16*	5	
4	16	12	[0]	67	
				√	

Step 3 Consider each row; if only one zero is there, enrectangle it and cross the zeroes in that column. Similarly consider each column, if only one zero is there, enrectangle it and cross the zeros in that row.

Step 4 Consider the row to which assignment is not made; mark and put a tick mark against that row.

Step 5 Look for zeroes in that ticked row. Put a tick mark in the corresponding column.

Step 6 Observe in that column; if there is an assigned cell, put a tick mark in that row.

Step 7 Draw lines through the unticked rows and the ticked column.

	101	102	103	104	
1	16	12	~~0~~	~~0~~*	√
2	24	20	6	[0]	√
3	[0]*	~~0~~*	10*	38	
4	16	12	[0]	100*	√
			√	√	

Step 8 Determine the minimum value out of those, through which lines are not drawn. Deduct this minimum value from all the values through which lines are not drawn.

Step 9 Repeat Steps 3 through 8.

	101	102	103	104
1	4	[0]	~~0~~	~~0~~*
2	12	8	6	[0]*
3	[0]*	~~0~~*	28*	50
4	4	~~0~~	[0]	100*

	Layover time
Flight 102 comes back as Flight 1–Jaipur-based	21.15 hours
Flight 2 comes back as Flight 104–Delhi-based	8.30 hours
Flight 3 comes back as Flight 101–Delhi-based	17.30 hours
Flight 103 comes back as Flight 4–Jaipur-based	5.15 hours
Total layover time = 52.30 hours	

Example 6.14

There are five jobs to be assigned, one each to five machines. The associated cost is as follows

		Machine				
		1	2	3	4	5
	A	11	17	8	16	20
	B	9	7	12	6	15
Job	*C*	13	16	15	12	16
	D	21	24	17	28	26
	E	4	10	12	11	15

Solve this problem by Bunch and Bound method.

Solution

Step 1 Consider each row. Deduct the smallest value in each row from all the values in that row.

Job	Machine 1	2	3	4	5
A	3	9	0	8	12
B	3	1	6	0	9
C	1	4	3	0	4
D	4	7	0	11	9
E	4	0	2	1	5

Step 2 Consider each column. Deduct the smallest value in each column from all the values in that column.

	1	2	3	4	5
A	2	9	0	8	8
B	2	1	6	0	5
C	0	4	3	0	0
D	3	7	0	11	5
E	3	0	2	1	1

Step 3 Consider each row. Determine penalty cost for each of the zero cells.

Penalty cost = Minimum in the corresponding row + Minimum in the corresponding column.

	1	2	3	4	5
A	2	9	0\2	8	8
B	2	1	6	0\1	5
C	0\2	4	3	0\0	0\1
D	3	7	0\3	11	5
E	3	0\2	2	1	1

D to 3

Step 4 Allocate to that cell which has maximum penalty cost. Here it is *D*-3. Delete the corresponding row and column.

	1	2	4	5
A	2	9	8	8
B	2	1	0\	5
C	0\	4	0\	0\
E	3	0\	1	1

Step 5 First row does not have zero. Therefore, deduct the minimum value 2 from all the values in that row.

	1	2	4	5
A	0\6	7	6	6
B	2	1	0\1	5
C	0\0	4	0\0	0\1
E	3	0\2	1	1

A to 1

Step 6 Repeat Steps 3 and 4.

	2	4	5
B	1	0\1	5
C	4	0\0	0\1
E	0\2	1	1

E to 2

Step 7 Repeat Steps 3 and 4.

	4	5
B	0\5	5
C	0\0	0\5

B to 4

C to 5

Assign	Cost
Job D to machine 3	17
Job A to machine 1	11
Job E to machine 2	10
Job B to machine 4	6
Job C to machine 5	16
Total cost = Rs. 60	

6.5 TRAVELLING SALESMAN PROBLEMS

This is similar to the assignment problems. A travelling salesman wants to minimise the total distance travelled during his visit to 'n' cities. A similar problem arises when 'n' items say $A_1 = 1, 2, 3, ..., n$ are to be produced on a machine continuously which has set-up costs of C_{ij} where $i = 1, 2, 3, ..., n$ and $j = 1, 2, 3..., n$.

Note that $C_{ij} = \infty$ when $i = j$, i.e., we do not produce A_i again after A_i. The matrix is given below

	1	2	3	.	.	.	n
1	∞	C_{12}	C_{13}	.	.	.	C_{1n}
2	C_{21}	∞	C_{23}				C_{2n}
3	.		.				.
.	.			.			.
.	.				.		.
.	.					.	.
n	C_{n1}	C_{n2}	.	.	.	.	∞

This problem is very similar to the assignment problem with two more restrictions.

1. We cannot select an element along the leading diagonal as we do not follow A_1 again by A_1. That is why we use the cost ∞.
2. We do not produce an item again until all the items are produced once.

A travelling salesman problem is said to be symmetric if the cost for $A_{ij} = A_{ji}$. Otherwise it is unsymmetric.

Example 6.15

Given is the matrix of set-up cost. Show how to sequence production so as to minimise the set-up cost per cycle.

	A_1	A_2	A_3	A_4	A_5
A_1	∞	2	5	7	1
A_2	6	∞	3	8	2
A_3	8	7	∞	4	7
A_4	12	4	6	∞	5
A_5	1	3	2	8	∞

Solution

Step 1 Consider each row. Deduct the smallest value in each row from all the values in that row.

	A_1	A_2	A_3	A_4	A_5
A_1	∞	1	4	6	0
A_2	4	∞	1	6	0
A_3	4	3	∞	0	3
A_4	8	0	2	∞	1
A_5	0	2	1	7	∞

Step 2 Consider each column. Deduct the smallest value in each column from all the values in that column.

	A_1	A_2	A_3	A_4	A_5
A_1	∞	1	3	6	0
A_2	4	∞	0	6	0
A_3	4	3	∞	0	3
A_4	8	0	1	∞	1
A_5	0	2	0	7	∞

Step 3 Determine the penalty cost for each zero cell.

	A_1	A_2	A_3	A_4	A_5
A_1	∞	1	3	6	0\1
A_2	4	∞	0\0	6	0\0
A_3	4	3	∞	0\9	3
A_4	8	0\2	1	∞	1
A_5	0\4	2	0\0	7	∞

A_3 - A_4

Step 4 Allocate to that cell which has the maximum penalty cost. Here, it is A_3-A_4. Delete corresponding row and column.

$$\therefore A_4\text{-}A_3 = \infty$$

	A_1	A_2	A_3	A_5	
A_1	∞	1	3	0\1	
A_2	4	∞	0\0	0\0	
A_4	8	0\2	∞	1	A_5 - A_1
A_5	0\4	2	0\0	∞	

Step 5 Allocate to the cell A_5-A_1. Delete the corresponding row and column.

$\therefore$ A_1-A_3 is ∞

	A_2	A_3	A_5
A_1	1	3	∞
A_2	∞	0\	0\
A_4	0\	∞	1

Step 6 Row A_1 does not have zero. Deduct the least value from all the values in that row.

	A_2	A_3	A_5	
A_1	0\2	2	∞	
A_2	∞	0\2	0\1	A_2 - A_3
A_4	0\1	∞	1	

Step 7 Allocate to the cell A_2-A_3. Delete the corresponding row and column.

	A_2	A_5
A_1	0\	∞
A_4	0\	1

Step 8 Column A_5 does not have zero. Deduct the least value from all the values in that column.

	A_2	A_5	
A_1	0\∞	∞	
A_4	0\0	0\∞	A_1-A_2

Allocate A_1 to A_2 and A_4 to A_5. Sequence of production is A_1-A_2-A_3-A_4-A_5-A_1

Set-up cost

A_1-A_2	2
A_2-A_3	3
A_3-A_4	4
A_4-A_5	5
A_5-A_1	1
Total cost = Rs. 15	

Example 6.16

A machine operator processes four types of items on his machine each week and must have a sequence for them. The set-up cost per change depends on the item presently on the machine and the item is to be made according to the following table.

	A	*B*	*C*	*D*
A	∞	4	7	3
B	4	∞	6	3
C	7	6	∞	7
D	3	3	7	∞

If he produces each type of item only once on his machine in order to minimise the total set-up cost, determine the optical assignment.

Solution

Step 1 Consider each row. Deduct the smallest value in each row from all the values in that row.

	A	*B*	*C*	*D*
A	∞	1	4	0
B	1	∞	3	0
C	1	0	∞	1
D	0	0	4	∞

Step 2 Consider each column. Deduct the smallest value in each column from all the values in that column.

	A	*B*	*C*	*D*
A	∞	1	1	0
B	1	∞	0	0
C	1	0	∞	1
D	0	0	1	∞

Step 3 Determine the penalty cost for zero cell.

	A	*B*	*C*	*D*
A	∞	1	1	0\1
B	1	∞	0\1	0\0
C	1	0\1	∞	1
D	0\1	0\0	1	∞

A-D

Step 4 Allocate to the cell with highest penalty cost. Allocate *A* to *D*. ∴ *D* to *A* is ∞.

	A	*B*	*C*
B	1	∞	0\
C	1	0\	∞
D	∞	0\	1

Step 5 Column A has no zero cell. Deduct the smallest value from all the values in that column.

	A	B	C	
B	0\0	∞	0\1	
C	0\0	0\0	∞	B - C
D	∞	0\1	1	

Step 6 Allocate to the cell B to C. ∴ C to B is ∞.

	A	B	
C	0\∞	∞	C-A
D	∞	0\∞	D-B

Step 7 Allocate to the cell C to A and D to B.

Sequence of items processed is A-D-B-C = A

Total set-up cost = 3 + 3 + 6 + 7 = Rs. 19

Example 6.17

A salesman wants to visit four cities B, C, D and E and then come back to his headquarters A. The distance matrix for these cities is given below. Find the optimal route for the salesman and the corresponding total distance he has to cover in one such cycle.

	A	B	C	D	E
A	0	50	90	20	140
B	50	0	40	80	70
C	90	40	0	120	110
D	20	80	120	0	65
E	140	70	110	65	0

Solution

Salesman can come back to the same city only after visiting all other cities. Therefore, the values of diagonals are taken as ∞.

	A	B	C	D	E
A	∞	50	90	20	140
B	50	∞	40	80	70
C	90	40	∞	120	110
D	20	80	120	∞	65
E	140	70	110	65	∞

Step 1 Consider each row. Deduct the smallest value in each row from all the values in that row.

	A	B	C	D	E
A	∞	30	70	0	120
B	10	∞	0	40	30
C	50	0	∞	80	70
D	0	60	100	∞	45
E	75	5	45	0	∞

Step 2 Consider each column. Deduct the smallest value in each column from all the values in that column.

	A	B	C	D	E	
A	∞	30	70	0\30	90	
B	10	∞	0\45	40	0\15	
C	50	0\45	∞	80	40	B-C
D	0\25	60	100	∞	15	
E	75	5	45	0\	∞	

Step 3 Determine the penalty cost for zero cells. Allocate to the cell B to C. Delete row B and Column C. Therefore, C to B is ∞.

	A	B	D	E
A	∞	30	0\	90
C	50	0\	80	40
D	0\	60	∞	15
E	75	5	0\	∞

Step 4 Deduct the smallest value in column E from all the values in that column and repeat Step 3.

	A	B	D	E	
A	∞	30	0\30	75	
C	50	0\30	80	25	D-A
D	0\50	60	∞	0\25	
E	75	5	0\5	∞	

Step 5 Delete Row D and Column A. Allocate to the cell D to A.

Therefore, A to D is ∞.

	B	D	E
A	30	∞	75
C	0\	80	25
E	5	0\	∞

Step 6 Deduct the smallest value in row A from all the values in that row.

	B	*D*	*E*
A	0\	∞	45
C	0\	80	25
E	5	0\	∞

Step 7 Deduct the smallest value in column E from all the values in that column and determine the penalty cost for all zero cells.

	B	*D*	*E*	
A	0\20	∞	20	
C	0\0	80	0\20	*E-D*
E	5	0\85	∞	

Step 8 Delete E and D. Allocate to the cell E to D.

	B	*E*	
A	0\20	20	*A-B*
C	0\0	0\20	*C-E*

Step 9 Allocate A to B and C to E.

Optimal route for the salesman is *A-B-C-E-D-A*

Total distance covered 50 + 40 + 110 + 65 + 20 = 285 km

Example 6.18

Solve the transportation problem using the given data. $C_{12} = 20$, $C_{13} = 4$, $C_{14} = 10$ $C_{23} = 5$, $C_{24} = 6$, $C_{25} = 10$, $C_{35} = 6$, $C_{45} = 20$. The value of C_{ij} is not given in the problem if there is no route between cities i and j.

Solution

$C_{ij} = C_{ji}$. Therefore, $C_{21} = 20$, $C_{31} = 4$, $C_{41} = 10$, $C_{32} = 5$, $C_{42} = 6$, $C_{52} = 10$, $C_{53} = 6$, $C_{54} = 20$.

	1	2	3	4	5
1	∞	20	4	10	∞
2	20	∞	5	6	10
3	4	5	∞	∞	6
4	10	6	∞	∞	20
5	∞	10	6	20	∞

Step 1 Consider each row. Deduct the smallest value in each row from all the values in that row.

	1	2	3	4	5
1	∞	16	0	6	∞
2	15	∞	0	1	5
3	0	1	∞	∞	2
4	4	0	∞	∞	14
5	∞	4	0	14	∞

Step 2 Consider each column. Deduct the smallest value in each column from all the values in that column. Determine penalty cost for each zero cell.

	1	2	3	4	5	
1	∞	16	0\5	5	60	
2	15	∞	0\0	0\5	3	
3	0\4	1	∞	∞	0\3	4 to 2
4	4	0\5	∞	∞	12	
5	∞	4	0\4	13	∞	

Step 3 Allocate to the cell with highest penalty cost. Delete row 4 and column 2. Therefore, 2 to 4 is ∞.

	1	3	4	5	
1	∞	0\5	5	∞	
2	15	0\0	0\5	3	
3	0\15	∞	∞	0\3	3 to 1
5	∞	0\13	13	∞	

Step 4 Allocate to the cell 3 to 1. Therefore, 1 to 3 is ∞.

	3	4	5
1	∞	5	∞
2	0\	0\	3
5	0\	13	∞

Step 5 Row 1 has no zero. Deduct the minimum in that row.

	3	4	5
1	∞	0\	∞
2	0\	0\	3
5	0\	13	∞

Step 6 Column 5 has no zero. Deduct the minimum in that column.

	3	4	5	
1	∞	0\0	∞	
2	0\0	0\0	0\∞	2 to 5
5	0\13	13	∞	

Step 7 Allocate to the cell 2 to 5. Therefore, 5 to 2 is ∞.

	3	4	
1	∞	0\∞	1 to 4
5	0\∞	13	5 to 3

Step 8 Allocate to the cell 1 to 4 and 5 to 3.

Sequence 1-4-2-5-3-1

Total cost = 10 + 6 + 10 + 6 + 4 = Rs. 36

Example 6.19

Solve the following travelling salesman problem.

	A	*B*	*C*	*D*	*E*	*F*	*G*
A	∞	6	12	6	4	8	1
B	6	∞	10	5	4	3	3
C	8	7	∞	11	3	11	8
D	5	4	11	∞	5	8	6
E	5	2	7	∞	∞	4	7
F	6	3	11	5	4	∞	2
G	2	3	9	7	4	2	∞

Solution

Step 1 Consider each row. Deduct the smallest value in each row by all the values in that row.

	A	*B*	*C*	*D*	*E*	*F*	*G*
A	∞	5	11	5	3	7	0
B	3	∞	7	2	1	0	0
C	5	4	∞	8	0	8	5
D	1	0	7	∞	1	4	2
E	3	0	5	∞	∞	2	5
F	4	1	9	3	2	∞	0
G	0	1	7	2	2	1	∞

Step 2 Consider each column. Deduct the smallest value in each column by all the values in that column and determine the penalty cost for each zero cell.

	A	*B*	*C*	*D*	*E*	*F*	*G*
A	∞	5	6	3	3	7	0 (3)
B	3	∞	2	0 (1)	1	0 (1)	0 (0)
C	5	4	∞	6	0 (5)	8	5
D	1	0 (2)	2	∞	1	4	2
E	3	0 (0)	0 (2)	∞	∞	2	5
F	4	1	4	1	2	∞	0 (1)
G	0 (2)	1	2	3	2	1	∞

C to *E*

Step 3 Allocate *C* to *E*. Therefore *E* to *C* is ∞.

	A	*B*	*C*	*D*	*F*	*G*
A	∞	5	6	3	7	0 (3)
B	3	∞	2	0 (1)	0 (1)	0 (0)
D	1	0 (1)	2	∞	4	2
E	3	0 (0)	∞	∞	2	5
F	4	1	4	1	∞	0 (1)
G	0 (2)	1	2	3	1	∞

A to *G*

Step 4 Allocate *A* to *G*. *G* to *A* is ∞. Therefore,

	A	*B*	*C*	*D*	*F*
B	3	∞	2	0	0
D	1	0	2	∞	4
E	3	0	∞	∞	2
F	4	1	4	1	∞
G	∞	1	2	3	1

Step 5 Deduct the smallest value from row *F* and row *G*.

	A	*B*	*C*	*D*	*F*
B	3	∞	2	0	0
D	1	0	2	∞	4
E	3	0	∞	∞	2
F	3	0	3	0	∞
G	∞	0	1	2	0

Step 6 Deduct the smallest value from column *A* and column *C*.

	A	*B*	*C*	*D*	*F*	
B	2	∞	1	0 \ 0	0 \ 0	
D	0 \ 2	0 \ 0	1	∞	4	*D* to *A*
E	2	0 \ 2	∞	∞	2	
F	2	0 \ 2	2	0 \ 0	∞	
G	∞	0 \ 0	0 \ 1	2	0 \ 0	

Step 7 Allocate *D* to *A*.

	B	C	D	F	
B	∞	1	0 \ 0	0 \ 0	
E	0 \ 2	∞	∞	2	*E* to *B*
F	0 \ 0	2	0 \ 0	∞	
G	0 \ 0	0 \ 1	2	0 \ 0	

Step 8 Allocate *E* to *B*.

	C	*D*	*F*	
B	1	0 \ 0	0 \ 1	
F	2	0 \ 2	∞	*F* to *D*
G	0 \ 1	2	0 \ 0	

Step 9 Allocate *F* to *D*.

	C	*F*	
B	1	0 \ 1	*B* to *F*
G	0 \ 1	0 \ 0	*G* to *C*

Step 10 Allocate *B* to *F* and *G* to *C*.

Sequence *A-G-C-E-B-F-D-A*

Distance travelled 1 + 9 + 3 + 2 + 3 + 5 + 5 = 28 km

Example 6.20

A toy manufacturer processes five types of dolls on his machine each week and he must choose a sequence for them. The set-up cost for change depends on the item presently on the machine and the item to be made, according to the following table.

	A	*B*	*C*	*D*	*E*
A	∞	14	17	13	14
B	14	∞	16	13	14
C	17	16	∞	17	15
D	13	13	17	∞	17
E	14	14	15	17	∞

If he produces each type of item only once each week, how should he sequence the items on his machine in order to minimise the total set-up cost.

Solution

Step 1 Deduct the smallest value in each row from all the elements in that row.

	A	B	C	D	E
A	∞	1	4	0	1
B	1	∞	3	0	1
C	2	1	∞	2	0
D	0	0	4	∞	4
E	0	0	1	3	∞

Step 2 Deduct the smallest value in each column from all the values in that column. Determine penalty cost at zero cell.

	A	B	C	D	E
A	∞	1	3	0^{1}	1
B	1	∞	2	0^{1}	1
C	2	1	∞	2	0^{2}
D	0^{0}	0^{0}	3	∞	4
E	0^{0}	0^{0}	0^{2}	3	∞

C to E

Step 3 Allocate C to E. E to C is ∞. Therefore,

	A	B	C	D
A	∞	1	3	0
B	1	∞	2	0
D	0	0	3	∞
E	0	0	∞	3

Step 4 Deduct smallest value in column C.

	A	B	C	D
A	∞	1	1	0^{1}
B	1	∞	0^{1}	0^{0}
D	0^{0}	0^{0}	1	∞
E	0^{0}	0^{0}	∞	3

A to D

Step 5 Allocate A to D. Therefore, D to A is ∞.

	A	B	C
B	1	∞	0^{2}
D	∞	0^{1}	1
E	0^{1}	0^{0}	∞

B to C

Step 6 Allocate *B* to *C*.

	A	*B*	
D	∞	0 ∞	*D* to *B*
E	0 ∞	0 0	*E* to *A*

Step 7 Allocate *D* to *B* and *E* to *A*.

Sequence of items *A-D-B-C-E-A*

Total set-up cost = 13 + 13 + 16 + 15 + 14 = Rs. 71

EXERCISES

1. Solve the assignment problem given in the table below for optimal solution using Hungarian method. The time taken by each worker is given in minutes in the table.

		Job			
		A	*B*	*C*	*D*
Worker	1	45	40	51	67
	2	57	42	63	55
	3	49	52	48	64
	4	41	45	60	55

2. Using the following cut matrix, determine optional job assignment and the cost associated with it.

		Job				
		1	2	3	4	5
Machine	*A*	10	3	3	2	8
	B	9	7	8	2	7
	C	7	5	6	2	4
	D	3	5	8	2	4
	E	9	10	9	6	10

3. Solve the following assignment problem and obtain the minimum cost at which all the jobs can be processed.

		Job				
		1	2	3	4	5
Worker	*A*	25	18	32	20	21
	B	34	25	21	12	17
	C	20	17	20	32	16
	D	20	28	20	16	27

4. A Solicitor firm employs typists on hourly page rate basis for their daily work. There are five typists and their charges and speed are different. According to an earlier understanding, only one job is given to one typist and the typist is paid for a full hour even when he works for a fraction of an hour. Find the least cost allocation for the following data.

Typist	Rate/hour Rs.	Number of pages typed per hour	Job	Number of pages
A	5	12	*P*	199
B	6	14	*Q*	175
C	3	8	*R*	145
D	4	10	*S*	298
E	4	11	*T*	178

5. A company has four sales representatives who are to be assigned to four different sales territories. The monthly sales increase estimated for each sales representative, for different sales territories (in lakhs of rupees) are shown in the following table.

Sales representative	Sales territory: I	II	III	IV
A	200	150	170	220
B	160	120	150	140
C	190	195	190	200
D	180	175	160	190

Suggest optimal assignment and the total maximum sales increase per month.

6. A firm produces four products. There are four operators who are capable of producing any of these four products. The firm records 8 hours a day and allows 30 minutes for lunch. The processing time in minutes and the profit for each of the products are given below.

Operator	Products: *A*	*B*	*C*	*D*
1	15	9	10	6
2	10	6	9	6
3	25	15	15	9
4	15	9	10	10
Profit Rs./unit	8	6	5	4

Find the optimal assignment of products to operators.

7. The owner of a small machine shop has four machinists available. To assign the jobs for the day, five jobs are offered with an expected profit in rupees for each machinist on each job as follows

		Job				
		A	*B*	*C*	*D*	*E*
Machinist	1	62	78	50	101	82
	2	71	84	61	73	59
	3	87	92	111	71	81
	4	48	64	87	77	80

Determine the assignment of machinists to jobs, which will result in a maximum profit and which job should be declined.

8. A company has four districts I, II, III and IV to sell its product, and four salesmen *A*, *B*, *C* and D. The district-wise sales record of each salesman is as given in the table. Determine the area allocation so as to maximise the sales.

		District			
		I	II	III	IV
Salesmen	*A*	420	350	280	210
	B	300	250	200	150
	C	300	250	200	150
	D	200	200	160	120

9. MCS Inc is a software company which has three projects of *Y2K* with the departments of health, education and housing for Maharashtra government. Based on the background and experiences of the project leader, they differ in terms of their performance at various projects. The performance score matrix is given below.

		Projects		
		Health	Education	Housing
Project tenders	P_1	20	26	42
	P_2	24	32	50
	P_3	32	34	44

Help the management by determining the optimal assignment which maximises the total performance score.

10. Obtain the optimal assignment from the following table.

	1	2	3	4	5	6	7
A				0			
B			0				
C		0		0			0
D	0	0		0	0		
E		0					
F	0	0	0	0			0
G		0			0		

11. Solve the following problem when the time table is as follows and minimun layover time is one hour.

Flight No.	Delhi Departure	Srinagar Arrival	Flight No.	Srinagar Departure	Delhi Arrival
1	7.30 AM	9.00 AM	2	7.00 AM	10 AM
3	8.15 AM	9.45 AM	4	7.45 AM	10.45 AM
5	2.00 PM	3.30 PM	6	11.00 AM	2.00 PM
7	5.45 PM	7.15 PM	8	6.00 PM	9.00 PM
9	7.00 PM	8.30 PM	10	7.30 PM	10.30 PM

12. ABC airline operating 7 days a week has given the following time table. Crews are meant to have a minimum layover of 5 hours between flights. Obtain the pairing flights that minimises layover time away from home. For any given pairing, the crew will be based at the city that results in the smallest layover.

Flight No.	Hyderabad Departure	Delhi Arrival	Flight No.	Delhi Departure	Hyderabad Arrival
A_1	6.00 AM	8.00 AM	B_1	8.00 AM	10.00 AM
A_2	8.00 AM	10.00 AM	B_2	9.00 AM	11.00 AM
A_3	2.00 PM	4.00 PM	B_3	2.00 PM	4.00 PM
A_4	8.00 PM	10.00 PM	B_4	7.00 PM	9.00 PM

13. A transport company provides daily bus service from Mysore to Chikmagalur and back. It takes 6 hours for a single trip. The drivers and conductors are provided with residential facilities at both the places (i.e., Mysore and Chikmagalur). They require at least 4 hours for rest before the return trip. From the following time table, prepare an assignment schedule for the minimum waiting time showing the place where the transport staff should be based.

Mysore Departure	Chikmagalur Arrival	Chikmagalur Departure	Mysore Arrival
5.00 AM	11.00 AM	4.00 AM	10.00 AM
7.00 AM	1.00 PM	5.00 AM	11.00 AM
9.00 AM	3.00 PM	10.00 AM	4.00 PM
4.00 PM	10.00 PM	1.00 PM	7.00 PM
5.00 PM	11.00 PM	3.00 PM	9.00 PM

14. A trip from Chennai to Bangalore takes six hours by bus. A typical time table of the bus service in both directions is given below.

Route No.	Chennai Departure	Bangalore Arrival	Route No.	Bangalore Departure	Chennai Arrival
A	06.00 hours	12.00 hours	1	05.30 hours	11.30 hours
B	07.30 hours	13.30 hours	2	9.00 hours	15.00 hours
C	11.30 hours	17.30 hours	3	15.00 hours	21.00 hours
D	19.00 hours	01.00 hours	4	18.30 hours	00.30 hours
E	00.30 hours	06.30 hours	5	00.00 hours	6.00 hours

The bus staff should be provided with at least 4 hours of rest before the return trip. Find which route.

15. The captain of a cricket team has to allot five middle batting positions to five batsmen. The average runs scored by each batsman at this position are as follows

		Batting position				
		1	2	3	4	5
	A	40	40	35	25	50
	B	42	30	16	25	27
Batsman	C	50	48	40	60	50
	D	20	19	20	18	25
	E	58	60	59	55	53

Solve this problem by Branch and Bound method and find the batsmen position which would give maximum number of score.

16. A small garment-making unit has five tailors stitching five different types of garments. All the five tailors are capable of stitching all the five types of garments. The output per tailor and the profit (Rs.) for each type of garment are given below.

		Garments				
		1	2	3	4	5
	A	7	9	4	8	6
	B	4	9	5	7	8
Tailor	C	8	5	7	9	8
	D	6	5	8	10	10
	E	7	8	10	9	9
Profit (Rs.) per garment		2	3	2	3	4

Find which type of garment should be assigned to which tailor in order to maximise the profit, using Branch and Bound method.

17. A machine operator processes five types of items on his machine each week and must choose a sequence for them. The set-up cost per change depends on the item presently on the machine and the set-up to be made according to the following table.

		Setup Cost				
		A	B	C	D	E
	A	∞	4	7	3	4
	B	4	∞	6	3	4
Machine operator	C	7	6	∞	7	5
	D	3	3	7	∞	7
	E	4	4	5	7	∞

If he processes each type of item only once each week, how should he sequence the items in order to minimise the total set-up cost.

18. Solve the following travelling salesman problem so as to minimise the cost per cycle.

		To				
		A	*B*	*C*	*D*	*E*
From	*A*	–	3	6	2	3
	B	3	–	5	2	3
	C	6	5	–	6	4
	D	2	2	6	–	6
	E	3	3	4	6	–

19. A salesman must travel from city to city to maintain his accounts. This week he has to leave his home base and visit all other cities and return home. The table shows the distance between the various cities in kilometres. The home city is city *A*. Use the assignment method to determine the tour that will minimise the total distance for visiting all the cities and returning home.

		To				
		A	*B*	*C*	*D*	*E*
City	*A*	–	375	600	150	190
	B	375	–	300	350	175
	C	600	300	–	350	500
	D	160	350	350	–	300
	E	190	175	500	300	–

20. Distance between the six islands in the Indian Ocean are given in the table below.

		Inlands					
		A	*B*	*C*	*D*	*E*	*F*
Island	*A*	–	120	90	100	110	160
	B	120	–	190	200	210	270
	C	90	190	–	30	40	90
	D	100	200	30	–	30	80
	E	110	210	40	30	–	70
	F	160	270	90	80	70	–

It is proposed to lay submarine cable of smallest length to connect all the entire islands by telephone connection. Find the minimum spanning tree for the six inlands mentioned. Use travelling salesman method.

Chapter 7

GAME THEORY

7.1 INTRODUCTION

Competition generally prevails in any business. Conflicts of interest among the competitors is a common issue. To overcome such situations, a certain course of action is followed by the competitors. The conflicting situation is called the game and the competitors are known as players. The course of action is known as the strategy.

The situations having the following properties are termed as games in competition.

1. The number of competitors usually referred to as players is finite.
2. There is a conflict of interest between the competitors.
3. A finite number of possible courses of action is available with each competitor.
4. The rules governing the choices of action are known to all competitors. Each competitor chooses one of these courses of action. These choices are assumed to be made simultaneously so that no one knows his opponent's choice until he has decided his own course of action.
5. The outcome of the game is affected by the choice made by the other competitors.
6. All combinations of courses of action are associated with an outcome—negative, zero or positive.

7.2 TWO-PERSON ZERO-SUM GAME

Here, there are exactly two competitors having conflicting interests. One competitor gains whatever the other loses. The sum of the net gain is zero. Therefore, they are called zero sum games. Each competitor knows the outcome for all possible strategies that he and his opponent may use during the course of the game.

In a two-person zero-sum game, the resulting gain can be represented in the form of a matrix called the pay-off matrix of the game.

7.3 PAY-OFF MATRIX

Let 'G' be an $m \times n$ matrix with entries g_{ij} for $i = 1, 2, 3, \ldots, m$ and $j = 1, 2, 3, \ldots, n$. Then G can be interpreted as the pay-off matrix of the following matrix game.

$$G = \begin{array}{|ccc} 0 & 2 & 1 \\ 1 & 3 & 6 \\ 2 & 1 & -3 \end{array}$$

The row player chooses any row '*i*' and simultaneously, the column player chooses any column '*j*'. The outcome of the game is that the column player pays to a row player an amount equal to g_{ij}. If g_{ij} is less than zero, the column player gives the row player $-g_{ij}$.

7.4 THE MAXIMIN AND MINIMAX PRINCIPLE

	B_1	B_2	B_3
A_1	1	3	6
A_2	2	1	–3
A_3	0	2	1

Consider a pay-off matrix for a two-person zero-sum game between two players *A* and *B*. A_1, A_2, A_3, B_1 B_2, B_3 are the strategies, e.g., paper, radio or TV advertisements, design features, reduction in selling price, credit sales, etc. If '*A*' chooses a particular row to gain the maximum, player *B* would like to choose a strategy in such a way that pay-off *A* is minimised for that particular strategy. Thus, if player *A* selects strategy A_1, *B* would select strategy B_1. Similarly, if *A* selects strategy A_2, *B* would select strategy B_3.

Here, *A* can gain 2 or lose 3. However he is guaranteed of the minimum of (2, 1, –3) regardless of strategies chosen by *B*. Player *A* will try to make the best use of the situation for which minimum assured is maximum. Hence he will identify the maximum among minimum pay-offs. This decision rule is called maximin principle and the strategy is called maximin strategy.

In a similar way, player *B* would like to minimise his losses; if he selects strategy B_2, his worst possible loss is 3. He tries to select that strategy where his loss is minimal. Therefore he will try to identify the minimum among the maximum possible losses. The corresponding decision is called the minimax principle. The strategy is minimax strategy.

When maximin value is equal to minimax value, the strategy is called the optimum strategy.

7.5 SADDLE POINT

Let '*G*' be an $m \times n$ matrix with payoff g_{ij}, where $i = 1, 2, 3, \ldots, m$ and $j = 1, 2, 3, \ldots, n$. Then g_{ij} is said to be the saddle point of *G* if g_{ij} is simultaneously the minimum of *i*th row and the maximum of *j*th column. If the matrix game *G* has a saddle point, it is said to be fully determined. The value of the game $v = g_{ij}$. When the value of the game is zero, the game is said to be fair.

Example 7.1

Consider the game having the following pay-off matrix. Determine whether it has a saddle point. If it does, determine the optimum strategy for each player and find the value of the game.

		Player *B* I	II	III	IV	V	Minimum of each row
Player *A*	I	3	(–1)	4	6	7	–1
	II	(–1)	8	2	4	[12]	–1
	III	[16]	8	[(6)]	[14]	[12]	6
	IV	1	[11]	(–4)	2	1	–4
Maximum of each column		16	11	6	14	12	

Solution

Step 1 Find the minimum in each row and encircle them.

Step 2 Find the maximum in each column and enrectangle them. Any value with circle and rectangle gives the saddle point value of the game = 6.

Player *A* plays strategy 3 and *B* also plays strategy 3.

Player *B* has to pay player *A*, 6.

Here maximin and minimax principle is used to find the saddle point.

Example 7.2

Solve the following game.

		Player *B* I	II	III	IV	V	Minimum of each row
Player *A*	I	4	0	1	7	−1	−1
	II	0	−3	−5	−7	5	−7
	III	3	2	3	4	3	2
	IV	−6	1	−1	0	5	−6
	V	0	0	0	0	0	0
Maximum of each column		4	2	3	7	5	

Solution

Player *A* plays strategy 3 and *B* plays strategy 2. Player *B* pays player *A* Rs. 2.

Value of the game = 2.

Example 7.3

Solve the following game.

		Player *B* I	II	III	Minimum of each row
Player *A*	I	−2	15	−2	−2
	II	−5	−6	−4	−6
	III	−5	20	−8	−8
Maximum of each column		−2	20	−2	

Solution

A has strategy 1 and *B* has either strategy 1 or 3. The value of the game is –2, i.e., player *A* has to pay player *B* Rs. 2.

Example 7.4

Solve the following game.

		Player *B* 1	2	3	Minimum of each row
Player *A*	1	2	–1	3	–1
	2	2	–1	2	–1
	3	–1	0	0	0
	4	2	0	4	0
Maximum of each column		2	0	4	

The value of the game is zero. Therefore, it is a fair game. Player *A* plays strategy 3 and player *B* plays strategy 2. Here neither player *B* pays to player *A* nor player *A* pays to player *B*.

Example 7.5

Solve the following game whose pay-off matrix is as follows

		Player *B* 1	2	3	Minimum of each row
Player *A*	1	1	3	1	1
	2	0	–4	–1	–4
	3	1	5	–1	–1
Maximum of each column		1	5	1	

Solution

Player *A* plays strategy 1. Player *B* plays either strategy 1 or 3. Value of the game is 1. Player *B* pays Rs. 1 to player *A*.

7.6 THEORY OF DOMINANCE

Consider a game with pay-off matrix as shown in this game. Player *B* observes that if player *A* plays strategy 1, he will lose 2; if he plays strategy 2 he gains 2. Similarly when player *A* plays strategy 2, player *B* will lose 4, if he plays strategy 1 and he loses 1 when he plays strategy 2. In both the cases, for player *B*, strategy 2 is the best. Therefore, he will not have any inclination to play strategy 1. Then we say that player *B*'s second action dominates over the first action.

		B I	II	III	IV
A	I	2	–2	–3	4
	II	4	1	2	2

In such cases, we can reduce the size of the pay-off matrix removing those actions that are dominated by some other action. Here, strategy 1 for player *B* can be eliminated. Now the matrix will be as follows

		B II	*B* III	*B* IV
A	I	–2	–3	4
	II	1	2	2

Actions 2 or 3 dominate over action 4 for player *B*. Therefore, column 4 can be removed. Then the matrix will be

		B II	*B* III
A	I	–2	–3
	II	1	2

If player *B* plays strategy 2, player *A* will lose 2 or gain 1. If player *B* plays strategy 3, player *A* will lose 3 or gain 2. For player *A*, strategy 2 is always advantageous. Therefore, strategy 2 dominates over strategy 1. The matrix will be as follows

		B II	*B* III
A	II	1	2

Column 2 dominates over column III. Thus the matrix will be as follows

		B II
A	II	1

Now the value of the game is 1.

Player *B* takes strategy 2 and player *A* takes strategy 2.

7.6.1 General Rule for Dominance

1. If all the elements of a column (*i*th column) are lesser than all the elements of another column (*j*th column), then, *i*th column dominates over *j*th column.
2. If all the elements of *i*th row are greater than or equal to the corresponding elements in *j*th row, it means that *i*th row dominates *j*th row.
3. A pure strategy of a player may also be dominated, if it is inferior to some convex combination of two or more pure strategies or as a particular case, inferior to the average of two or more strategies.

Example 7.6

Solve the game whose pay-off matrix is given below.

		Player *B* I	II	III	IV	Minimum of each row
Player *A*	I	1	7	3	4	1
	II	5	6	4	5	4
	III	1	2	0	3	0
Maximum of each column		5	7	4	5	

Solution

By using minimax and maximin principle, the value of the game is obtained. Saddle point = 4. Player *B* pays 4 to player *A*. Player *A* plays strategy 2 and player *B* plays strategy 3.

By using dominance principle

		B I	II	III	IV
A	I	1	7	3	4
	II	5	6	4	5
	III	1	2	0	3

Column player pays. Therefore, if all the elements of a column are less than all the elements of another column, that column dominates over the other column. Here, column 1 dominates over column 2. Therefore, remove column 2.

		B I	III	IV
A	I	1	3	4
	II	5	4	5
	III	1	0	3

Columns 1 dominates over column 4. Therefore, remove column 4.

		B I	III
A	I	1	3
	II	5	4
	III	1	0

The row player receives the money. Therefore, if all the elements in a row are greater than the corresponding elements in any other row that row dominates over the other rows.

Here, row 2 dominates row1 and row 3. Therefore, remove rows 1 and 3.

		B: I	B: III
A	II	5	4

Column 3 dominates over column 1. Therefore, remove column 1.

		B: III
A	II	4

Value of the game is 4. Player *A* plays strategy 2 and player *B* plays strategy 3. Player *B* pays player *A*, 4.

Example 7.7

Use dominance property to solve the following games.

		B: I	II	III	IV	V
A	I	3	5	4	9	6
	II	5	6	3	7	8
	III	8	7	9	8	7
	IV	4	2	8	5	3

Solution

Column 2 dominates over column 4. Therefore, delete column 4.

		B: I	II	III	V
A	I	8	5	4	6
	II	5	6	3	8
	III	8	7	9	7
	IV	4	2	8	3

Column 2 dominates over column 5. Therefore, delete column 5.

		B: I	II	III
A	I	3	5	4
	II	5	6	3
	III	8	7	9
	IV	4	2	8

Row 3 dominates over rows 1, 2, and 4. Therefore, remove rows 1, 2 and 4.

		B I	B II	B III
A	III	8	7	9

Column 2 dominates over columns 1 and 3. Therefore, remove columns 1 and 3.

		B II
A	III	7

Value of the game is 7. Player A plays strategy 3 and player B plays strategy 2. Player B pays player A, 7.

Example 7.8

Solve the game whose pay-off matrix is given below.

$$\begin{bmatrix} 9 & 3 & 1 & 8 & 1 \\ 6 & 5 & 4 & 6 & 7 \\ 2 & 4 & 3 & 3 & 8 \\ 5 & 6 & 2 & 2 & 1 \end{bmatrix}$$

By Minimax and Maximise Principle

		B 1	B 2	B 3	B 4	B 5	Minimum of each row
A	1	[9]	3	1	[8]	(1)	1
A	2	6	5	[(4)]	6	7	4
A	3	(2)	4	3	3	[8]	2
A	4	5	[6]	2	2	(1)	1
Maximum of each column		9	6	4	8	8	

Solution

Value of the game is 4. Player A plays strategy 2 and player B plays strategy 3. Player B pays player A, 4.

By Dominance Principle

		B 1	B 2	B 3	B 4	B 5
A	1	9	3	1	8	1
A	2	6	5	4	6	7
A	3	2	4	3	3	8
A	4	5	6	2	2	1

Column 3 dominates column 4. Therefore, delete column 4.

		B			
		1	2	3	5
A	1	9	3	1	1
	2	6	5	4	7
	3	2	4	3	8
	4	5	6	2	1

Column 3 dominates column 2. Therefore, delete column 2.

		B		
		1	3	5
A	1	9	1	1
	2	6	4	7
	3	2	3	8
	4	5	2	1

Average of rows 1 and 2 dominate row 4. Therefore, delete row 4.

		B		
		1	3	5
A	1	9	1	1
	2	6	4	7
	3	2	3	8

Column 3 dominates column 5. Therefore, delete column 5.

		B	
		1	3
A	1	9	1
	2	6	4
	3	2	3

Row 2 dominates row 3. Therefore, delete row 3.

		B	
		1	3
A	1	9	1
	2	6	4

Column 3 dominates column 1. Therefore, delete column 1.

		B
A	1	1
	2	4

Row 2 dominates row 1. Therefore, delete row 1.

		B
		3
A	2	4

Value of the game is 4. Player A plays strategy 2 and player B plays strategy 3. Player B pays player A, 4.

Example 7.9

Solve the game whose pay-off matrix is given below.

$$\begin{bmatrix} -2 & 0 & 0 & 5 & 3 \\ 3 & 2 & 1 & 2 & 2 \\ -4 & -3 & 0 & -2 & 6 \\ 5 & 3 & -4 & 2 & 3 \end{bmatrix}$$

By Minimax and Maximin Principle

		B: 1	2	3	4	5	Minimum of each row
A	1	(−2)	0	0	[5]	3	−2
	2	3	2	[(1)]	2	2	1
	3	(−4)	−3	0	−2	[6]	−4
	4	[5]	[3]	(−4)	2	(3)	−4
Maximum of each column		5	3	1	5	6	

Solution

Value of the game is 1. Player A plays strategy 2 and player B plays strategy 3. Player B pays player A, 1.

By Dominance Principle

		B: 1	2	3	4	5
A	1	−2	0	0	5	3
	2	3	2	1	2	2
	3	−4	−3	0	−2	6
	4	5	3	−4	2	3

Column 3 dominates column 5. Therefore, delete column 5.

		B 1	2	3	4
A	1	–2	0	0	5
	2	3	2	1	2
	3	–4	–3	0	–2
	4	5	3	–4	2

Row 2 dominates row 3. Therefore, delete row 3.

		B 1	2	3	4
A	1	–2	0	0	5
	2	3	2	1	2
	4	5	3	–4	2

Column 3 dominates column 2. Therefore, delete column 2.

		B 1	3	4
A	1	–2	0	5
	2	3	1	2
	4	5	–4	2

Column 3 dominates column 4. Therefore, delete column 4.

		B 1	3
A	1	–2	0
	2	3	1
	4	5	–4

Row 2 dominates row 1. Therefore, delete row 1.

		B 1	3
A	1	3	1
	2	5	–4

Column 3 dominates column 1. Therefore, delete column 1.

		B 3
A	2	1
	4	–4

Row 2 dominates row 4. Therefore, delete row 4.

		B
		3
A	2	1

Value of the game is 1. Player *A* plays strategy 2 and player *B* plays strategy 3. Player *B* pays player *A*, 1.

7.7 MIXED STRATEGY GAME WITH 2 × 2 PAY-OFF MATRIX AND WITHOUT SADDLE POINT

		B: I	B: II
A	I	a_{11}	a_{12}
	II	a_{21}	a_{22}

Let player *A* play strategy 1 with probability X_1 and strategy 2 with probability X_2. If player *B* plays strategy 1, the net gain to player *A* is $X_1a_{11} + X_2a_{21}$. If player *B* plays strategy 2, player *A*'s net gain will be $X_1a_{12} + X_2a_{22}$.

Let the value of game be *v*. As player *A* is receiving the amount, the net gain should be greater than or equal to ϑ.

i.e.,

$$X_1a_{11} + X_2a_{21} \geq \vartheta$$
$$X_1a_{12} + X_2a_{22} \geq \vartheta$$

Let player *B* play strategy 1 with probability Y_1 and strategy 2 with probability Y_2. If player *A* plays strategy 1, player *B*'s net loss is $a_{11}Y_1 + a_{12}Y_2$. If player *A* plays strategy 2, player *B*'s net loss will be $a_{21}Y_1 + a_{22}Y_2$. As player *B* is losing, he expects the net loss to be less than or equal to ϑ.

Therefore,

$$a_{11}Y_1 + a_{12}Y_2 \geq \vartheta$$
$$a_{21}Y_1 + a_{22}Y_2 \geq \vartheta$$

Bringing inequality to equality

$$X_1a_{11} + X_2a_{12} = \vartheta = X_1a_{12} + X_2a_{22}$$
$$X_1a_{11} + X_2a_{21} = X_1a_{12} + X_2a_{22}$$
$$X_1a_{11} - X_1a_{12} = X_2a_{22} - X_2a_{21}$$
$$X_1(a_{11} - a_{12}) = X_2(a_{22} - a_{21})$$
$$\frac{X_1}{X_2} = \frac{a_{22} - a_{21}}{a_{11} - a_{12}}$$
$$Y_1a_{11} + Y_2a_{12} = Y_1a_{21} + Y_2a_{22}$$
$$Y_1(a_{11} - a_{21}) = Y_2(a_{22} - a_{12})$$

$$\frac{Y_1}{Y_2} = \frac{a_{22} - a_{21}}{a_{11} - a_{12}}$$

$$X_1 + X_2 = 1.$$

Because the total probability is 1.

$$\therefore X_1 = 1 - X_2$$

$$X_2 = 1 - X_1$$

$$X_1 a_{11} + X_2 a_{21} = X_1 a_{12} + X_2 a_{22}$$

Substituting $X_2 = 1 - X_1$ in the above equation we get,

$$X_1 a_{11} + a_{21}(1 - X_1) = X_1 a_{12} + a_{22}(1 - X_1)$$

$$X_1 a_{11} + a_{21} - X_1 a_{21} = X_1 a_{12} + a_{22} - X_1 a_{22}$$

$$X_1(a_{11} - a_{21} - a_{12} + a_{22}) = a_{22} - a_{21}$$

$$\therefore X_1 = \frac{a_{22} - a_{21}}{a_{11} + a_{22} - (a_{21} + a_{12})}$$

Similarly,

$$X_2 = \frac{a_{11} - a_{12}}{a_{11} + a_{22} - (a_{21} + a_{12})}$$

$$Y_1 = \frac{a_{22} - a_{12}}{a_{11} + a_{22} - (a_{21} + a_{12})}$$

$$Y_2 = \frac{a_{11} - a_{21}}{a_{11} + a_{22} - (a_{21} + a_{12})}$$

We know that,

$$a_{11} X_1 + a_{22} X_2 = \vartheta$$

Substituting the value of X_1 and X_2, we get

$$\vartheta = \frac{a_{11} \times a_{22} - a_{12} \times a_{21}}{a_{11} + a_{22} - (a_{21} + a_{12})}$$

***Example* 7.10**

Solve the game whose pay-off matrix is given by

		B I	B II	B III
A	I	1	7	2
A	II	6	2	7
A	III	5	1	6

Solution

Column 1 dominates column 3.

		B: I	B: II
A	I	1	7
	II	6	2
	III	5	1

Row 2 dominates row 3. Therefore, delete row 3.

		B: I	B: II
A	I	1	7
	II	6	2

There is no dominance in this 2 × 2 matrix. Here both the players have to play mixed strategies, i.e., Strategies 1 and 2.

For player A

Probability of playing strategy 1,

$$X_1 = \frac{a_{22} - a_{21}}{a_{11} + a_{22} - (a_{21} + a_{12})}$$

$$X_1 = \frac{2-6}{1+2-(6+7)}$$

$$X_1 = \frac{-4}{-10}$$

$$X_1 = \frac{2}{5}$$

Probability of playing strategy 2,

$$X_2 = 1 - X_1 = 1 - \frac{2}{5} = \frac{3}{5}$$

$$X_2 = \frac{3}{5}$$

For player B

Probability of playing strategy 1,

$$Y_1 = \frac{2-7}{1+2-(6+7)}$$

$$Y_1 = \frac{-5}{-10}$$

$$Y_1 = \frac{1}{2}$$

Probability of playing strategy 2,

$$Y_2 = 1 - Y_1 = 1 - \frac{1}{2}$$

$$Y_2 = \frac{1}{2}$$

Value of game

$$\vartheta = \frac{a_{11} \times a_{22} - a_{12} \times a_{21}}{a_{11} + a_{22} - (a_{21} + a_{12})}$$

$$= \frac{2 \times 1 - 6 \times 7}{1 + 2 - (6 + 7)}$$

$$= \frac{-40}{-10}$$

$$\vartheta = 4$$

The value of the game is 4. Strategy for A is $\left[\frac{2}{5}, \frac{3}{5}, 0\right]$.

Strategy for B is $\left[\frac{1}{2}, \frac{1}{2}, 0\right]$.

Example 7.11

Use the concept of dominance to solve the following game.

		B			
		I	II	III	IV
A	I	3	2	4	0
	II	3	4	2	4
	III	4	2	4	0
	IV	0	4	0	8

Solution

Row 3 dominates row 1. Therefore, delete row 1.

		B			
		I	II	III	IV
A	II	3	4	2	4
	III	4	2	4	0
	IV	0	4	0	8

Column 3 dominates column 1. Therefore, delete column 1.

		B		
		II	III	IV
A	II	4	2	4
	III	2	4	0
	IV	4	0	8

Average of columns 3 and 4 dominates column 2. Therefore, delete column 2.

		B: III	IV
A	II	2	4
	III	4	0
	IV	0	8

Average of rows 3 and 4 dominates row 2. Therefore, delete row 2.

		B: III	IV
A	II	4	0
	III	0	8

Player A plays strategies 3 and 4.

Player B plays strategies 3 and 4.

$$X_1 \text{ probability of playing strategy 3} = \frac{8-0}{4+8-0}$$

$$X_1 = \frac{8}{12} = \frac{2}{3}$$

$$X_2 = 1 - \frac{2}{3} = \frac{1}{3}$$

Similarly,

$$Y_1 = \frac{8-0}{4+8-0} = \frac{8}{12} = \frac{2}{3}$$

$$= 1 - \frac{2}{3} = \frac{1}{3}$$

Value of the game

$$\vartheta = \frac{4 \times 8 - 0}{4 + 8 - 0} = \frac{32}{12}$$

$$\vartheta = \frac{8}{3}$$

Strategies for A are $\left[0, 0, \frac{2}{3}, \frac{1}{3}\right]$.

Strategies for B are $\left[0, 0, \frac{2}{3}, \frac{1}{3}\right]$.

***Example* 7.12**

Solve the following game.

		B: I	II	III	IV
A	I	6	8	3	13
	II	4	1	5	3
	III	8	10	4	12
	IV	3	6	7	12

Column 2 dominates column 4. Therefore, delete column 4.

		B		
		I	II	III
A	I	6	8	3
	II	4	1	5
	III	8	10	4
	IV	3	6	7

Row 3 dominates row 1. Therefore, delete row 1.

		B		
		I	II	III
A	I	4	1	5
	III	8	10	4
	IV	3	6	7

Average of Rows 3 and 4 dominate row 2. Therefore, delete row 2.

		B		
		I	II	III
A	III	8	10	4
	IV	3	6	7

Column 1 dominates column 2. Therefore, delete column 2.

		B	
		I	III
A	III	8	4
	IV	3	7

It is a 2 × 2 matrix with mixed strategies.

$$X_1 = \frac{7-3}{15-7} = \frac{4}{8}$$
$$= \frac{1}{2}$$

$$X_2 = 1 - \frac{1}{2} = \frac{1}{2}$$

$$Y_1 = \frac{7-4}{8} = \frac{3}{8}$$

$$Y_2 = 1 - \frac{3}{8} = \frac{5}{8}$$

Value of the game

$$\vartheta = \frac{56-12}{8} = \frac{44}{8}$$
$$= \frac{11}{2}$$

Strategies for A are $\left[0, 0, \frac{1}{2}, \frac{1}{2}\right]$

Strategies for B are $\left[\frac{3}{8}, 0, \frac{5}{8}, 0\right]$.

Example 7.13

Solve the following game.

		B 1	2	3
	1	–5	–1	–1
A	2	4	0	2
	3	–5	2	0

Row 3 dominates row 1. Therefore, delete row 1.

		B 1	2	3
A	2	4	0	2
	3	–5	2	0

Average of columns 1 and 2 dominates column 3. Therefore, delete column 3.

		B 1	2
A	2	4	0
	3	–5	2

$$X_1 = \frac{a_{22} - a_{21}}{a_{11} + a_{22} - (a_{21} + a_{12})}$$

$$X_1 = \frac{2 - (-5)}{4 + 2 - (-5 + 0)}$$

$$X_1 = \frac{2 + 5}{6 + 5} = \frac{2}{5}$$

$$X_2 = 1 - X_1 = 1 - \frac{7}{11}$$

$$X_2 = \frac{4}{11}$$

$$Y_1 = \frac{2 - 0}{6 + 5} = \frac{2}{11}$$

$$Y_2 = 1 - \frac{2}{11}$$

$$Y_2 = \frac{9}{11}$$

Value of the game

$$\vartheta = \frac{a_{11} \times a_{22} - a_{12} \times a_{21}}{a_{11} + a_{22} - (a_{21} + a_{12})}$$

$$= \frac{4 \times 2 - 0(-5)}{4 + 2 - (-5 + 0)}$$

$$\vartheta = \frac{8}{11}$$

Strategies for A are $\left[0, \frac{7}{11}, \frac{4}{11}\right]$.

Strategies for B are $\left[\frac{2}{11}, \frac{9}{11}, 0\right]$.

7.8 GRAPHICAL SOLUTION

Any game with pay-off matrix $m \times n$ can be reduced to $2 \times n$ or $m \times 2$ matrix by using dominance principle. This rectangular matrix with either two rows or two columns can be solved by simple graphical method.

Considering $2 \times n$ game as follows

		B						
		1	2	3	.	.	.	n
A	1	a_{11}	a_{12}	a_{13}	.	.	.	a_{1n}
	2	a_{21}	a_{22}	a_{23}	.	.	.	a_{2n}

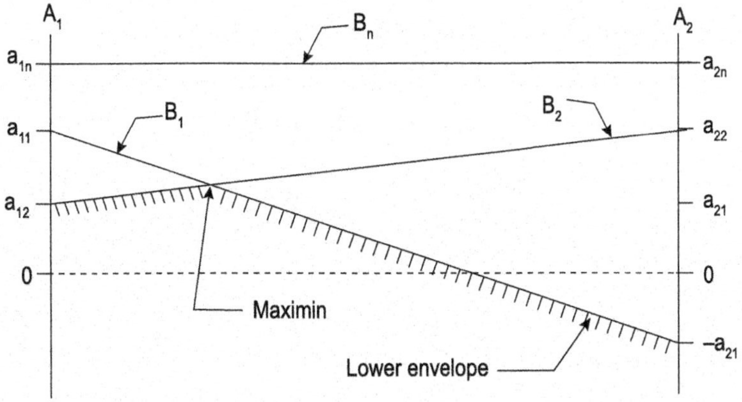

Figure 7.1 Graphical solution for row matrix

Strategies A_1 and A_2 are fitted on two lines at certain distance apart with suitable scale. Player A is receiving money. His minimum gain should be maximum pay-off. Values corresponding to each strategy of Player B are plotted as shown in the Figure 7.1. Lower envelope is drawn. Maximum value on the lower envelope is determined. Strategic lines intersecting are determined. Hence $2 \times n$ matrix is converted to 2×2 matrix with mixed strategies. Finally 2×2 matrix is solved by usual procedure.

Considering $m \times 2$ game as follows

		B: 1	B: 2
A	1	a_n	a_{12}
	2	a_{21}	a_{22}
	.	.	.
	.	.	.
	m_1	a_{m1}	a_{m2}

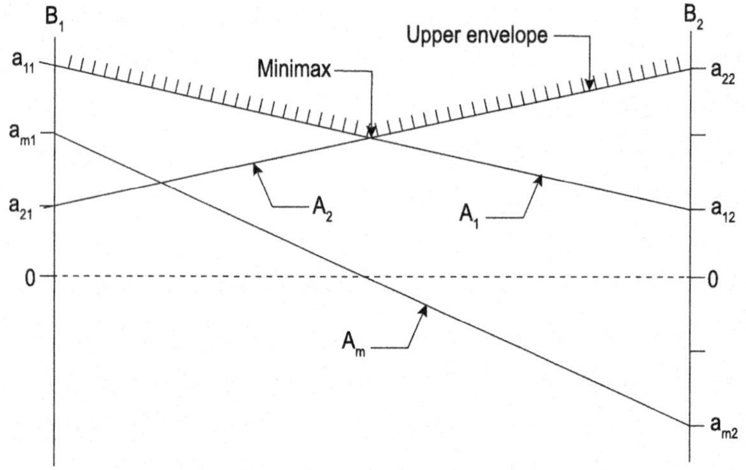

Figure 7.2 Graphical solution for column matrix

Strategies B_1 and B_2 are fitted on two lines at a certain distance apart with a suitable scale. Player B is paying money. His maximum loss should be minimum. Pay-off values corresponding to each strategy of A are plotted as shown in Figure 7.2.

Upper envelope is drawn. Minimax value on the upper envelope is determined. Strategic lines intersecting are determined. Hence $m \times 2$ matrix is converted to 2×2 matrix with mixed strategies. Finally 2×2 matrix is solved by the usual procedure.

Example 7.14

Solve the following game graphically.

		B		
A	1	3	−3	7
	2	5	4	−6

Step 1 Fit the strategies of A on two lines at a certain distance apart.

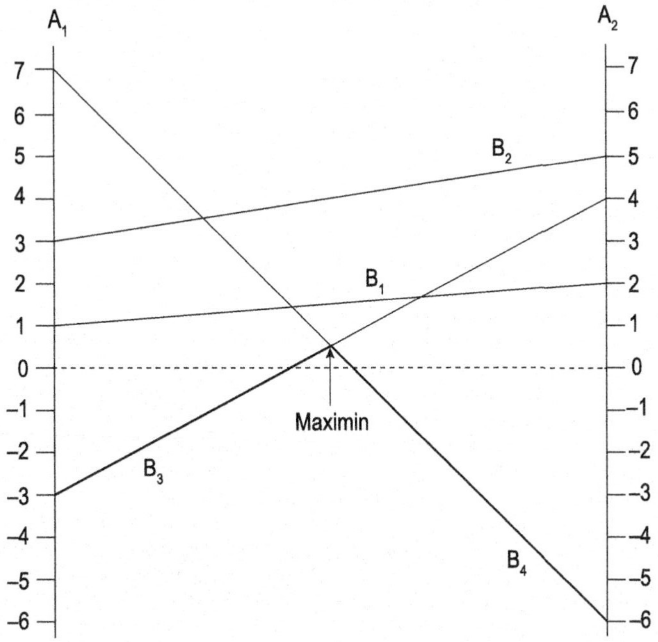

Figure 7.3 Strategies for player B

Step 2 Plot the values corresponding to B's strategies.

Step 3 Draw the lower envelope and determine the strategies for B from the maximum point of intersection from Figure 7.3. Intersecting strategies are B_3 and B_4.

2 × 2 matrix is as follows

		B 3	B 4
A	1	–3	7
	2	4	–6

Probability of playing strategy A_1

$$X_1 = \frac{a_{22} - a_{21}}{a_{11} + a_{22} - (a_{21} + a_{12})}$$

$$= \frac{6 - 4}{-3 - 6 - (4 + 7)}$$

$$= \frac{-10}{-20}$$

$$X_1 = \frac{1}{2}$$

$$X_2 = 1 - X_1 = 1 - \frac{1}{2}$$

Probability of playing strategy B_1

$$Y_1 = \frac{a_{22} - a_{12}}{a_{11} + a_{22} - (a_{21} + a_{12})}$$

$$= \frac{-6-7}{-20}$$

$$= \frac{-13}{-20}$$

$$Y_1 = \frac{13}{20}$$

$$Y_2 = 1 - Y_1 = 1 - \frac{13}{20}$$

$$Y_2 = \frac{7}{20}$$

Value of the game

$$\vartheta = \frac{a_{11} \times a_{22} - a_{12} \times a_{21}}{a_{11} + a_{22} - (a_{21} + a_{12})}$$

$$\vartheta = \frac{-3(-6) - 7(4)}{-20}$$

$$= \frac{18-28}{-20}$$

$$= \frac{-10}{-20}$$

$$\vartheta = \frac{1}{2}$$

Strategies for A are $\left[\frac{1}{2}, \frac{1}{2}\right]$.

Strategies for B are $\left[0, 0, \frac{13}{20}, \frac{7}{20}\right]$.

Example 7.15

Use graphical method of solving the following game.

		B			
		1	2	3	4
A	1	2	2	4	−2
	2	4	3	2	6

Step 1 Find the strategies of A on two lines at certain distance apart.

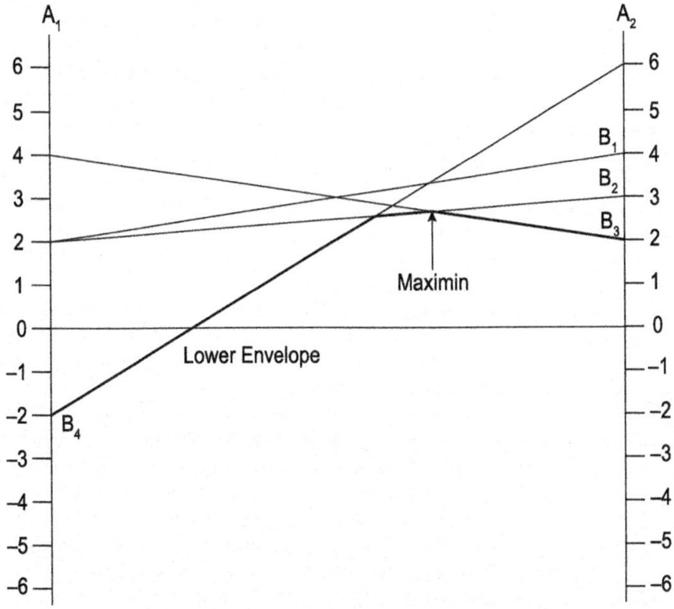

Figure 7.4 Strategies for player B

From Figure 7.4 maximum value is the intersection of strategies B_2 and B_3. 2×2 matrix is as follows

		B 2	B 3
A	1	2	4
	2	3	2

Probability of playing strategy A_1,

$$X_1 = \frac{a_{22} - a_{21}}{a_{11} + a_{22} - (a_{21} + a_{12})}$$

$$X_1 = \frac{2-4}{2+2-(3+4)}$$

$$= \frac{-2}{-3}$$

$$X_1 = \frac{2}{3}$$

$$X_2 = 1 - X_1$$

$$= 1 - \frac{2}{3}$$

$$X_2 = \frac{1}{3}$$

Probability of playing strategy B_2

$$Y_1 = \frac{a_{22} - a_{12}}{a_{11} + a_{22} - (a_{21} + a_{12})}$$

$$= \frac{2-4}{-3}$$

$$= \frac{-2}{-3}$$

$$Y_1 = \frac{2}{3}$$

$$Y_2 = 1 - \frac{2}{3}$$

$$Y_2 = \frac{1}{3}$$

Value of the game

$$\vartheta = \frac{a_{11} \times a_{22} - a_{12} \times a_{21}}{a_{11} + a_{22} - (a_{21} + a_{12})}$$

$$\vartheta = \frac{2(2) - 3(4)}{-20}$$

$$= \frac{4-12}{-3}$$

$$= \frac{-8}{-3}$$

$$\vartheta = \frac{8}{3}$$

Strategies for player A are $\left[\frac{2}{3}, \frac{1}{3}\right]$.

Strategies for player B are $\left[0, \frac{2}{3}, \frac{1}{3}, 0\right]$

Example 7.16

Solve the following game graphically.

		B 1	B 2
A	1	1	2
	2	5	4
	3	–7	9
	4	–4	–3
	5	2	1

Step 1 Fit the strategies of B on two lines at certain distance apart

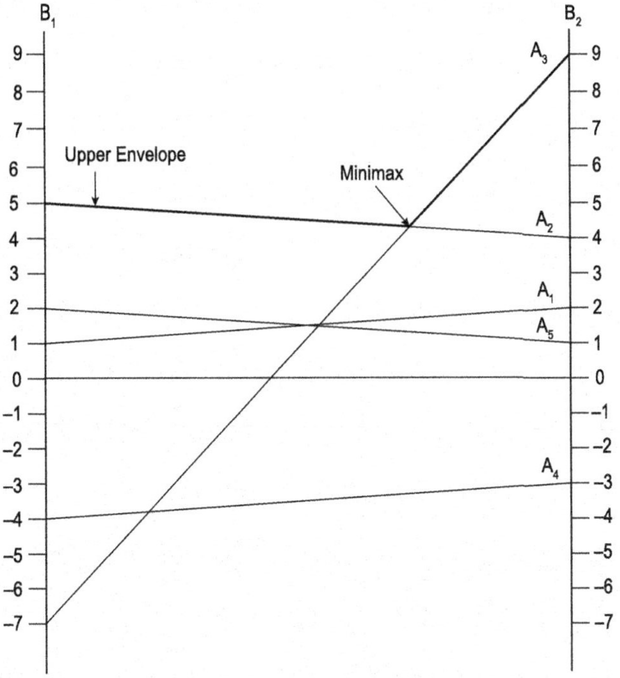

Figure 7.5 Strategies for player A

Step 2 Plot the values corresponding to A's strategies

Step 3 Draw the upper envelope and determine the strategies for A from the minimax point of intersection. From Figure 7.5 intersecting strategies are A_2 and A_3. 2×2 matrix is as follows

		B: 1	B: 2
A	2	5	4
	3	–7	9

Probability of playing strategy A_1

$$X_1 = \frac{a_{22} - a_{21}}{a_{11} + a_{22} - (a_{21} + a_{12})}$$

$$X_1 = \frac{9+7}{5+9-(-7+4)}$$

$$= \frac{16}{14+3}$$

$$X_1 = \frac{16}{17}$$

$$X_2 = 1 - X_1$$
$$= 1 - \frac{16}{17}$$
$$X_2 = \frac{1}{17}$$

Probability of playing strategy B_1.

$$Y_1 = \frac{a_{22} - a_{12}}{a_{11} + a_{22} - (a_{21} + a_{12})}$$
$$= \frac{9-4}{5+9-(-7+4)}$$
$$Y_1 = \frac{5}{17}$$
$$Y_2 = 1 - Y_1$$
$$= 1 - \frac{5}{17}$$
$$Y_2 = \frac{12}{17}$$

Strategies for A are $\left[0, \frac{16}{17}, \frac{1}{17}, 0, 0\right]$.

Strategies for B are $\left[\frac{5}{17}, \frac{12}{17}\right]$.

Value of game

$$\vartheta = \frac{a_{11} \times a_{22} - a_{12} \times a_{21}}{a_{11} + a_{22} - (a_{21} + a_{12})}$$
$$= \frac{5(9) - 4(-7)}{17}$$
$$= \frac{45 + 28}{17}$$
$$\vartheta = \frac{73}{17}$$

Example 7.17

Use graphical method in solving the following game.

		B	
		1	2
A	1	–4	3
	2	–7	1
	3	–2	–4
	4	–5	–2
	5	–1	–6

Step 1 Fit the strategies of B on two lines at certain distance apart.

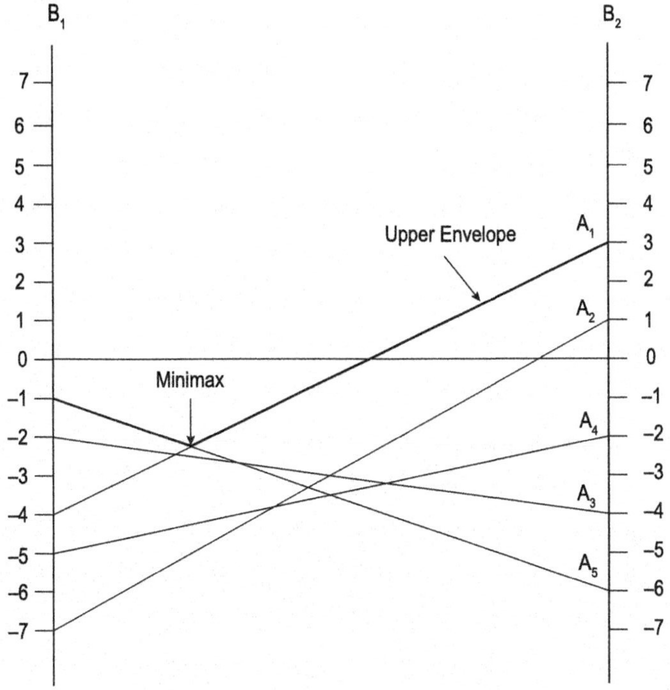

Figure 7.6 Strategies for player A

Step 2 Plot the values corresponding to A's strategies.

Step 3 Draw the upper envelope and determine the strategies for A from the minimax point of intersection. From Figure 7.6 intersecting strategies are A_1 and A_5. The 2 × 2 matrix with mixed strategies is as follows

		B 1	B 2
A	1	–4	3
	5	–1	–6

Probability of playing strategy A_1,

$$X_1 = \frac{a_{22} - a_{21}}{a_{11} + a_{22} - (a_{21} + a_{12})}$$

$$X_1 = \frac{-6+1}{-4+(-6)-(-1+3)}$$

$$= \frac{-5}{-10-(2)}$$

$$X_1 = \frac{-5}{12}$$

$$X_2 = 1 - X_1$$
$$= 1 - \frac{5}{12}$$
$$X_2 = \frac{7}{12}$$

Probability of playing strategy B_1

$$Y_1 = \frac{a_{22} - a_{12}}{a_{11} + a_{22} - (a_{21} + a_{12})}$$
$$= \frac{-6 - 3}{-4 + (-6) - (-1 + 3)}$$
$$Y_1 = \frac{-9}{-12}$$
$$Y_2 = 1 - \frac{3}{4}$$
$$Y_2 = \frac{1}{4}$$

Value of the game

$$\vartheta = \frac{a_{11} \times a_{22} - a_{12} \times a_{21}}{a_{11} + a_{22} - (a_{21} + a_{12})}$$
$$\vartheta = \frac{-4(-6) - 3(-1)}{-12}$$
$$= \frac{27}{-12}$$
$$\vartheta = \frac{-9}{4}$$

Strategies for A are $\left[\frac{5}{12}, 0, 0, 0, \frac{7}{12}\right]$.

Strategies for B are $\left[\frac{3}{4}, \frac{1}{4}\right]$.

EXERCISES

1. Find the saddle point for the following game.

		B 1	2	3	4
A	1	5	8	2	4
	2	2	6	1	3

2. Given the following pay-off matrix of a zero-sum game, determine the optimal strategies for the players and the value of the game.

		B1	B2	B3	B4
	1	5	–4	5	8
A	2	6	2	0	–5
	3	7	15	8	7
	4	2	8	–6	5

3. Solve the game with the following pay-off matrix.

		B1	B2	B3	B4	B5	B6
	1	18	8	18	8	18	8
A	2	15	6	15	6	15	6
	3	18	8	18	8	18	8
	4	–15	–6	–15	–6	–15	–6

4. Solve the following game.

		B1	B2	B3	B4	B5
	1	8	10	–3	–8	–12
	2	3	6	0	6	12
A	3	7	5	–2	–8	17
	4	–11	12	–10	10	20
	5	–7	0	0	6	2

5. Solve the following two-person zero-sum game.

		B1	B2	B3
	1	5	9	3
A	2	6	–12	–11
	3	8	16	10

6. In a game of matching coins with 2 players, suppose player *A* wins one unit of value when there are two heads, wins nothing when there are two tails and loses $\frac{1}{2}$ unit of value when there is one head and one tail, determine the pay-off matrix and the optimum strategies for both the players.
7. Two players *A* and *B* without showing to each other toss a coin on a table with head or tail up. If the coin shows the same side, the player A takes both the coins, otherwise *B* gets them. Construct the matrix of the game and solve it.

8. *A* and *B* play one or two matches. Guess how many matches the opponent has taken. If one of the players guesses correctly, the opponent has to pay him as many rupees as the sum of the number of matches held by both the players, otherwise the payout is zero. Write down the pay-off matrix and obtain the optimum strategies for both the players.

9. Consider a matching coins game. The matching players are paid Rs. 8, if the two coins are both heads and Re. 1 if both are tails. The non-matching player is paid Rs. 3. When the coins do not match, both the players are given the chance to toss. What would be your strategy for the players.

10. Solve the following game whose pay-off matrix is given below.

(a)

		B 1	2	3
	1	−3	−2	6
A	2	2	0	2
	3	5	−2	−4

(b)

		B 1	2	3
	1	−2	15	−2
A	2	−5	−6	−4
	3	−5	20	−8

11. Firm *X* is fighting for its life against the determination of firm *Y* to it out of the industry. Firm *X* has the option of increasing the price, leaving it unchanged or lowering it. Firm *Y* has the same three options. Firm *X*'s gross sales in the event of the pairs of choices are shown below.

		Firm *Y*'s pricing strategy		
		Increase price	Do not change	Reduce price
	Increase price	90	80	110
Firm *X*'s pricing strategies	Do not change	110	100	90
	Reduce price	120	70	80

Solve the game and find its value.

12. There are two companies *A* and *B* in a certain city. Both companies have similar reputation and the total number of customers is equally divided between the two companies. Both the companies want to attract a greater number of customers by using different media of advertisement. By seeing the market trend, company *A* constructed the following pay-off matrix, where the number in the matrix indicates a gain or a loss of customer to it.

		Company *B*		
		Newspaper	Radio	TV
	Newspaper	40	50	−17
Company *A*	Radio	10	25	−10
	TV	100	30	60

Find the optimal strategies for both the companies and also the value of the game.

13. Two leading firms A and B are planning to make fund allocation for advertising their product. The matrix given below shows the percentage of market shares of the firms A and B for their various advertising policies.

		Firm B		
		No advertising	Medium advertising	Heavy advertising
	No advertising	60	50	40
Firm A	Medium adverting	70	70	50
	Heavy advertising	80	60	75

Find the optimum strategies for the two firms and the expected outcome when both the firms follow their optimum strategies.

14. Even though there are several manufacturers of scooters, two firms with branch names Janta and Praja, control their market in Western India. If both manufacturers make model changes of the same type for their market segment in the same year, their respective market shares remain constant. Likewise, if neither makes model changes, then also their market shares remain constant. The pay-off matrix in terms of increased/decreased percentage market share under different possible conditions is as follows

		Praja		
		No change	Minor change	Major change
	No change	0	–4	–10
Janta	Minor change	3	0	5
	Major change	8	1	0

i. Find the value of the game.

ii. What change should Janta consider, if this information is available only to itself.

15. Solve the following 2-person zero-sum game.

		Player B		
		B_1	B_2	B_3
	A_1	15	10	4
Player A	A_2	10	–3	10
	A_3	12	23	0

16. Solve the following 2-person zero-sum game.

		Player B		
		B_1	B_2	B_3
	A_1	10	5	7
Player A	A_2	6	7	5
	A_3	7	6	7

17. The following matrix represents the pay-off to P_1 in a rectangular game between two persons P_1 and P_2.

	P_2			
P_1	8	15	–4	–2
	19	15	17	16
	0	20	15	5

By the notation of dominance, reduce the game to a 2×4 game and solve it graphically.

18. Solve the following game.

		Player *B*			
		I	II	III	IV
Player *A*	I	3	2	4	0
	II	3	4	2	4
	III	4	2	4	0
	IV	0	4	0	8

19. Two firms are competing for business under the condition that one firm's gain is another firm's loss. Firm *A*'s pay-off matrix is given below.

		Firm *B*		
		No advertising	Medium advertising	Heavy advertising
Firm *A*	No advertising	10	5	–2
	Medium advertising	13	12	15
	Heavy advertising	16	14	10

20. Use the notion of dominance to simplify the rectangular game with the following pay-off.

		Player *A*			
		1	2	3	4
Player *B*	1	18	4	6	4
	2	6	2	13	7
	3	11	5	17	3
	4	7	6	12	2

Find its graphical solution.

Chapter 8

NETWORK ANALYSIS

8.1 INTRODUCTION

The graphical representation of a project plan is called the network. It shows the interrelationship between activities. Network analysis plays a major role in project management. Today customers expect the products at lower cost for minimum lead time. Network analysis is the mathematical tool that helps the project manager to determine the sequence of activities to be performed and the optimum project cost and optimum project time.

Activity An activity is part of a network which consumes certain time and resource. It should have a definite beginning and ending. It is represented by an arrow (→). The name and duration of the activity are written above and below the arrow respectively.

Node or Event The beginning and ending of an activity is known as node. It is represented by a circle having a number.

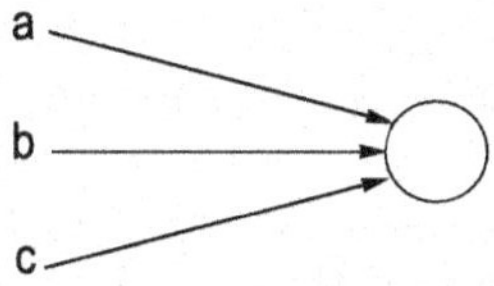

Figure 8.1 Merge event

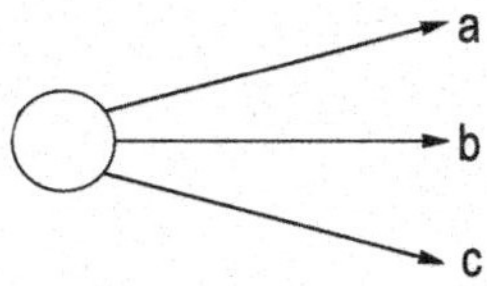

Figure 8.2 Burst event

If more than one activity is entering a node, it is called a **merge event**. If more than one activity is going away from a node, it is called a **burst event**.

8.2 BASIC RULES FOR DRAWING NETWORKS

Rule 1 Before an activity can start, all activities preceding it should be completed.

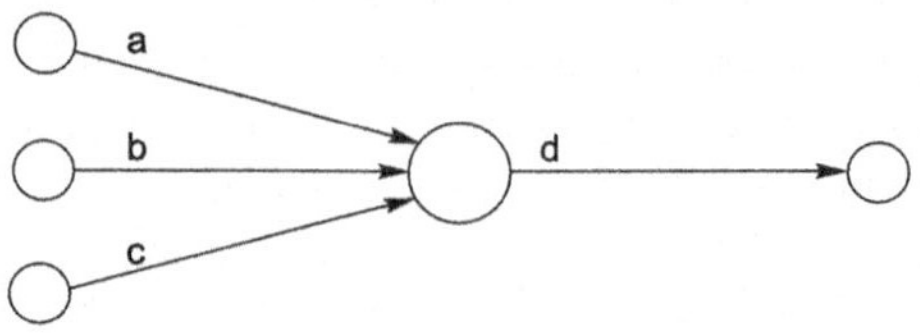

Figure 8.3 Network

For example, before activity d can start, activities a, b and c should have been completed.

Rule 2 Arrows indicate only the logical relationship. Neither the length of the arrow nor its direction has any significance.

Rule 3 Event numbers should not be duplicated.

Rule 4 No two events should be connected by more than one activity.

Rule 5 A network should have only one initial event and only one final or terminal event.

Rule 6 Arrows should not cross each other.

8.3 TYPES OF ACTIVITIES

Predecessor activity An activity that must be completed before other activities can start is called predecessor activity.

Successor activity An activity that cannot start until another activity is completed is called successor activity.

Concurrent activity Activities that can start simultaneously are called concurrent activities.

Dummy activity According to Rule 4, no two events should be connected by more than one activity. To avoid this, we use a dummy activity. A dummy activity is one which consumes no time or resource. It only shows the logical relationship and is represented by a dotted arrow.

Example 8.1

The following table gives the description of the jobs and their precedence. Construct an arrow diagram.

		Job description	Predecessor	Duration in days
	a	Forecast unit sales	—	14
	b	Pricing	a	3
Job	c	Preparing production schedule	a	7
	d	Costing production	c	4
	e	Preparing budget	b and d	10

Solution

Network for the given project is as follows

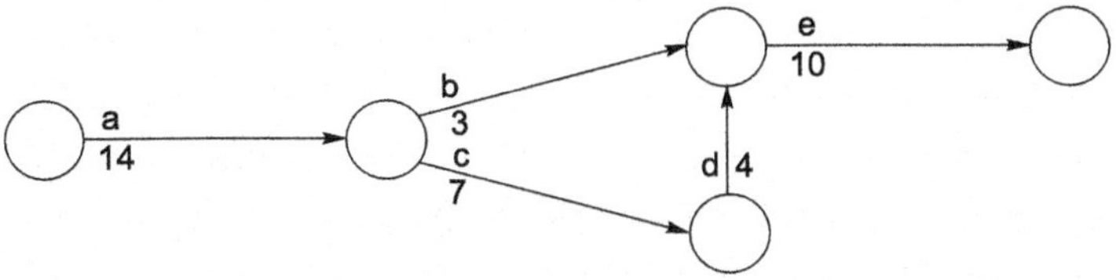

8.4 FULKERSON'S RULE

Rule 1 An initial event is one which has no arrows entering into it but has arrows going out of it. Number this as 1.

Rule 2 Remove all arrows coming out of the initial event. It will create at least one initial event. Number it and repeat the procedure till the end.

Example 8.2

A project consists of 11 activities. Draw the network and number the events as per Fulkerson's rule.

Activity	Predecessor
a	Start activity
b	*a*
c	*b*
d	*b*
e	*d*
f	*b*
g	*e*
h	*g*, *f*
j	*h*
k	*c*, *j*
l	*k*

Solution

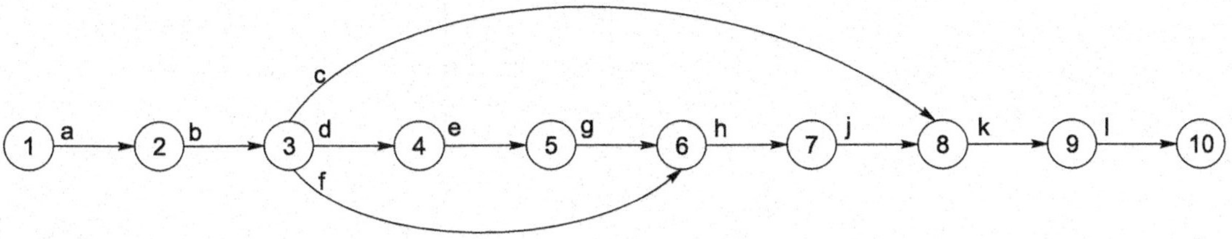

Example 8.3

A project consists of 16 activities having their predecessor relation as follows.

- * *a* is the first activity.
- * *b*, *c*, *d* follow *a* and can start concurrently.
- * *e*, *g* cannot begin until *c* is completed and starts simultaneously.

* f is the immediate successor to activities b and e.
* h and k run in parallel and both succeed g.
* l succeeds f and h.
* i and j are immediate successor activities to activity d.
* m and n are immediate successors to i and k. However both m and n cannot start at the same time.
* Activities o and p are last activities and immediate successors to n and l.
* Activity p is the immediate successor to m and j.

Draw the network, and number the events as per Fulkerson's rule.

Solution

The following is the precedence table.

Activity	Predecessor
a	-
b	a
c	a
d	a
e	c
f	b, e
g	c
h	g
i	d
j	d
k	g
l	f, h
m	i, k
n	i, k
o	n, l
p	n, l, m, j

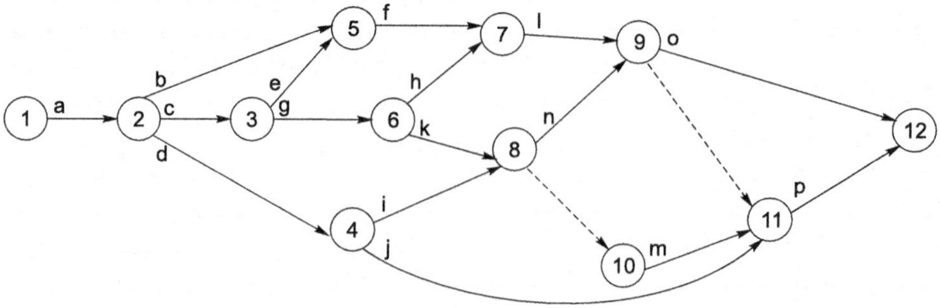

***Example* 8.4**

A small project is as follows

* Jobs *a*, *b* and *c* can start concurrently.
* Job *b* precedes job *e*.
* Job *d* is the successor of job *a*.
* Job *f* can start after job *a*.
* Job *g* depends on job *c*.
* Job *i* cannot start until jobs *e* and *f* are completed.
* Job *h* is the successor of job *e*.
* Job *j* depends on jobs *e* and *g*.
* Job *k* is the successor of jobs *i* and *h*.
* Job *m* depends on jobs *j* and *h*.
* Job *l* depends upon *i*, *j* and *h*.

Prepare a precedence table using Fulkerson's rule.

Solution

The precedence table is as follows.

Activity	Predecessor
a	–
b	–
c	–
d	*a*
e	*b*
f	*a*
g	*c*
h	*e*
i	*e*, *f*
j	*e*, *g*
k	*i*, *h*
l	*i*, *j*, *h*
m	*j*, *h*

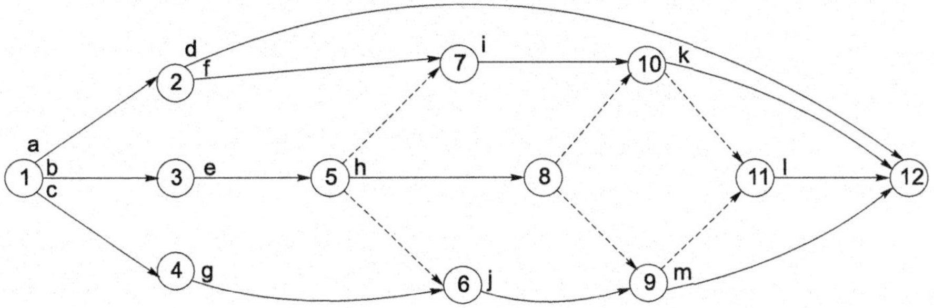

Example 8.5

Draw the network for the following project and number the events.

Activity	Predecessor
A	–
B	*A*
C	*A*
D	*A*, *C*
E	*B*, *D*
F	*B*, *D*, *E*
G	*B*, *D*
H	*E*, *F*, *G*

s

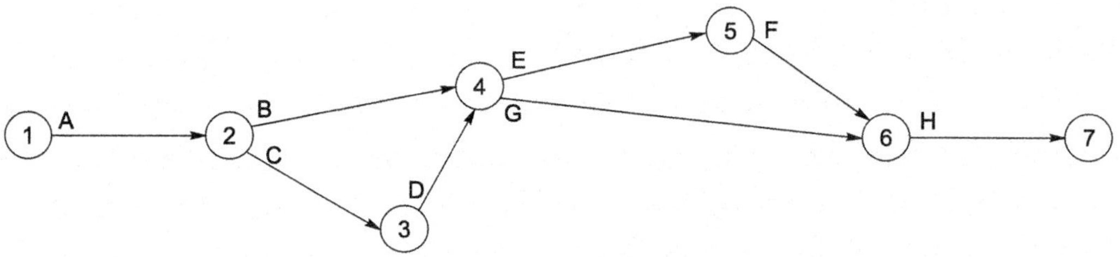

Example 8.6

Draw the network and have the events for the following project.

Activity	Immediate Predecessor
A	–
B	–
C	–
D	*A*, *B*
E	*B*, *C*
F	*A*, *B*
G	*C*
H	*D*, *E*, *F*
I	*D*
J	*G*
K	*G*
L	*H*, *J*
M	*K*
N	*I*, *L*

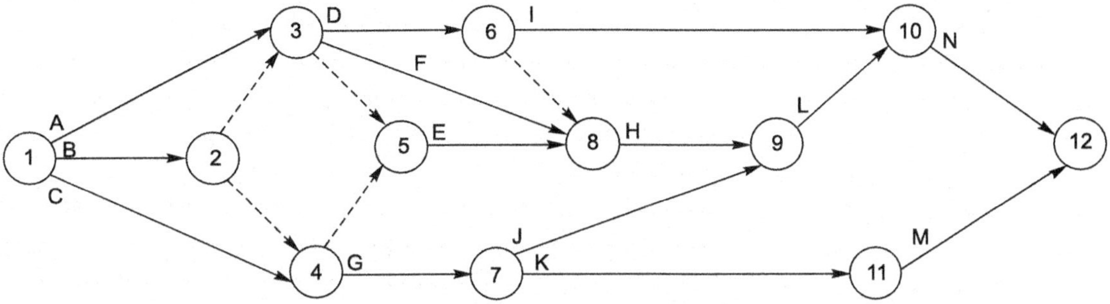

8.5 CRITICAL PATH IN NETWORK ANALYSIS

A path is a series of activities in the network that starts with the initial node and terminates with the last node. Any network consists of a number of such paths consisting of activities to be carried out in sequence to complete the project. The path having the maximum length of duration, determines the total duration of the project. The path with the longest duration is called the critical path.

8.6 SCHEDULING OF ACTIVITIES

When the project network is ready and the activity times are estimated, the questions that a rise are as follows

* How long would the project take to complete?
* What is the early start time for each activity?
* What is the early finish time for each activity?
* What is the late start time for each activity?
* What is the late finish time for each activity?
* What is the delay time for each activity?

As we have discussed in Section 8.5, the duration of the critical path determines the duration of the project. The forward pass calculation for activities along different patterns of the network gives the early start and early finish time for each activity. Early start time of an activity is the time by which an activity can be started after the completion of all the preceding activities. Early finish time of an activity is the sum of the early start time and the duration of that activity.

The backward pass calculations along different paths of the network gives the late start and the late finish time for each activity. Late start time of an activity is the latest time of starting that activity without delaying the completion of the project. Late finish time of an activity is the sum of the late start time and the duration of that activity.

Total float is the time by which the completion of an activity can be delayed without affecting the overall project duration. Free float is the time by which the completion of an activity can be delayed without affecting the early start time of a succeeding activity.

Example 8.7

The following table gives a list of jobs and their duration in days.

Job	Duration (days)
1 – 2	2
2 – 5	3
1 – 3	4
3 – 5	5
5 – 8	5
3 – 6	5
3 – 4	2
4 – 6	2
6 – 7	6
4 – 7	3
7 – 8	1

i. Draw the arrow diagram for the project.

ii. List all the paths and identify the critical path.

iii. The due date given by the management is 15 days. Compute for each job Early Start (ES), Early Finish (EF), Late Start (LS), Late Finish (LF), Total Slack (TS), Fress Slack (FS).

Solution

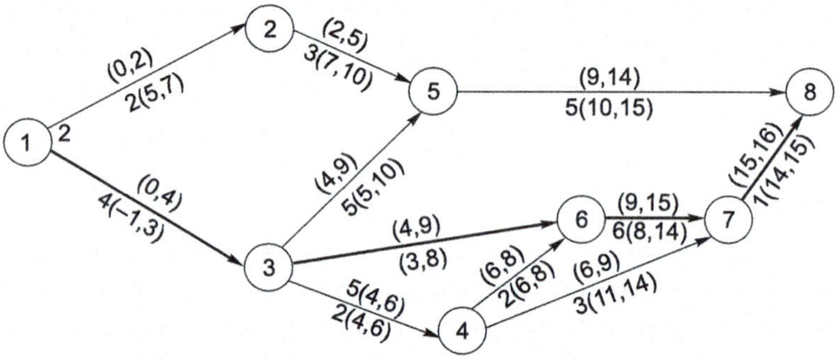

The above network shows the logical relationship between the activities in a project. The different paths in the network are as follows

Path	Total time consumed for activities
1 – 2 – 5 – 8	2 + 3 + 5 =10 days
1 – 3 – 5 – 8	4 + 5 + 5=14 days
1 – 3 – 6 – 7 – 8	4 + 5 + 6 + 1 = 16 days
1 – 3 – 4 – 6 – 7 – 8	4 + 2 + 2 + 6 + 1 = 15 days
1 – 3 – 4 – 7 – 8	4 + 2 + 3 + 1 = 10 days

The path taking maximum duration is called the critical path. In this case, 1 – 3 – 6 – 7 – 8 is the critical path with 16 days duration. Duration along the critical path is the duration of the project. In the above network, thick arrows represent the activities along the critical path.

Consider activity 1 – 2 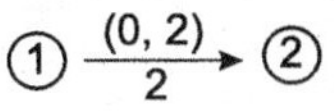

Numbers above the arrow represent Early Start (ES) and Early Finish (EF) of activity 1 – 2 respectively. If an activity is preceeded by only one activity, early finish of the previous activity becomes the early start of the next activity.

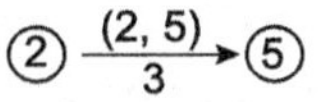

Early finish of activity 1 – 2, is 2. The same is the Early Start (ES) of the successor activity 2 – 5.

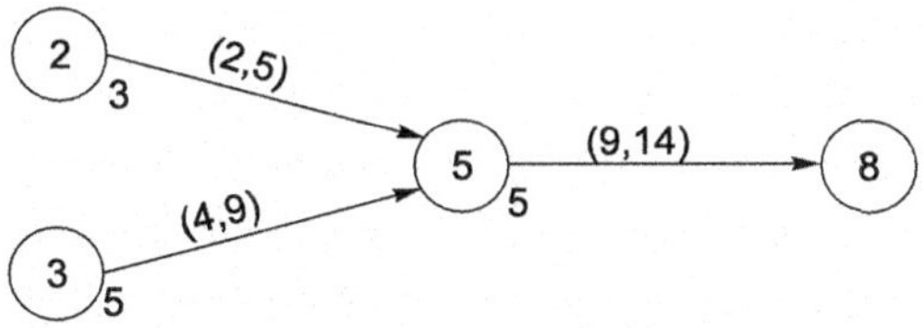

If an activity is preceded by more than one activity, the highest among the early finish times of preceding activities is taken as Early Start (ES) of that activity.

The due date for the completion of the project is considered as the project duration. Here it is 15 days. Back pass calculation starting with project duration of 15 days yield the Late Start (LS) and Late Finish (LF) times for each activity. The numbers below the arrows show the Late Start (LS) and Late Finish (LF) time respectively for an activity.

Consider last activities 5 – 8 and 7 – 8.

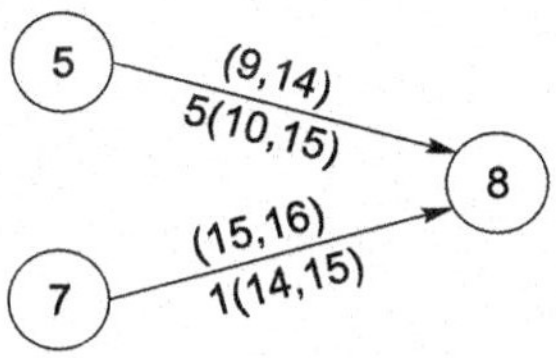

15 is the Late Finish for both activities 5 – 8 and 7 – 8. Late Finish time minus duration of an activity gives the Late Start for that activity. If an activity is succeeded by only one activity, Late Start of the successor activity becomes the Late Finish of that activity.

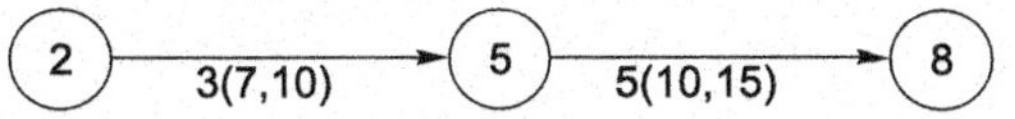

LF of activity 5 – 8 is 10. Same is the LF for the activity 2 – 5. If an activity is succeeded by more than one activity, the least among the LS time of the successor activities is taken as the LF of that activity.

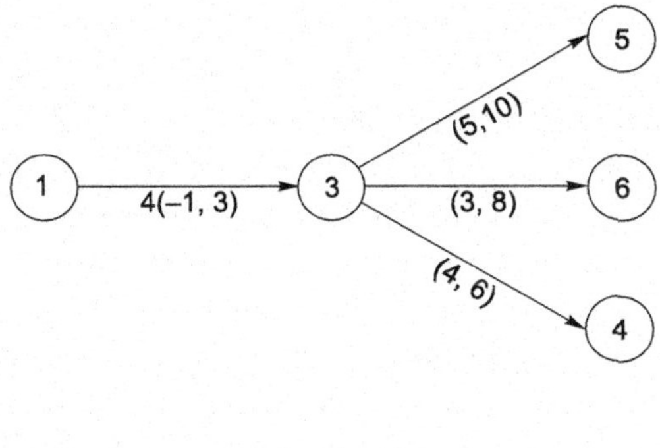

Activity 1 – 3 is succeeded by three activities. Least Late Start (LS) time is 3. Therefore Late Finish (LF) for activity 1 – 3 is 3.

Total slack of an activity is the difference between the Late Start (LS) and Early Start (ES) times of that activity. Free slack is the difference between the Early Start (ES) of successor activity and Early Finish (EF) of that activity.

The ES, EF, LS, LF, TS and FS for the given project are tabulated below.

Activity	ES	EF	LS	LF	TS	FS
1 – 2	0	2	5	7	5	0
2 – 5	2	5	7	10	5	4
1 – 3	0	4	–1	3	–1	0
3 – 5	4	9	5	10	1	0
5 – 8	9	14	10	15	1	2
3 – 6	4	9	3	8	–1	0
3 – 4	4	6	4	6	0	0
4 – 6	6	8	6	8	0	1
6 – 7	9	15	8	14	–1	0
4 – 7	6	9	11	14	5	6
7 – 8	15	16	14	15	–1	0

Example 8.8

The following table gives the details of the project.

Job	Immediate predecessor	Duration days
A	–	5
B	–	10
C	*A*	6
D	*B*	8
E	*C*, *D*	7
F	*C*	4
G	*C*	8
H	*E*, *F*, *B*	7
I	*G*, *H*, *D*	6

i. Draw the network.
ii. List all the paths and identify the critical path.
iii. The due date given by the management is 35 days. Compute for each job Early Start (ES), Early Finish (EF), Late Start (LS), Late Finish (LF), Total Slack (TS), Fress Slack (FS).

Solution

Network for the given project is as follows

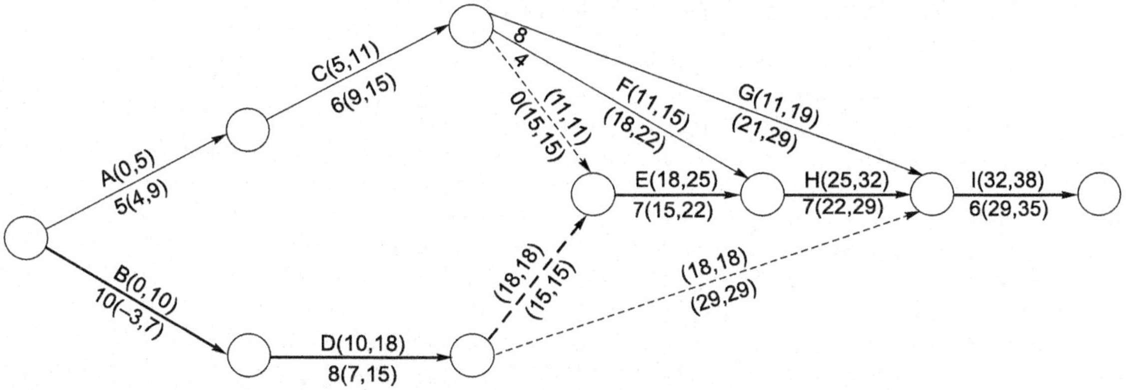

Path	Total duration	
$A-C-G-I$	25 days	
A − C − F − H − I	28 days	
A − C − E − H − I	31 days	
B − D − E − H − I	38 days	Critical path
B − D − I	24 days	

$B-D-E-H-I$ with 38 days duration is the critical path ES, EF, LS, LF, TS and FS are tabulated below.

Activity	ES	EF	LS	LF	TS	FS
A	0	5	4	9	4	0
B	0	10	−3	7	−3	0
C	5	11	9	15	4	0
D	10	18	7	15	−3	0
E	18	25	15	22	−3	0
F	11	15	18	22	7	10
G	11	19	21	29	10	13
H	25	32	22	29	−3	0
I	32	38	29	35	−3	0

ES and EF for each activity are calculated using forward pass calculation. LS and LF for each activity are calculated using backward pass calculation for an activity. Total slack for an activity is the difference between the LS and ES of the same activity. Free slack for an activity is the difference between the ES of the successor activity and EF of that activity.

8.7 PERFORMANCE EVALUATION AND REVIEW TECHNIQUE (PERT)

PERT is used for research and development work. Organisations involved in the research and development activities related to satellite, space vehicles, missiles, ship-building, etc., the time taken is not certain. This uncertainty is taken into account by PERT. Here, we use three time estimates.

t_o – optimistic time

t_p – pessimistic time

t_m – most probable time

Optimistic time is the time taken by an activity, if everything goes right. For example, in an industry, a product can be finished in the stipulated time, if there is no power interruption, no machine breakdown, if the sufficient material and men are available at the right time, this time is called optimistic time.

Pessimistic time is the time taken by an activity, if everything goes wrong. For example, a student leaves his house and goes on a two-wheeler to attend the class. He is confident that he can reach the college and enter the classroom in 15 minutes within which he is supposed to be present in the class. The 15 minutes time is the optimistic time for him. Unfortunately, a mob on strike blocks the road from morning till evening on that day. The student could not attend the class on that day. Everything went wrong. This is the pessimistic time.

Most probable time is the time taken by an activity under normal conditions. For example, completion of a certain job is delayed by one hour because of power interruption. This is most likely time.

The expected time t_e is calculated using the formula

$$t_e = \frac{t_o + 4t_m + t_p}{6}$$

Here, we assume t_o and t_p are equally likely to occur t_m is 4 times more likely to occur than t_0 and t_p. If an activity is repeated a number of times, on an average the time taken by that activity is the estimated time t_e. What is important to us is to know how reliable t_e is?

For example $t_o = 5$, $t_p = 17$ and $t_m = 8$ for a project.

then $$t_e = \frac{5 + 4(8) + 17}{6} = 9$$

Let another time estimate be $t_o = 8$ $t_p = 10$ and $t_s = 9$

then $$t_e = \frac{8 + 4(9) + 10}{6} = 9$$

Between these two time estimates, 2nd one is reliable. It is because the deviation from the values is less. Therefore, to know how reliable the estimated time is, we must determine the standard deviation of these values. One measure of deviation is variance.

Variance for an activity $$= V_t = \sqrt{S_t} = \left(\frac{t_p - t_o}{6}\right)$$

How does the management interpret t_e and S_T. What is important to the management is to determine the probability of completing the project within the due date. This is determined by using the formula

$$Z = \frac{D - T_e}{S_T}$$

where,

D = The due date

t_e = The total expected time

S_T = Standard deviation of all the activities along the critical path

Example 8.9

The following table gives a list of jobs along with their time estimates.

Job	Optimistic time (t_o)	Most likely time (t_m)	Pessimistic time (t_p)
1 – 2	3	6	15
1 – 6	2	5	14
2 – 3	6	12	30
2 – 4	2	5	8
3 – 5	5	11	17
4 – 5	3	6	15
6 – 7	3	9	27
5 – 8	1	4	7
7 – 8	4	19	28

i. Draw the project network.

ii. What is the approximate probability that jobs in the critical path will be completed by the due date of 38 days?

iii. What is the approximate probability that jobs on the next most critical path will be completed by the due date?

iv. What is your estimate of the probability that the entire project will be completed by the due date?

v. Which due date has 90% chance of being met?

Solution

Estimated time for an activity $t_e = \frac{t_o + 4t_m + t_p}{6}$

For example, for activity 1– 2, $t_e = \frac{3 + (4 \times 6) + 15}{6} = 7 \text{ days}$

Similarly t_e is calculated for all the activities and tabulated. Variance is calculated using the formula

$$\left(\frac{t_p - t_o}{6}\right)^2$$

For example, for activity 1 – 2, $\left(\frac{16-3}{6}\right)^2$

The estimated time t_e and variance of time V_T for all the jobs in the network is tabulated below.

Job	t_e	V_T
1 – 2	7	4
1 – 6	6	4
2 – 3	14	16
2 – 4	5	1
3 – 5	11	4
4 – 5	7	4
6 – 7	11	16
5 – 8	4	1
7 – 8	18	16

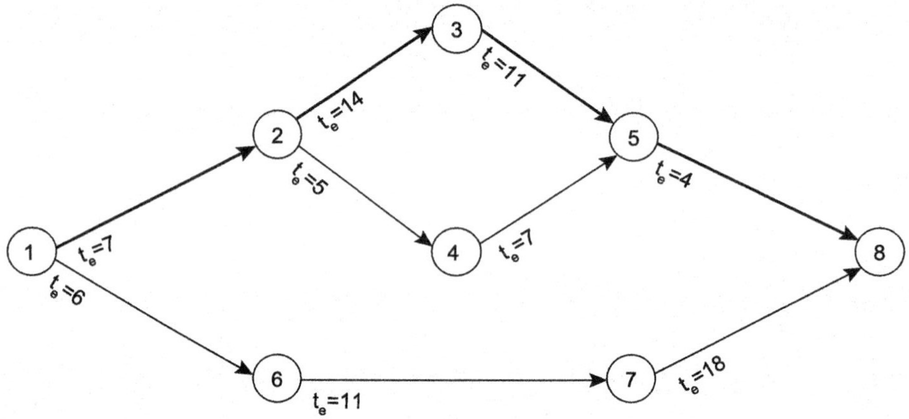

Different paths are as follows

Path	Duaration	
1 – 2– 3 – 5 – 8	36 Days	Critical path
1 – 2 – 4 – 5 – 8	23 Days	
1 – 6 – 7 – 8	35 Days	

From the above Figure 1 – 2 – 3 – 5 – 8 is the critical path.

$$V_T = \Sigma V_t \text{ along critical path}$$

$$V_T = 25$$

Standard deviation

$$S_T = \sqrt{V_T}$$

$$= \sqrt{25}$$

$$S_T = 5$$

We know that

$$Z = \frac{D - T_e}{S_T}$$

Due date $D = 38$ days

$$Z = \frac{38 - 36}{5}$$

$T_e = \Sigma t_e$ along critical path = 36 days

$$Z = 0.4$$

From the normal distribution table 'A' for $Z = 0.4$, the probability of completion of the project is 65.5%. Next most critical path is $1 - 6 - 7 - 8$.

t_e along next critical path = 35 days.

V_T along next critical path = 36 days

$$S_T = \sqrt{V_T} = \sqrt{25} = 6$$

$$\therefore \quad Z = \frac{D - T_e}{S_T} = \frac{38 - 35}{6} = 0.5$$

From normal distribution table for $Z = 0.5$. Probability is 64.15%

If the variance is more than the time taken T_e for an activity, it will not be reliable. Then the next most critical path may become the critical path. The product of these two probabilities gives the probability of completing the project.

Probability of completing the project $= P_{CP} \times P_{ncp}$

$= 0.6554 \times 0.6915$

$= 0.4532$

Probability $= 0.9$ (given)

From the table, for probability of 0.9, Z is 1.29.

$$Z = \frac{D - T_e}{S_T}$$

$$1.29 = \frac{D - 36}{5}$$

$$D = 42.45 \text{ days}$$

Example 8.10

The following table gives a list of jobs with their time estimates.

Job	Duration		
	t_o	t_m	t_p
1 – 2	1	1	7
1 – 3	1	4	7
2 – 4	2	2	8
3 – 4	1	1	1
2 – 5	2	5	14
4 – 5	2	5	8
4 – 6	3	6	15
6 – 8	6	15	13
6 – 7	5	14	17
4 – 7	3	12	21
5 – 8	4	6	8
7 – 8	2	4	6

Draw the project network and identify all paths through it. What is the probability that the project will be completed at least three days earlier than what due date has about 95% of chance of being met.

Solution

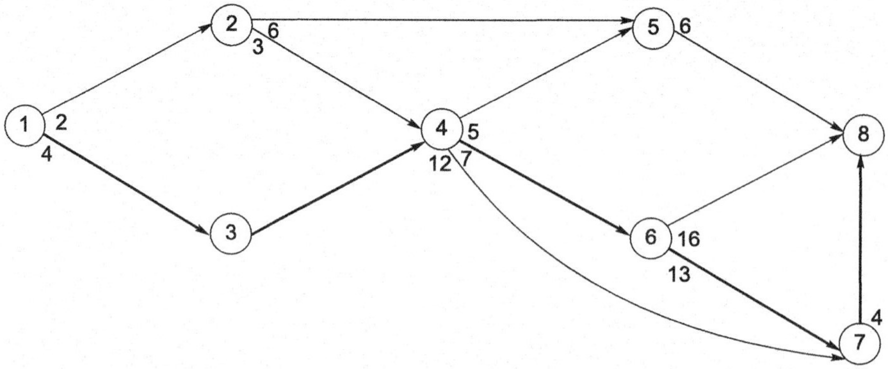

The different paths are as follows

Path	Duration	
1 – 2 – 5 – 8	14 days	
1 – 2 – 4 – 5 – 8	16 days	
1 – 2 – 4 – 6 – 8	28 days	
1 – 2 – 4 – 7 – 8	21 days	
1 – 2 – 4 – 6 – 7 – 8	29 days	Critical path
1 – 3 – 4 – 6 – 8	28 days	
1 – 3 – 4 – 7 – 8	21 days	
1 – 3 – 4 – 5 – 8	16 days	
1 – 3 – 4 – 6 – 7 – 8	29 days	Critical path

Job	$t_e = \frac{t_o + 4t_m + t_p}{6}$	V_T
1 – 2	2	1
1 – 3	4	1
2 – 4	3	1
3 – 4	1	0
2 – 5	6	4
4 – 5	5	1
4 – 6	7	4
6 – 8	16	1
6 – 7	13	4
4 – 7	12	9
5 – 8	6	0.44
7 – 8	4	0.44

V_T along 1 – 2 – 4 – 6 – 7 – 8 = 10.44

V_T along 1 – 3 – 4 – 6 – 7 – 8 = 9.44

Critical path with least standard deviation gives the optimum solution.

∴ Critical path 1 – 3 – 4 – 6 – 7 – 8 is taken to determine the probability of completion of the project.

Standard deviation

$$S_T = \sqrt{V_T} = \sqrt{9.44}$$

$$S_T = 3.072$$

$$Z = \frac{D - T_e}{S_T}$$

$$T_e = \Sigma t_e \text{ along the path } 1 - 3 - 4 - 6 - 7 - 8$$

$$T_e = 29 \text{ days}$$

If $D = 32$ days,

$$Z = \frac{32 - 29}{3.072} = 0.976$$

From table 'A', for $Z = 0.976$, probability of completion of the project = 83.40%.

If $D = 26$ days,

$$Z = \frac{26 - 29}{3.072} = -0.976$$

For $Z = -0.97$ from the tables, probability of completion of the project is 16.6%. If the probability of completion is 95%, from the table 'A' $Z = 1.65$

$$Z = 1.65 = \frac{D - T_e}{S_T} = \frac{D - 29}{3.072}$$

$$\therefore D = 34.06 \text{ days}$$

Example 8.11

A project has the following 3 time estimates.

Activity	Duration (months)		
	t_o	t_m	t_p
1 – 2	7	8	9
1 – 3	5	7	8
2 – 6	6	9	12
3 – 4	4	4	4
3 – 5	7	8	10
3 – 6	10	13	49
4 – 5	3	4	6
5 – 6	4	5	7
5 – 7	7	9	11
6 – 7	3	4	8

(a) Based on the above estimates calculate the expected time for each activity.

(b) Draw the project network and determine the critical path.

(c) What is the probability that the project will be completed at least 2 months earlier than the expected time?

(d) Which due date has 84% chance of being met?

Solution

Time estimates are listed below.

Activity	t_e	V_T
1 – 2	8	0.11
1 – 3	6.83	0.25
2 – 6	9	1
3 – 4	4	0
3 – 5	8.16	0.25
3 – 6	55	42.25
4 – 5	4.16	0.25
5 – 6	5.16	0.25
5 – 7	9	0.44
6 – 7	4.5	0.69

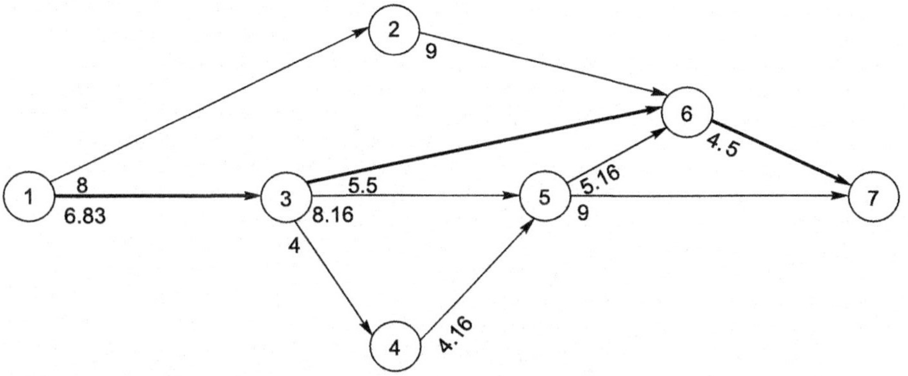

Paths	Duration	
1 – 2 – 6 – 7	23.5 months	Critical path
1 – 3 – 6 – 7	66.33 months	
1 – 3 – 5 – 6 – 7	24.69 months	
1 – 3 – 5 – 7	28 months	
1 – 3 – 4 – 5 – 7	24 months	
1 – 3 – 4 – 5 – 6 – 7	24.66 months	

$$V_T \text{ along } 1-3-6-7 = 43.19$$

$$\therefore S_T = 6.57$$

$$Z = \frac{D - T_e}{S_T}$$

$T_e = \Sigma t_e$ along $1 - 3 - 6 - 7 = 66.33$ months

$$D = 68.33 \text{ months}$$

$$\therefore Z = \frac{68.33 - 66.33}{6.57}$$

$$Z = 0.304$$

From table 'A' for $Z = 0.304$, probability of completion of the project = 63.31%

If probability of completion is 84%, from tables,

$$Z = 1.0$$

$$\therefore Z = 0.995 = \frac{D - 66.33}{6.57}$$

$$\therefore \text{Due date } D = 72.867 \text{ months}$$

Example 8.12

The following table lists the jobs of a network along with the time estimates.

Job	Duration (days)		
	t_o	t_P	t_m
1 – 2	3	5	7
1 – 3	1	2	3
2 – 5	6	8	12
3 – 4	8	12	17
4 – 5	0	0	0
4 – 6	6	9	12
4 – 7	3	6	8
7 – 10	8	15	20
6 – 9	1	2	3
5 – 8	5	7	9
8 – 9	3	6	8
9 – 10	2	4	6

Determine the critical path and the probability of finishing the project within the scheduled time

(a) When due date = 34 days

(b) When due date = 36 days

Solution

Following is the list of activities with t_e and V_T.

Activity	t_e	V_T
1 – 2	5	0.11
1 – 3	2	0.027
2 – 5	8.33	0.11
3 – 4	12.16	0.44
4 – 5	0	0
4 – 6	9	0.25
4 – 7	5.83	0.25
7 – 10	14.66	1.36
6 – 9	2	0.027
5 – 8	7	0.11
8 – 9	5.83	0.25
9 – 10	2.66	0.11

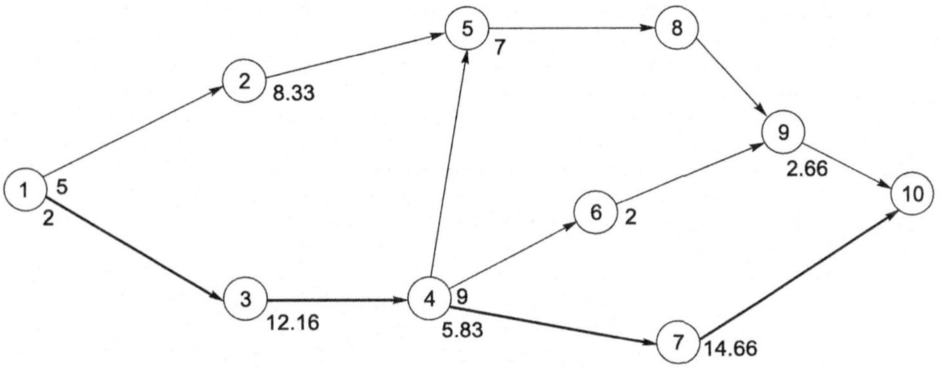

Path	Duration	
1 – 2 – 5 – 8 – 9 – 10	28.32 days	
1 – 3 – 4 – 5 – 8 – 9 – 10	29.15 days	
1 – 3 – 4 – 6 – 9 – 10	27.82 days	
1 – 3 – 4 – 7 – 10	34.65 days	Critical path

$T_e = \Sigma\, t_e$ along critical path $=$ 34.65 days

$V_T = \text{SV}_t$ along critical path $=$ 2.077

$$S_T = \sqrt{V_T} = 1.44$$

We know that
$$Z = \frac{D - T_e}{S_T}$$

where,

$$D = 34 \text{ days}$$

$$Z = \frac{34 - 34.65}{1.44} = -0.45$$

From the table A, for $Z = -0.45$, probability of completion is 32.64%. When $D = 36$ days,

$$Z = \frac{36 - 34.65}{1.44} = 0.9375$$

For $Z = 0.937$, probability of completion is 82.64%.

8.8 CRITICAL PATH METHOD (CPM)

Critical path method is used to determine the optimum duration and optimum project cost. This technique is used to arrive at the minimum lead time within which the product can be delivered to the customer. Present-day manufacturing activity is customer-oriented. CPM plays a major role in determining the optimum lead time and to see that the customer's requirements are fulfilled within the stipulated time.

Example 8.13

Given below is the network data for the time–cost trade-off data for a small project. Indirect cost is Rs. 150/day. Determine the following

(a) Normal project length

(b) Optimum project length

(c) Draw graph of duration vs total cost

Job	Normal		Crash	
	Time	Cost	Time	Cost
1 – 2	5	600	3	800
1 – 3	3	800	1	900
2 – 4	4	500	3	800
1 – 4	8	1200	5	1800
4 – 5	1	400	1	400
3 – 5	4	300	2	400
5 – 6	6	1000	4	1100
3 – 6	7	800	5	950

Solution

Crash time is the time up to which the duration of an activity can be reduced. Cost associated with it is known as crash cost.

For example, for activity 1 – 2, normal time required is 5 days and the associated cost is Rs. 600. The duration of this activity can be reduced from 5 days to 3 days. But here we have to pay Rs. 200 extra to reduce the duration by two days. Then what is the extra amount to be paid for the reduction of duration by one day? Here it is Rs. 100 per day.

This extra amount to be paid to reduce the duration by one day is called cost slope.

i.e., $$\text{Cost slope} = \frac{\text{Crash cost} - \text{Normal cost}}{\text{Normal time} - \text{Crash time}}$$

For example, for activity 1 – 2,

$$\text{Cost Slope} = \frac{800-600}{5-3} = \text{Rs.}100/\text{day}$$

Job	Cost slope	No. of days an activity can be crashed
1 – 2	100	~~2~~ ~~1~~
1 – 3	50	2
2 – 4	300	~~1~~
1 – 4	200	~~3~~ ~~2~~ 1
4 – 5	–	–
3 – 5	50	2
5 – 6	50	~~2~~
3 – 6	75	2

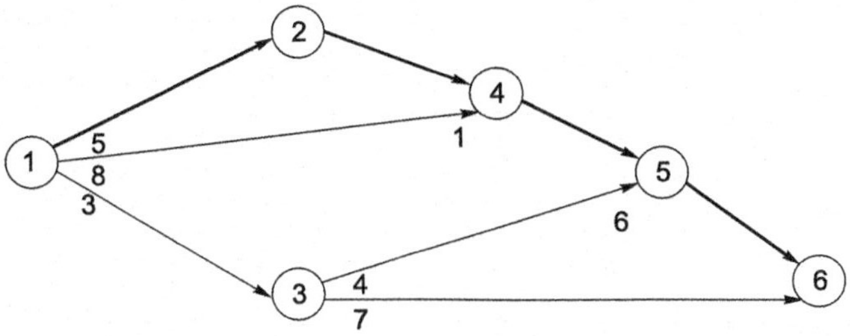

Path	Duration	
1 – 2 – 4 – 5 – 6	16 days	Critical path
1 – 4 – 5 – 6	15 days	
1 – 3 – 5 – 6	13 days	
1 – 3 – 6	10 days	

Normal project length is 16 days.

$$\begin{aligned}\text{Normal cost of the project} &= \Sigma\text{Normal costs of all activities} + \text{Indirect cost}\\ &= 5600 + 150\times16\\ &= 5600 + 2400\\ &= \text{Rs. } 8000\end{aligned}$$

To reduce the duration of the project, we must crash the activities along the critical path. Here they are 1 – 2, 2 – 4, 4 – 5 and 5 – 6.

Crash the activities whose cost slope is minimum. Here, for activity 5 – 6, the cost slope is 50 per day.

First crash It can be crashed only for two days.

$$\begin{aligned}\text{Duration} &= 14 \text{ days}\\ \text{Crashing cost} &= \text{Rs. } 50/\text{day} \times 2 = \text{Rs. } 100\\ \therefore \text{Total cost} &= \text{Normal cost} + \text{Overhead cost} + \text{Crashing cost}\\ &= 5600 + (150 \times 14) + 100\\ \text{Total cost} &= \text{Rs. } 7800\end{aligned}$$

Crashing of project duration is shown in the following figure.

Second crash

Crash activities 1 – 2 or 2 – 4. Crashing cost of 1 – 2 is less. Therefore, we crash activity 1 – 2. Activity 1 – 2 can be crashed by 2 days. Then 1 – 4 – 5 – 6 becomes the critical path. Therefore, we can crash activity 1 – 2 by one day only.

$$\begin{aligned}\text{Duration} &= 13 \text{ days}\\ \text{Crashing cost} &= \text{Rs. } 100 \text{ for 1st crash} + (100 \times 1) \text{ for second crash}\\ &= \text{Rs. } 200\\ \therefore \text{Total cost} &= 5600 + (150 \times 13) + 200\\ &= 5600 + 1950 + 200\\ &= \text{Rs. } 7750\end{aligned}$$

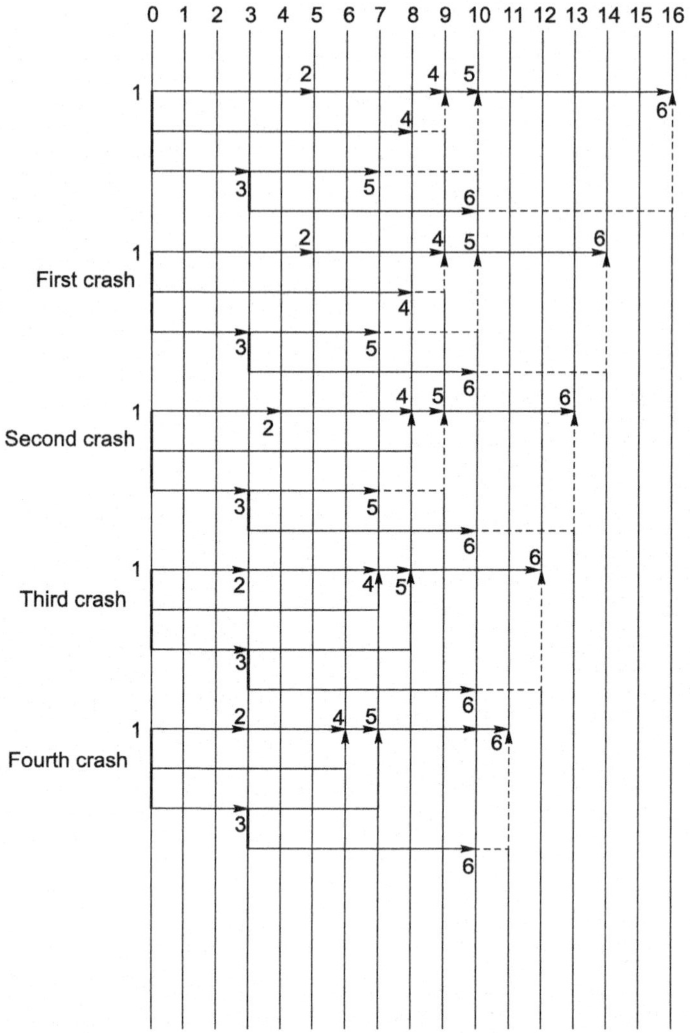

Third crash

Now there are two critical paths 1 – 2 – 4 – 5 – 6 and 1 – 4 – 5 – 6. If we want to reduce the duration, we must crash activities along both critical paths.

We can crash 1 – 2 and 1 – 4 or 2 – 4 and 1 – 4.

We crash 1 – 2 and 1 – 4 because crashing cost is minimum. 1 – 2 can be crashed by one day, 1 – 4 can be crashed by 3 days. Put together, we can crash them by one day.

Now, duration = 12 days

$$\text{Crashing cost} = 200 + 300$$
$$= 500$$
$$\text{Total cost} = 5600 + (150 \times 12) + 500$$
$$= \text{Rs. } 7900$$

Fourth crash

Crash 2 – 4 and 1 – 4 by one day

Duration = 11 days

Crashing cost = 500 + 500

= Rs. 1000

Total cost = 5600 + (150 × 11) + 1000

= Rs. 8250

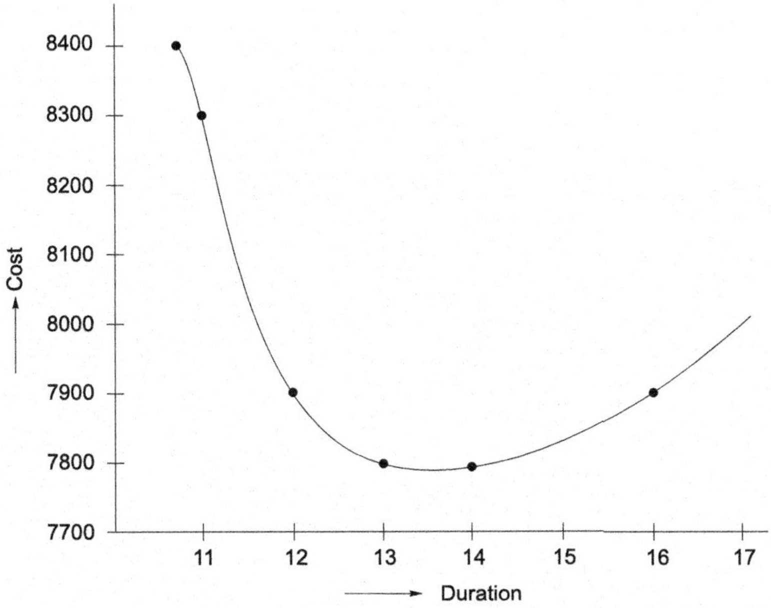

From the graph, it is found that the minimum duration is 11 days and optimum duration is 13 days.

Example 8.14

The following table gives a list of jobs, normal duration and cost, and crash duration and cost.

(a) Draw a network.

(b) What is the normal duration and minimum duration of the project.

(c) Plot a graph of cost–time relationship. Indirect cos t is Rs. 100/day.

Job	Normal		Crash	
	Time days	Cost Rs.	Time days	Cost Rs.
1 – 2	4	180	2	200
2 – 4	6	300	4	400
1 – 3	2	200	1	300
4 – 6	8	520	5	700
3 – 5	5	500	3	600
5 – 6	10	800	6	1200

Solution

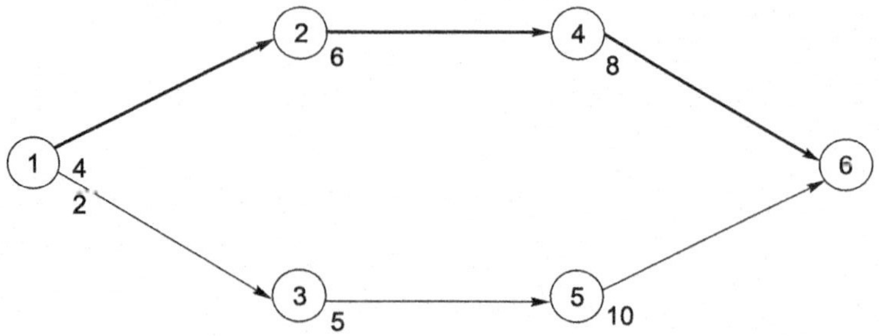

Path	Duration	
1 – 2 – 4 – 6	18 days	Critical path
1 – 3 – 5 – 6	17 days	

$$\text{Cost slope} = \frac{\text{Crash cost} - \text{Normal cost}}{\text{Normal time} - \text{Crash time}}$$

Job	Cost slope	No. of days an activity can be crashed
1 – 2	10	~~2~~ ~~1~~
2 – 4	50	~~2~~ ~~1~~
1 – 3	100	1
4 – 6	60	3
3 – 5	50	~~2~~ ~~1~~
5 – 6	100	4

Critical path duration is the normal duration of the project

$\therefore$ Normal duration = 18 days

Normal cost of the project = (180 + 300 + 200 + 520 + 500 + 800) + (18 × 100)

= Rs. 4300

To reduce the duration of the project, we must crash the activities along the critical path. Here they are 1 – 2, 2 – 4 and 4 – 6.

Crash activities whose cost slope is minimum. Hence for activity 1 – 2 it is Rs. 10/day.

First crash

Crash 1 – 2 by one day

Duration = 17 days

Crashing cost = Rs.10

Total cost = 2500 + (17 × 100) + 10

= Rs. 4210

The crashing of project duration is shown in the following figure.

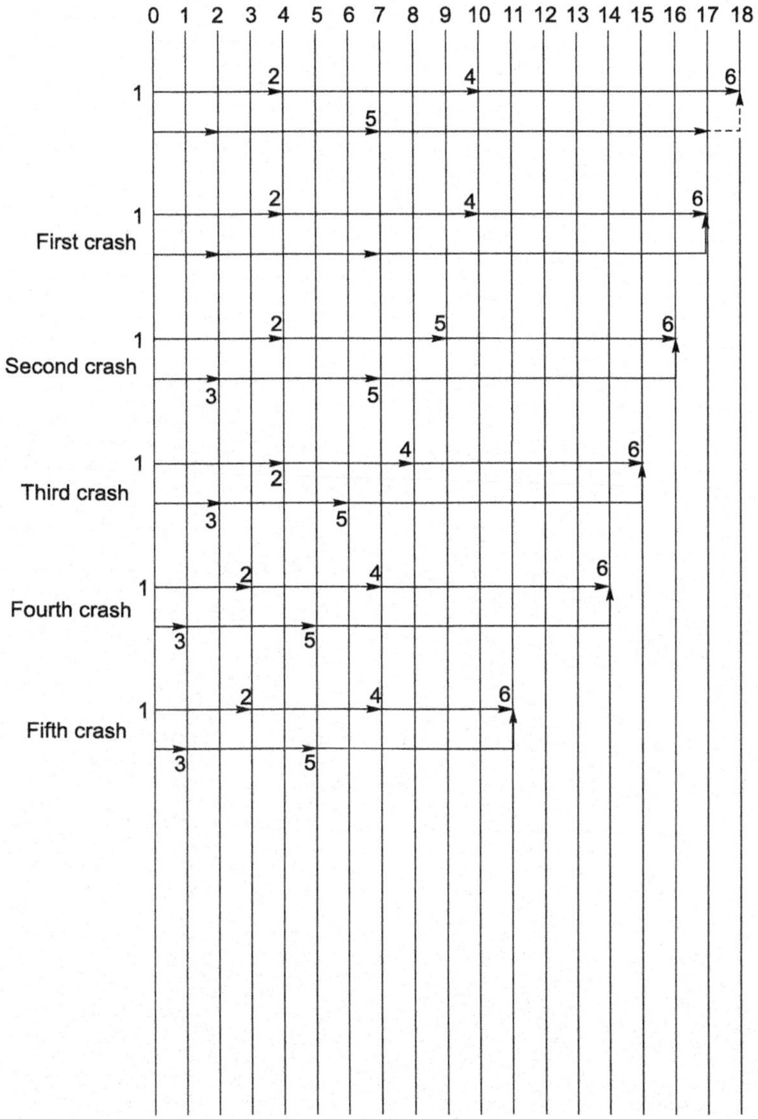

Second crash

Path 1 – 3 – 5 – 6 becomes the critical path.

Therefore, crash activities 1 – 2 and 3 – 5 by one day.

Duration = 16 days

Crashing cost = 10 + 10 + 50

= Rs. 70

Total cost = 2500 + (16 × 100) + 70

= Rs. 4170

Third crash

Crash activities 2 – 4 and 3 – 5 by one day

$$\begin{aligned} \text{Duration} &= 15 \text{ days} \\ \text{Crashing cost} &= 70 + 50 + 50 \\ &= \text{Rs. } 170 \\ \text{Total cost} &= 2500 + (15 \times 100) + 170 \\ &= \text{Rs. } 4170 \end{aligned}$$

Fourth crash

Crash activities 2 – 4 and 1 – 3 by one day

$$\begin{aligned} \text{Duration} &= 14 \text{ days} \\ \text{Crashing cost} &= 170 + 50 + 100 \\ &= \text{Rs. } 320 \\ \text{Total cost} &= 2500 + (14 \times 100) + 320 \\ &= \text{Rs. } 4220 \end{aligned}$$

Fifth crash

Crash activities 4 – 6 and 5 – 6 by 3 days

$$\begin{aligned} \text{Duration} &= 11 \text{ days} \\ \text{Crashing cost} &= 320 + 3(60 + 100) \\ &= \text{Rs. } 800 \\ \text{Total cost} &= 2500 + (11 \times 100) + 800 \\ &= \text{Rs. } 4400 \end{aligned}$$

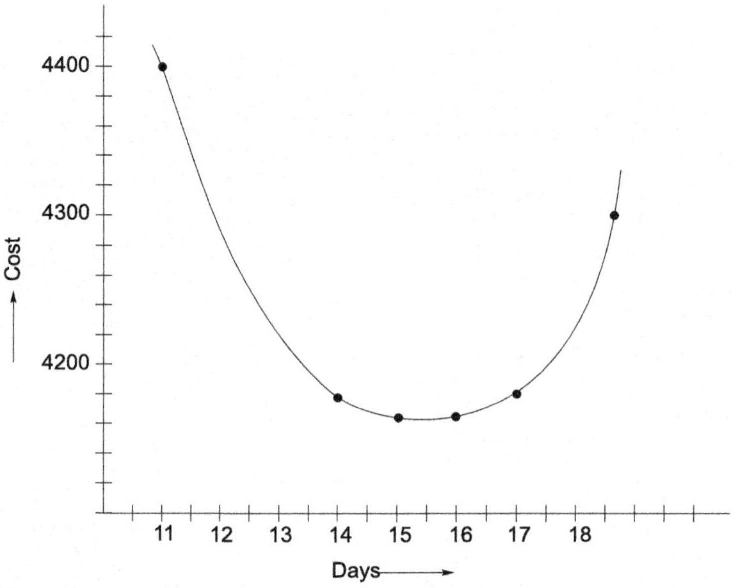

From the graph, it is found that the minimum duration is 11 days and optimum duration is 15 days.

Example 8.15

A maintenance foreman has given the following estimates of time and cost for jobs of a project.

(a) What is the normal duration, minimum duration and the corresponding costs of the project?

(b) What is the optimum duration and minimum cost?

(c) Plot a graph of cost–time relationship. Indirect cost is Rs. 250/week.

Job	Predecessor	Normal		Crash	
		time	cost	time	cost
A	–	8	500	6	1000
B	*A*	7	400	4	940
C	*A*	12	1000	5	1840
D	*A*	9	700	5	1020
E	*B, C, D*	6	500	6	500

Solution

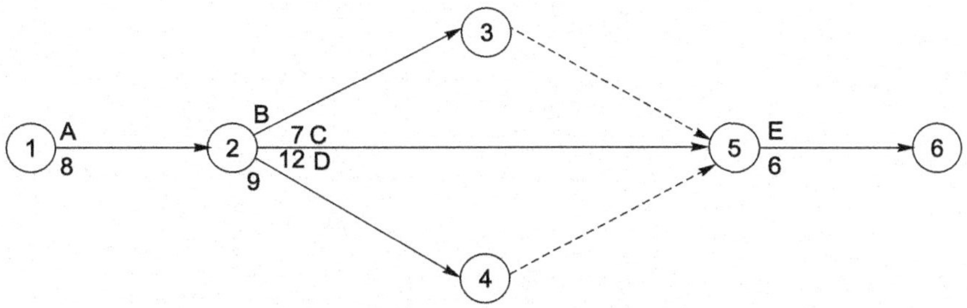

Activity	Cost slope	No. of days an activity can be crashed
1 – 2 (*A*)	100	~~2~~
2 – 3 (*B*)	180	3
2 – 5 (*C*)	120	~~7~~ ~~4~~ 2
2 – 4 (*D*)	80	4
5 – 6 (*E*)	–	–

Path	Duration
1 – 2 – 3 – 5 – 6	21 days
1 – 2 – 5 – 6	26 days
1 – 2 – 4 – 5–6	23 days

Normal duration of the project = Duration of critical path

= 26 days

Normal cost = 3400 + 26(250)

= Rs. 9900

First crash

Crash 1 – 2 by 2 days

Duration = 24 days

Crashing cost = 2(100) = Rs. 200

Total cost = 3400 + 24 (250) + 2(100)

= Rs. 9600

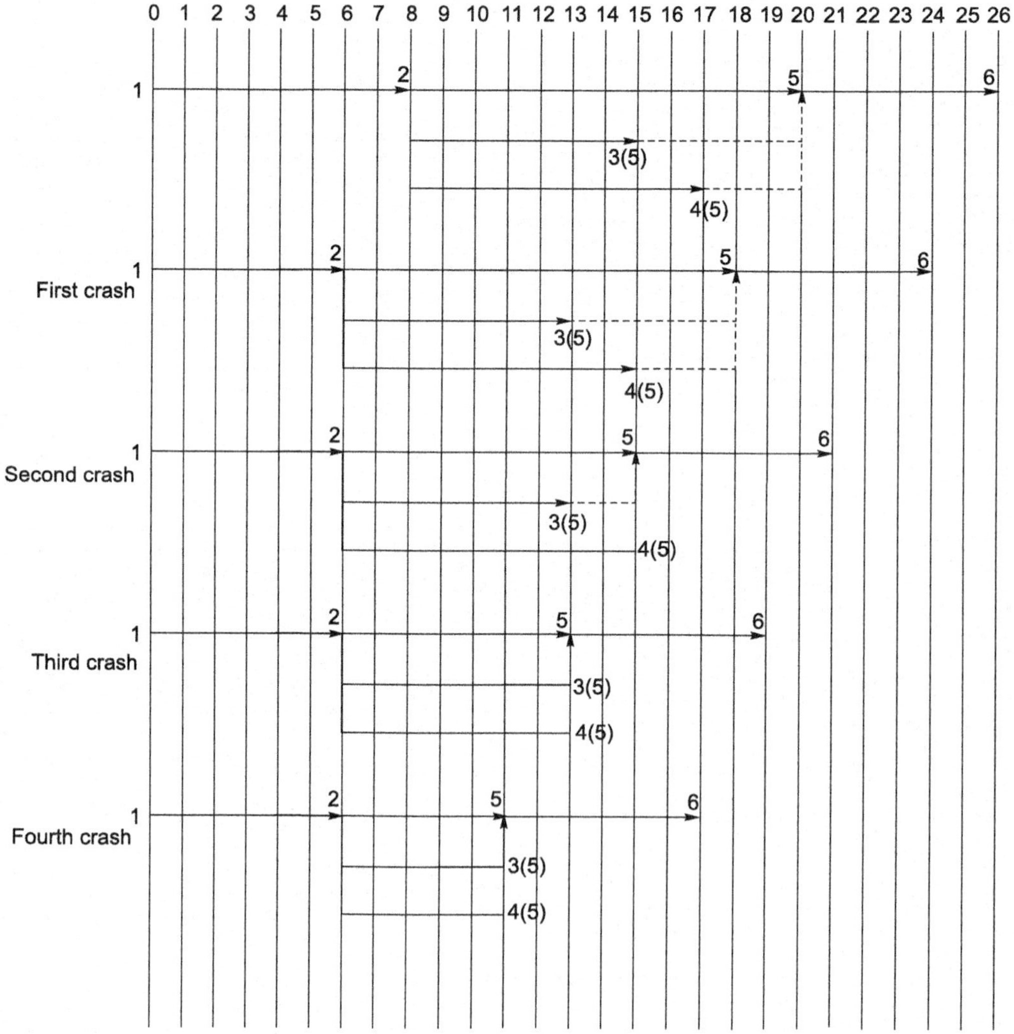

Second crash

Crash 2 – 5 by 3 days

$$\begin{aligned}\text{Duration} &= 21 \text{ days}\\ \text{Crashing cost} &= 200 + (120 \times 3)\\ &= \text{Rs. } 560\\ \text{Total cost} &= 3400 + 21(250) + 560\\ &= 3400 + 5250 + 560\\ &= \text{Rs. } 9210\end{aligned}$$

The crashing of project duration is shown in the following figure.

Third crash

Crash activities 2 – 4 and 2 – 5 by 2 days

$$\begin{aligned}\text{Duration} &= 19 \text{ days}\\ \text{Crashing cost} &= 560 + (80 \times 2 + 120 - 2)\\ &= 560 + 160 + 240\\ &= \text{Rs. } 960\\ \text{Total cost} &= 3400 + 19(250) + 960\\ &= 3400 + 4750 + 960\\ &= \text{Rs. } 9110\end{aligned}$$

Fourth crash

Crash activities 2 – 5, 2 – 3 and 2 – 4 by 2 days

$$\begin{aligned}\text{Duration} &= 17 \text{ days}\\ \text{Crashing cost} &= 960 + (120 \times 2 + 180 \times 2 + 80 \times 2)\\ &= 960 + 240 + 360 + 160\\ &= \text{Rs. } 1720\\ \text{Total cost} &= 3400 + 17(250) + 1720\\ &= \text{Rs. } 9120\end{aligned}$$

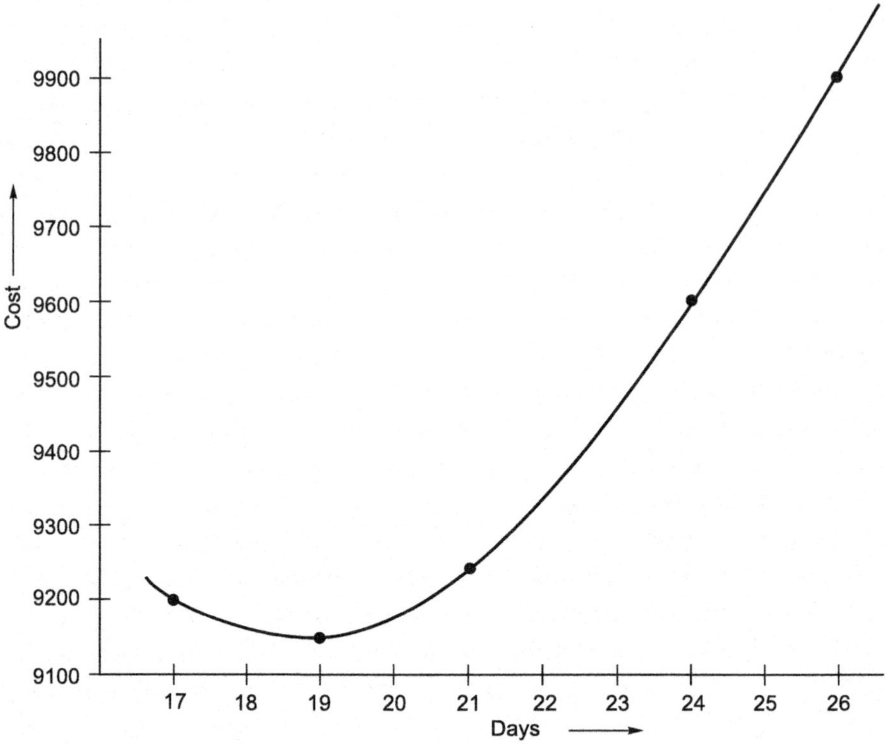

Minimum duration = 17 days

Optimum duration = 19 days

EXERCISES

1. Draw a network for the following project.

Activity	Immediate predecessors	Activity	Immediate predecessors
A	–	*H*	*F*
B	–	*I*	*C, D, G, H*
C	–	*J*	*I*
D	–	*K*	*J, K*
E	*A*	*L*	*J, K*
F	*B*	*M*	*J, K*
G	*E*	*N*	*M*

2. The precedence relationship for the activities in a project is given below. Draw a network.

Activity	Immediate predecessors	Activity	Immediate predecessors
A	–	*I*	*F, G, H*
B	–	*J*	*O, E, N*
C	–	*K*	*B, C, D*
D	–	*L*	*K*
E	*B, C, D*	*M*	*B, C, D*
F	*A, B, C, D*	*N*	*B, C, D*
G	*A, B, C, D*	*O*	*A, B, C, D*
H	*A, B, C, D*	*P*	*I, J, M, L*

3. Draw a network for the following project.

Activity	Description	Preceding Activity
A	Drawing prepared and approved	–
B	Cost analysis	*A*
C	Tool feasibility (economics)	*A*
D	Tool manufactured	*C*
E	Favourable cost	*B, C*
F	Raw material procured	*D, E*
G	Sub-assemblies ordered	*E*
H	Sub-assemblies received	*G*
I	Parts manufactured	*D, F*
J	Final assembly	*I, H*
K	Testing and shipment	*J*

4. Write down the precedence table for the following network.

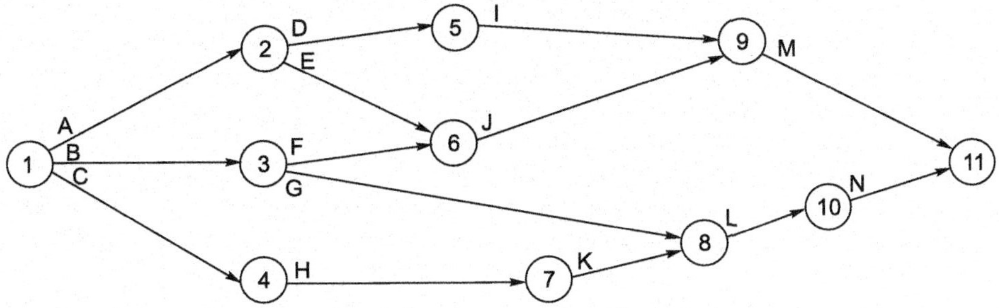

5. A project has the following time schedule.

Activity	Time in weeks	Activity	Time in week
1 – 2	4	5 – 7	8
1 – 3	1	6 – 8	1
2 – 4	1	7 – 8	2
3 – 4	1	8 – 9	1
3 – 5	6	8 – 10	8
4 – 9	5	9 – 10	7
5 – 6	4		

Construct the network and compute Early Start (ES), Early Finish (EF), Late Start (LS), Late Finish (LF), Free Slack (FS) and Total Slack (TS) for each activity.

6. A project has the following time estimates.

Activity	Time in weeks	Activity	Time in week
1 – 2	2	4 – 6	3
1 – 3	2	6 – 8	1
1 – 4	1	7 – 9	5
2 – 5	4	8 – 8	4
3 – 6	8	8 – 9	3
3 – 7	5		

Construct the network and compute ES, EF, LS, LF, FS and TS for all the activities.

7. Draw the network of the following activities and tabulate early start, early finish, late start, late finish, free slack and total slack for each activity.

Activity	Description	Duration (days)
1 – 2	Excavation	4
1 – 3	Order and delivery of steel	17
2 – 3	Frame work of steel	4
2 – 4	Foundation	5
3 – 4	Dummy	0
3 – 5	Concrete work	8
4 – 6	Placement of framer	2
5 – 6	Dummy	0
5 – 9	Back filling	3
6 – 7	Concrete stage - 2	8
7 – 8	Dummy	0
7 – 9	Dummy	0
8 – 10	Steel work stage - 2	10
9 – 10	Black filling stage - 2	5

8. Draw the network and compute ES, EF, LS, LF, FS and TS for all the activities.

Activity	*A*	*B*	*C*	*D*	*E*	*F*	*G*	*H*	*I*	*J*	*K*	*L*
Predecessor	–	–	–	*B*, *C*	*A*	*C*	*E*	*E*	*D*, *F*, *H*	*E*	*I*, *J*	*G*
Duration (days)	9	4	7	8	7	5	10	8	6	9	10	2

9. The data for a PERT network is given in the following table.

Activity	Time (t_o)	Duration (t_m)	(Days) (t_p)
1 – 2	2	4	6
1 – 3	6	6	6
1 – 4	6	12	24
2 – 3	2	5	8
2 – 5	11	14	28
3 – 4	15	24	45
3 – 6	3	6	9
4 – 6	9	15	27
5 – 6	4	10	16

i. Draw the network and find the critical path.

ii. What is the probability of completion of the project within 60 days?

iii. What is the probability of completing the project between 45 and 54 days?

10. Consider a project having the following activities and their time estimates. Draw the network and determine the duration to complete the project with 99% chance.

Activity	Immediate predecessors	Expected time (Days)		
		Most optimistic	Most likely	Most pessimistic
A	–	2	2	2
B	–	1	3	7
C	*A*	4	7	8
D	*A*	3	5	7
E	*B*	2	6	9
F	*B*	5	9	11
G	*C*, *D*	3	6	8
H	*E*	2	6	9
I	*C*, *D*	3	5	8
J	*G*, *H*	1	3	4
K	*F*	4	8	11
L	*J*, *K*	2	5	7

11. Consider a project having the following activities and their time estimates

Activity	Immediate predecessors	Expected time (Days)		
		Most optimistic	Most likely	Most pessimistic
A	–	3	4	5
B	–	4	8	10
C	*B*	5	6	8
D	*A*, *C*	9	10	15
E	*B*	4	6	8
F	*D*, *E*	3	4	5
G	*D*, *E*	5	6	8
H	*D*, *E*	1	3	4
I	*G*	2	4	5
J	*F*, *I*	7	8	10
K	*G*	4	5	6
L	*H*	8	9	13
M	*J*, *K*, *L*	6	7	8

i. Draw the network and find the critical path.

ii. Determine the due date on which 84% chance of project being met.

12. Following table lists the activities with time estimates for a project.

Activity	t_o	t_m	t_p	Activity	t_o	t_m	t_p
1 – 2	6	9	11	4 – 9	6	8	9
1 – 3	7	9	10	5 – 8	5	7	8
2 – 4	8	12	13	5 – 9	10	12	13
2 – 5	9	11	12	6 – 9	11	12	13
2 – 6	12	13	15	6 – 10	13	15	18
2 – 7	11	14	16	7 – 9	9	10	14
3 – 4	10	13	15	7 – 10	8	12	14
3 – 5	14	16	18	8 – 11	10	11	15
3 – 6	11	13	15	9 – 11	11	13	16
3 – 7	12	15	16	10 – 11	6	8	11
4 – 8	8	10	12				

i. Draw the project network and find the critical path.

ii. What is the probability of completion of the project within 47 days?

13. The following table gives data on normal time and cost, and crash time and cost for a project.

Activity	Duration (Weeks)		Total cost (Rs.)	
	Normal	Crash	Normal	Crash
1 – 2	3	2	300	450
2 – 3	3	3	75	75
2 – 4	5	3	200	300
2 – 5	4	4	120	120
3 – 4	4	1	100	190
4 – 6	3	2	90	130
5 – 6	3	1	60	110

i. Draw the network and find out the critical path and the normal project duration.

ii. If the indirect costs are Rs. 100 per week, find out the optimum duration by crashing and the corresponding project costs.

14. In planning a project to introduce a new project in the market, a company lists the various activities, their normal times and costs, and their crash times and costs. They are

Activity	Immediate predecessor	Normal time (Weeks)	Crash	Total cost (Rs.)	
			Cost (Rs.)	Time (Weeks)	Cost (Rs.)
A	-	5	10,000	4	12,000
B	-	2	6000	2	6000
C	*A*	4	8000	3	10,000
D	*A*	4	10,000	3	15,000
E	*A*	3	11,000	1	16,000
F	*C*	1	7000	1	7000
G	*D*	4	8000	2	12,000
H	*B, E*	5	9000	3	12,000
I	*H*	2	8500	2	8500
J	*F, G, I*	3	7500	2	10,000

15. The normal and crash points for each activity are given in the following table. Find the different minimum crash schedules (32 days schedule) that can occur between normal and crash times.

Activity	Normal		Crash	
	Duration days	Cost Rs.	Duration days	Cost Rs.
1 – 2	20	2000	15	3000
1 – 3	10	1500	7	2400
2 – 5	15	1000	10	1500
3 – 4	16	3000	12	4000
3 – 5	22	4500	16	5700
4 – 5	14	1500	10	2100

16. The following table gives the various activities, their duration and their direct costs. The indirect costs are Rs. 2000 per week. Find the minimum cost schedule using CPM technique.

Activity	Time (Weeks)		Cost (Rs.)	
	Normal	Crash	Normal	Crash
1 – 2	8	4	3000	6000
1 – 3	5	3	4000	8000
2 – 4	9	6	4000	5500
3 – 5	7	5	2000	3200
2 – 5	5	1	8000	12,000
4 – 6	3	2	10,000	11,200
5 – 6	6	2	4000	6800
6 – 7	10	7	6000	8700
5 – 7	9	5	4200	9000

17. The relevant data for a project are as follows

Activity	Preceding activity	Time (Weeks)		Cost (Rs.)	
		Normal	Crash	Normal	Crash
A	–	12	10	5000	7400
B	*A*	9	8	4300	5900
C	*A*	6	4	3700	4900
D	*C*	3	3	2500	2500
E	*B*	7	6	3900	4800
F	*D*	8	5	4600	6100
G	*D*, *E*	4	4	2800	2800
H	*F*	6	3	4100	6200
I	*G*	7	5	4800	7800

This project must be completed in 34 weeks. Suggest the least cost schedule for completing the project.

Chapter 9

QUEUING THEORY

9.1 INTRODUCTION

In mass production industry, the material flows through a sequence of processing stages and is finally converted to saleable product. Automobiles form a line at the fuel service station for filling the fuel. The material flowing through a sequence of stages in industry and automobile at fuel station are the examples for queue.

In a railway station, people seeking train tickets form a queue. They will be served through several counters.

In a supermarket, customers form several queues at different counters to purchase things of their interest.

These examples illustrate the queuing system. How long does an item or a customer wait in the queue to get service? What is the length of the line? What is the service time/unit? What is the arrival rate? These are the questions that arise from the queuing systems. Queuing theory, a mathematical technique, answers all these questions and helps to determine optimum waiting time.

9.2 CHARACTERISTICS OF QUEUING SYSTEM

Customers are generated from a source. These customers in the queuing system need not be humans or even physical objects. A customer could be a message, a transaction, an order or work. As soon as the customer arrives at the facility, they have to wait in a queue if the facility is busy. When a facility completes a service, it moves a waiting customer in the queue for service. If the queue is empty, the facility becomes idle till a new customer arrives.

A server at the facility could be a person or crew of workers attending a single customer or a machine, etc.

A queuing system refers to the whole system consisting of the customers waiting in the queue and the customers in service. The number of customers in the queuing system varies with time.

Queue size may be finite or infinite. The customers may arrive at the facility one by one or in batches of fixed size or variable size. If the facility is busy, customers arrived will form a queue. The time gap between successive arrivals is uncertain. The arrival pattern may be probabilistic or deterministic.

In common queuing models, we assume the arrival rate in a Poisson distribution and the time gap between arrivals in an exponential distribution.

Selection of customers plays a major role in the analysis of queuing models. Commonly used selection methods are as follows

1. First Come First Served (FCFS)
2. Last Come First Served (LCFS)
3. Selection of Service in Random Order (SIRO)

Behaviour of the customer in the queue affects the analysis of the queuing system. For instance, if the queue is too long, the customer arrived may decide not to enter the queue. He is said to balk from joining a queue. In another case, a customer may enter the queue but after sometime, he may come out. He is said to renege from a queue. There may be two or more queues. In such case, customer may move from one queue to another. He is said to jockey from one queue to another.

Variations in the characteristics of a queuing system lead to variety of queuing models.

9.3 PURE BIRTH MODEL

In a queuing system, arrival of additional customers is referred to as birth. In pure birth model, only arrival of the customer into the queuing system is considered. But the departure of serviced customer from the queuing system is ignored.

Let $P_n(t)$ be the probability of n arrivals during time t and $P_n(t+h)$ be the probability of n arrivals during time $(t+h)$.

$$P_n(t+h) = P_n(t) \cdot P_0(h) + P_{n-1}(t) \cdot P_1(h) \quad (9.1)$$

$$P_0(t) = P\{\text{time between arrivals} \geq t\}$$

$$= 1 - P\,\{\text{time between arrivals} \leq t\}$$

$$= 1 - (1 - e^{-\lambda t})$$

where λ is the arrival rate in unit time and time gap between arrivals is exponential.

$$P_0(t) = e^{-\lambda t}$$

$$\therefore \quad P_0(h) = e^{-\lambda h} = 1 - + \ldots$$

$$P_0(h) = 1 - \lambda h + 0(h)^2$$

Assumption During a small time h, atmost one arrival can take place.

$$\therefore P_0(h) = 1 - \lambda h \quad (9.2)$$

$$P_1(h) = 1 - P_0(h)$$

$$= 1 - (1 - \lambda h)$$

$$P_1(h) = \lambda h \quad (9.3)$$

Substituting Equations (9.2) and (9.3) in (9.1), we get

$$P_n(t+h) = P_n(t).(1 - \lambda h) + P_{n-1}(t) \,.\, \lambda h \quad (9.4)$$

If $n = 0$

$$P_0(t+h) = P_0(t)(1 - \lambda h) \quad (9.5)$$

$$P_n(t+h) = P_n(t) - P_n(t)\lambda h + P_{n-1}(t)\,\lambda h$$

$$P_n(t+h) - P_n(t) = -\lambda h\,P_n(t) + P_{n-1}(t) \,.\, \lambda h$$

Dividing both the sides by h and taking the limit as h tends to zero, we get

$$\lim_{h \to 0} \frac{P_n(t+h) - P_n(t)}{h} = -\lambda P_n(t) + \lambda P_{n-1}(t)$$

$$\frac{d}{dt} P_n(t) = -\lambda P_n(t) + \lambda P_{n-1}(t) \quad (9.6)$$

If $n = 0$
$$\frac{d}{dt} P_0(t) = -\lambda P_0(t) \quad (9.7)$$

The solution of the differential Equations (9.5) and (9.6) gives

$$P_n(t) = \frac{(\lambda t)^n \cdot e^{-\lambda t}}{n!}$$

where, $n = 0, 1, 2, 3$, etc.

9.4 PURE DEATH MODEL

In a queuing system, departure of served customer is referred to as death. In pure death model, only departure of served customer is considered, but additional arrivals of customers are ignored. We start with N customers at $t = 0$. μ is the rate of departure of served customers which is very similar to pure birth model.

Let $P_n(t)$ be the probability of n departures during time t and $P_n(t + h)$ be the probability of 4 departures during time $(t + h)$.

Then
$$P_n(t+h) = P_n(t).\,(1 - \mu h) + P_{n+1}(t).\,\mu h \quad (9.8)$$

If $n = 0$
$$P_0(t+h) = P_0(t)\,(1 - \mu h) \quad (9.9)$$

$$P_n(t+h) = P_n(t) - P_n(t)\mu h + P_{n-1}(t)\mu h$$

$$P_n(t+h) - P_n(t) = -\mu h P_n(t) + P_{n-1}(t).\mu h$$

Dividing both the sides by h and taking limits as h tends to zero, we get

$$\lim_{h \to 0} \frac{P_n(t+h) - P_n(t)}{h} = -\mu P_n(t) + \mu P_{n+1}(t)$$

$$\frac{d}{dt} P_n(t) = -\mu P_n(t) + \mu P_{n+1}(t) \quad (9.10)$$

If $n = 0$,
$$\frac{d}{dt} P_0(t) = +\mu P_1(t) \quad (9.11)$$

If $n = N$,
$$\frac{d}{dt} P_N(t) = -\mu P_N(t) \quad (9.12)$$

The solution of the differential Equations (9.10), (9.11) and (9.12) gives

$$P_n(t) = \frac{(\lambda t)^n \cdot e^{-\lambda t}}{n!} \quad (9.13)$$

where $N = 1, 2, 3, \cdots, N$
and

$$P_0(t) = 1 - \sum_{n=1}^{N} P_n(t) \quad (9.14)$$

9.5 STEADY-STATE PERFORMANCE ANALYSIS

A queuing system is said to be at steady state when the characteristics of the queuing system are independent of time.

Let

X_S be the number of customers in the queuing system

X_q be the number of customers in the queue

W_S be the waiting time in the queuing system

W_q be the waiting time in the queue

C be the number of service facilities

We know that the queuing system consists of both queue and service facility.

P_n is the steady-state probability of n customers in the queuing system.

$$X_S = \sum_{n=1}^{\infty} nP_n$$

$$X_q = \sum_{n=c-1}^{\infty} (n-c)P_n$$

Then

$$X_S = \lambda \cdot W_S$$

and

$$X_q = \lambda \cdot W_q$$

W_S and W_q are related as follows

Waiting time in the system = Waiting time on queue + Service time

i.e.,

$$W_S = W_q + \frac{1}{\mu} \quad (9.15)$$

Similarly,

$$X_S = X_q + \frac{\lambda}{\mu} \quad (9.16)$$

$$C_1 = X_S - X_q = \frac{\lambda}{\mu} \quad (9.17)$$

where C_1 denotes service facilities which are busy.

$$\text{Percentage Facility Utilization} = \frac{C_1}{C} \times 100 \quad (9.18)$$

9.6 SINGLE-SERVER QUEUING MODELS

Model 1

$m|m|1:\infty|FCFS$ is a single-server queuing model.

where,

m is the Poisson arrival or departure distribution

1 is the number of servers ($C = 1$)

∞ is the capacity of the queuing system

FCFS is the queue discipline (First Come First Served)

We know that λ is the arrival ratio of customer, and μ is the rate of departure of served customers.

ρ is the ratio of arrival rate to departure rate

$$\rho = \frac{\lambda}{\mu}$$

In generalised model,

$$P_n = \left(\frac{\lambda}{\mu}\right)^n P_0, \qquad n = 0,\ 1, 2, \ldots \tag{9.19}$$

To determine P_0, we use the boundary condition $\sum_{n=0}^{\infty} P_n = 1$

$$\sum_{n=0}^{\infty} P_n = 1 = \sum_{n=0}^{\infty}\left(\frac{\lambda}{\mu}\right)^n P_0$$

$$= P_0 \times \frac{1}{1-\frac{\lambda}{\mu}} \qquad \left[\text{Assuming } \frac{\lambda}{\mu} < 1\right]$$

$$P_0 = 1 - \frac{\lambda}{\mu}$$

$$\therefore \qquad P_0 = 1 - \rho$$

Substituting in Equation (9.13), we get

$$P_n = \left(\frac{\lambda}{\mu}\right)^n \cdot (1 < \rho),\ n = 1, 2, 3, \cdots, (\rho \le 1)$$

$$P_n = \rho^n.(1-\rho) \tag{9.20}$$

The measure of performance, the number of customers in the queuing system

$$X_S = \sum_{n=0}^{\infty} n\, P_n$$

$$= \sum_{n=0}^{\infty} n\,.\,\rho^n.\,(1-\rho)$$

$$= (1-\rho)\sum_{n=0}^{\infty} n\,.\rho^n$$

$$= (1-\rho).\,\rho.\sum_{n=0}^{\infty} n\,.\,\rho^{n-1}$$

$$= (1-\rho)\cdot\rho\cdot\frac{d}{d\rho}\sum_{n=0}^{\infty}\rho^n$$

$$= (1-\rho)\cdot \rho \cdot \frac{d}{d\rho}\left(\frac{1}{1-\rho}\right)$$

$$X_s = \left(\frac{\rho}{1-\rho}\right) \tag{9.21}$$

Waiting time in the queuing system,

$$W_S = \frac{X_S}{\lambda}$$

$$W_s = \frac{\rho}{(1-\rho)} \times \frac{1}{\lambda}$$

$$= \frac{\lambda}{\mu(1-\rho)} \cdot \frac{1}{\lambda}$$

$$W_s = \frac{1}{\mu - \rho} \tag{9.22}$$

Waiting time in the queue,

$$W_q = W_S - \frac{1}{\mu}$$

$$W_q = \frac{1}{\mu-\rho} - \frac{1}{\mu}$$

$$W_q = \frac{\rho}{\mu(1-\rho)} \tag{9.23}$$

The number of customers in the queue,

$$X_q = \lambda \times W_q$$

$$X_q = \frac{\rho^2}{(1-\rho)} \tag{9.24}$$

Service facilities which are busy,

$$C_1 = X_S - X_q$$

$$C_1 = \rho \tag{9.25}$$

Waiting time distribution For $m|m|1$ model, the waiting time distribution is based on *FCFS* discipline. The waiting time of a customer in the queuing system

$$T = t_1 + t_2 + \ldots, t_{n+1}$$

where t_1 is the service time for a customer taking services at the facility.

$t_2, t_3, \ldots, t_n$ are the time required for $n - 1$ customers in the queue.

t_{n+1} is the service time for a person just arriving.

Average waiting time of a customer in the queuing system

$$W_{av} = \sum_{n=0}^{\infty} W\left(\frac{T}{n+1}\right) \times P_n$$
$$= \sum_{n=0}^{\infty} \frac{\mu(\mu T)^n \cdot e^{-\mu T}}{n!} \times \rho^n (1-\rho)$$
$$= \mu \cdot e^{-\mu T} \cdot (1-\rho) \sum_{n=0}^{\infty} \frac{\mu^n \cdot T^n}{n!} \frac{\lambda^n}{\mu^n}$$
$$= \mu \cdot e^{-\mu T} \cdot (1-\rho) \sum_{n=0}^{\infty} \frac{(\lambda T)^n}{n!}$$
$$= \mu \cdot e^{-\mu T} \cdot (1-\rho) \cdot e^{\lambda T}$$
$$= \mu(1-\rho) e^{-\mu T} \cdot e^{\lambda T}$$
$$= \mu\,(1-e) e^{-(\mu T)} \cdot e^{\lambda T}$$
$$W_{av} = (\mu - \lambda) \cdot e^{-(\mu-\lambda)T}$$

where $T > 0$. (9.26)

Average waiting time of a customer is an exponential distribution.

Model 2

$m|m|1 : N|FCFS$ is also a single-server queuing model. In this model, the maximum number of customers is limited to N.

Using the boundary condition

$$\sum_{n=0}^{\infty} P_n = 1$$

and

$$P_n = \left(\frac{\lambda}{\mu}\right)^n \times P_0$$
$$P_n = \rho^n . P_0 \qquad n \leq N$$
$$P_n = 0 \qquad n > N$$

We determine

$$P_0(1 + \rho + \rho + ... + \rho^N) = 1$$
$$P_o \times \sum_{n=0}^{N} P^n = 1$$

Therefore,

$$1 = P_o \sum_{n=0}^{N} P_n = P_0 \times 1 - \rho^{N+1} \quad \text{if } \rho \neq 1$$
$$1 = P_0 \times (N+1)$$

if $\rho = 1$

Therefore, $$P_o = \frac{1-\rho}{1-\rho^{N+1}} \qquad \text{if } \rho \neq 1$$

$$P_0 = \frac{1}{N+1} \qquad \text{if } \rho = 1 \qquad n = 1, 2, 3, ..., N$$

The measure of performance, the number of customers in the queuing system

$$X_S = \sum_{n=0}^{N} n \cdot \rho^n$$

$$X_n = \sum_{n=0}^{N} n \sum_{n=0}^{N} \rho^n \cdot P_0$$

$$= \frac{1-\rho}{1-\rho^{N+1}} \sum_{n=0}^{N} n \cdot \rho^n$$

$$X_n = \left[\frac{1-\rho}{1-\rho^{N+1}}\right] P \frac{d}{dp} \sum_{n=0}^{N} \rho^n$$

$$= \frac{(1-\rho)}{1-\rho^{N+1}} \frac{\rho d}{d\rho}\left(\frac{1-\rho^{N+1}}{1-\rho}\right)$$

$$P_n = \frac{\lambda^n}{\mu(2\mu)(3\mu)\cdots(n\mu)} \times P_0 \qquad \rho = 1 \qquad (9.27)$$

9.7 MULTIPLE SERVER MODELS

In this type of queuing models, multiple facilities serve the customers in parallel. The arrival rate is λ and the service rate is μ.

Model 1

$m|m|c : \infty/FCFS$

In this model 'C' facilities serve the customers simultaneously.

If the number of customers n are less than the service facilities C (i.e., $n < C$)

$$P_n = \frac{\lambda^n}{\mu(2\mu)(3\mu)\cdots(n\mu)} \times P_0$$

$$= \frac{\lambda^n}{n!\mu^n} \times P_0$$

$$P_n = \frac{\rho^n}{n!} \times P_0$$

If the number of customers n are more than the service facilities C (i.e., $n > C$)

$$P_n = \frac{\lambda_n}{\left(\pi_{i=1}^{c} i\mu\right)(C\mu)^{n-c}} \times P_0$$

$$P_n = \frac{\lambda^n}{C!C^{n-c}\mu^n} \times P_0$$

$$P_n = \frac{p^n}{C!C^{n-c}\mu^n} \times P_o \tag{9.28}$$

We know that

$$\sum_{n=0}^{\infty} P_n = 1$$

Therefore,

$$\sum_{n=0}^{C-1} P_n + \sum_{n=C}^{\infty} P_n = 1$$

For $n < C$,

$$P_n = \frac{\rho^n}{n!} \times P_0$$

i.e.,

$$\sum_{n=0}^{C-1} \frac{\rho^n}{n!} \cdot P_0 + \sum_{n=C}^{\infty} P_n = 1$$

$$\sum_{n=0}^{C-1} \frac{\rho^n}{n!} \cdot P_0 + \sum_{n=C}^{\infty} \frac{\rho^n}{C!C^{n-c}} \times P_0 = 1$$

$$\left\{\sum_{n=0}^{C-1} \frac{\rho^n}{n!} \cdot P_0 + \sum_{n=C}^{\infty} \frac{\rho^n}{C!C^{n-c}}\right\} \times P_0 = 1$$

$$P_0 = \left\{\sum_{n=0}^{C-1} \frac{\rho^n}{n!} + \sum_{n=C}^{\infty} \frac{\rho^c \cdot \rho^{n-c}}{C!C^{n-c}}\right\}^{-1}$$

$$P_0 = \left\{\sum_{n=0}^{C-1} \frac{\rho^n}{n!} + \frac{\rho^C}{C!} \sum_{n=C}^{\infty} \left(\frac{\rho}{C}\right)^{n-C}\right\}^{-1}$$

If $\frac{\rho}{C} < 1$

$$P_0 = \left\{\sum_{n=0}^{C-1} \frac{\rho^n}{n!} + \frac{\rho^C}{C!}\left(\frac{1}{1-\frac{\rho}{C}}\right)\right\}^{-1} \tag{9.29}$$

Number of customers in the queue,

$$X_q = \sum_{n=C}^{\infty} (n-C).P_n$$

$$= \sum_{a=0}^{\infty} a \cdot P_{a+c}$$

$$= \sum_{a=0}^{\infty} a \cdot \frac{\rho^{a+c}}{CaC!} \cdot P_0$$

$$= \frac{\rho^{c+1}}{C!C!} \cdot P_0 \sum_{a=0}^{\infty} a \left(\frac{P}{C}\right)^{a-1}$$

$$= \frac{\rho^{c+1}}{C!C!} \cdot P_0 \frac{d}{dy} \sum_{a=0}^{\infty} Y^a$$

where $Y = \frac{\rho}{C}$

$$X_q = \frac{\rho^{C+1}}{(C-1)!(C-P)^2} P_0 \tag{9.30}$$

Number of customers in the queuing system,

$$X_S = X_q + \rho$$

Waiting time in the queue,

$$X_q = \frac{\rho^{C+1}}{(C-1)!(C-P)^2} P_0$$

Waiting time in the queuing system

$$W_s = \frac{X_s}{\lambda}$$

$$= \frac{X_q}{\lambda} + \frac{\rho}{\lambda}$$

$$= \frac{X_q}{\lambda} + \frac{1}{\mu}$$

$$W_s = \frac{\rho^{c+1}}{\lambda(C-1)!(C-P)^2} \times P_0 + \frac{1}{\mu}$$

$$= \frac{\rho^c \cdot \rho}{\lambda(C-1)!(C-P)^2} \times \frac{1}{\mu}$$

$$= \frac{\rho^c . \frac{\lambda}{\mu}}{\lambda(C-1)!(C-P)^2} \times P_0 + \frac{1}{\mu}$$

$$W_S = \frac{\rho^c}{\mu(C-1)!(C-P)^2} \times P_0 + \frac{1}{\mu} \quad (9.31)$$

***Example* 9.1**

A foreign bank is considering opening a drive-in window for customer service. The management estimates that customers will arrive for service at the rate of 12 per hour. The teller whom it is considering to staff the window can serve customers at the rate of one every three minutes. Assuming Poisson arrivals and exponential service, find

i. Utilization of teller,
ii. Average number in the system,
iii. Average waiting time in the line, and
iv. Average waiting time in the system.

Solution

Arrival rate = λ = 12 per hour
Service rate = μ = 1 in every three minutes
μ = 20 per hour

This problem is a single-server queuing model with infinite number of customers in the system.

i.e., $m|m|1 : \infty|FCFS$ model

Ratio of arrival rate to departure rate is ρ

$$\rho = \frac{\lambda}{\mu} = \frac{12}{20} = 0.6$$

Utilization of the teller is 60%

Idle time of teller $\quad P_0 = 1 - \rho = 1 - 0.6 = 0.4$

$$P_0 = 40\%$$

Average number of customers in the queuing system,

$$X_S = \frac{\rho}{1-\rho} = \frac{0.6}{1-0.6}$$

$$X_S = 1.5 \text{ (approximately 2 customers)}$$

Average waiting time in the line,

$$W_q = \frac{\rho}{\mu(1-\rho)}$$

$$W_q = \frac{0.6}{20(1-0.6)}$$

$$W_q = \text{hours} = 4.5 \text{ minutes}$$

Average waiting time in the system

$$W_S = \frac{X_S}{\lambda} = \frac{15}{12} = 0.125 \text{ hours}$$

$$W_S = 7.5 \text{ minutes}$$

Example 9.2

At a one-man barber shop, customers arrive according to Poisson distribution with a mean arrival rate of 5 per hour and his hair cutting was exponentially distributed with an average haircut taking 10 minutes. It is assumed that because of his excellent reputation, customers were always willing to wait. Calculate the following.

i. Average number of customers in the shop and the average number of customers waiting for a haircut.

ii. The percentage of time a customer can walk in without having to wait.

iii. The percentage of customers who have to wait prior to getting into the barber's chair.

Solution

$$\text{Arrival rate} = \lambda = 5 \text{ per hour}$$

$$\text{Service time} = 10 \text{ minutes}$$

$$\text{Service rate} = \mu = \text{customers/minutes}$$

$$\mu = 6 \text{ customers/hour}$$

Average number of customers in the shop (system)

$$X_S = \frac{\rho}{1-\rho} = \frac{5}{6\left(1-\frac{5}{6}\right)}$$

$$X_S = 5 \text{ customers}$$

Average number of customers waiting for haircut (queue),

$$X_q = \frac{\rho^2}{1-\rho} = \frac{\left(\frac{5}{6}\right)^2}{1-\frac{5}{6}}$$

$$X_q = 4.16 \text{ customer}$$

The percentage of customers who have to wait

$$= \frac{X_q}{X_s} \times 100$$

$$= \frac{4.16}{5} \times 100$$

$$= 83.2\%$$

Percentage of time a customer can walk in without waiting is

$$= \left(1 - \frac{W_q}{W_S}\right) \times 100 \qquad \text{(A)}$$

$$W_q = \frac{X_S}{\mu} = \frac{5}{6} \text{ hours} = 0.833 \text{ hours}$$

$$W_S = \frac{X_S}{\lambda} = \frac{5}{5} = 1 \text{ per hour}$$

Substituting W_q and W_s in Equation (A), we get

$$= \left(1 - \frac{0.833}{1}\right) \times 100 = 16.66\%$$

Percentage of time a customer can walk in without waiting is 16.66%.

Example 9.3

Customers arrive at a sales counter manned by a single person according to a Poisson process with a mean rate of 20 per hour. The time required to serve a customer has an exponential distribution with a mean of 100 seconds. Find the average waiting time of a customer.

Solution

This is a problem on $m|m|1 : \infty|FCFS$ model

$$\text{Arrival rate} = \lambda = 20 \text{ per hour}$$

$$\text{Service time} = \frac{1}{\mu} = 100 \text{ seconds}$$

$$\text{Service rate} = \mu = \frac{1}{100} \text{ per second}$$

$$= \frac{60 \times 60}{100} \text{ per hour}$$

$$\mu = 36 \text{ per hour}$$

Number of customers in the queuing system

$$X_S = \frac{\rho}{1-\rho}$$

$$\rho = \frac{\lambda}{\mu} = \frac{20}{36} = 0.555$$

Average waiting time in the queue

$$W_q = \frac{X_S}{\mu} = \frac{\rho}{\mu(1-\rho)}$$

$$= \frac{0.555}{36(1-0.55)} = 0.0347 \text{ hours}$$

$$W_q = 2.083 \text{ minutes}$$

Average waiting time in the system

$$W_S = \frac{X_S}{\lambda} = \frac{\rho}{\lambda(1-\rho)}$$

$$= \frac{0.555}{20(1-0.555)} = 0.0625 \text{ hours}$$

$$W_S = 3.75 \text{ minutes}$$

Example 9.4

Patients arrive at a clinic at an average of 6 patients per hour. Some patients require only the repeat prescription, some come for minor check-up while, some others require thorough inspection for diagnosis. This takes the doctor six minutes per patient on an average. It can be assumed that arrivals follow a Poisson distribution and the doctor's inspection time follows an exponential distribution. Calculate

i. The percentage of time a patient can walk right inside the doctor's cabin.
ii. The average number of patients in the clinic.
iii. The average number of patients waiting for their turn.
iv. The average time a patient spends in the clinic.

Solution

This is a problem on $m|m|1 : \infty|FCFS$ model

$$\text{Arrival rate} = \lambda = 6 \text{ patients/hour}$$

$$\text{Service rate} = \frac{1}{\mu} = 6 \text{ minutes/patient}$$

Service rate

$$\mu = \frac{1}{6} \text{ patient/minute}$$

$$\mu = \frac{60}{6} = 10 \text{ patients/hour}$$

$$\rho = \frac{\lambda}{\mu} = \frac{6}{10} = 0.6$$

Average number of patients waiting for their turn

$$X_q = \left(\frac{\rho^2}{(1-\rho)}\right)$$

$$= \left(\frac{(0.6)^2}{1-0.6}\right)$$

$$X_q = 0.9 \text{ patients (approximately one patient)}$$

Average number of patients in the clinic

$$X_S = \left(\frac{\rho}{(1-\rho)}\right) = \frac{0.6}{1-0.6}$$

$$X_S = 1.5 \text{ patients (approximately two patients)}$$

Average time a patient spends in the clinic

$$W_S = \frac{X_S}{\lambda}$$

$$= \frac{15}{6}$$

$$W_S = 0.25 \text{ hours}$$

$$W_S = 15 \text{ minutes}$$

Average time a patient waits for his turn

$$W_q = \frac{X_S}{\mu}$$

$$= \frac{15}{10}$$

$$W_q = 0.15 \text{ hour}$$

$$W_q = 9 \text{ minutes}$$

The percentage of times a patient can walk inside the doctor's cabin without having to wait.

$$= \left(1-\frac{W_q}{W_S}\right) \times 100$$

$$= \left(1-\frac{9}{15}\right) \times 100$$

$$= 40\%$$

Example 9.5

A supermarket has a single cashier. During the peak hours, customers arrive at the rate of 20 customers per hour. The average number of customers that can be processed by the cashier is 24 per hour. Calculate

i. The probability that the cashier is idle.

ii. The average number of customers in the queuing system.

iii. The average time a customer spends in the system.

iv. The average number of customers in the queue.

v. The average time a customer spends in the queue waiting for service.

Solution

This is a problem on $m|m|1 : \infty|FCFS$ model

$$\text{Arrival rate } \lambda = 20 \text{ customers per hour}$$

$$\text{Service rate } \mu = 24 \text{ per hour}$$

$$\rho = \frac{\lambda}{\mu} = \frac{20}{24} = 0.833$$

Probability that the cashier is idle is

$$P_0 = 1 - \rho = 1 - 0.833$$

$$P_0 = 0.1666 = 16.66\%$$

Average number of customers in the queuing system,

$$X_S = \frac{\rho}{1-\rho} = \frac{0.833}{1-0.833}$$

$$X_S = 5 \text{ customers}$$

Average time a customer spends in the system

$$W_S = \frac{X_S}{\lambda} = \frac{5}{20} = 0.25 \text{ hours}$$

$$W_S = 15 \text{ minutes}$$

Average number of customers in the queue

$$X_q = \frac{\rho^2}{1-\rho} = \frac{(0.833)^2}{(1-0.833)}$$

$$X_q = 4.16 \text{ customers}$$

The average time a customer spends in the queue waiting for service

$$W_q = \frac{X_S}{\mu} = \frac{5}{24}$$

$$= 0.208 \text{ hours}$$

$$W_q = 12.499 \text{ minutes}$$

Example 9.6

Cars arrive at a petrol pump with exponential inter-arrival times having mean $\frac{1}{2}$ minute. The attendant takes an average of $\frac{1}{5}$ minute per car to supply petrol, the service times being exponentially distributed. Determine

i. The average number of cars waiting to be served
ii. The average number of cars in the queue
iii. The proportion of time for which the pump attendant is idle

Solution

This is a problem on $m|m|1 : \infty|FCFS$ model

$$\text{Inter arrival time} = \frac{1}{\lambda} = \frac{1}{2} \text{ minute}$$

$$\text{Arrival rate} = \lambda = 2 \text{ cars/minute}$$

$$\text{Service time} = \frac{1}{\mu} = \frac{1}{5} \text{ minute}$$

$$\text{Service rate } \mu = 5 \text{ cars/minute}$$

$$\rho = \frac{\lambda}{\mu} = \frac{2}{5} = 0.4$$

Average number of cars waiting to be served

$$X_S = \frac{\rho}{1-\rho} = \frac{0.4}{1-0.4}$$

$$X_S = 0.66 \text{ cars}$$

Average number of cars in the queue

$$X_q = \frac{\rho^2}{1-\rho} = \frac{(0.4)^2}{1-0.4}$$

$$X_q = 0.266 \text{ cars}$$

The probability of time for which the pump attendant is idle is

$$P_0 = 1-\rho = (1-0.4)$$

$$P_0 = 0.6 = 60\%$$

Example 9.7

An airline organisation has one reservation clerk on duty in its local branch, at any given time. The clerk handles information regarding passenger reservation and flight timings. Assume that the number of customers arriving during any given period is Poisson-distributed with an arrival rate of eight per hour and that the reservation clerk can serve a customer in six minutes on an average, with an exponentially distributed service time.

i. What is the probability that the system is busy?
ii. What is the average time a customer spends in the system?
iii. What is the average length of the queue and what is the number of customers in the system?

Solution

This is a problem $m|m|1 : \infty|FCFS$ model

$$\text{Arrival rate } \lambda = 8 \text{ passenger/hour}$$

$$\text{Service time} = \frac{1}{\mu} = 6 \text{ minutes}$$

$$\text{Service rate} = \mu = \frac{1}{6} \text{ passengers/minute}$$

$$\mu = \frac{60}{6} = 10 \text{ passengers/hour}$$

$$\rho = \frac{\lambda}{\mu} = \frac{8}{10} = 0.8$$

Probability that the system is busy = 80%

Average time a customer spends in the system,

$$W_S = \frac{1}{(\mu - \rho)} = \frac{1}{10 - 0.8}$$

$$= 0.108 \text{ hours}$$

$$W_S = 6.52 \text{ minutes}$$

The average length of the queue,

$$X_q = \frac{\rho^2}{(1-\rho)} = \frac{0.8^2}{(1-0.8)}$$

$$X_q = 3.2 \text{ passengers}$$

Number of customers in the system,

$$X_S = \frac{\rho}{1-\rho} = \frac{0.8}{1-0.8}$$

$$X_S = 4 \text{ passengers}$$

Example 9.8

Consider a single server queuing system with Poisson input, exponential service times. Suppose the mean arrival ratio is 3 calling units per hour, the expected service time is 0.25 hours and the maximum permissible number calling of units in the system is two. Derive the steady-state probability distribution of the number of calling units in the system. Then calculate the expected numbers in the system.

Solution

$$\text{Arrival rate } \lambda = 3 \text{ units/hour}$$

$$\text{Service time } \frac{1}{\mu} = 0.25 \text{ hours}$$

$$\text{Service rate } \mu = \frac{1}{0.25} = 4 \text{ units/hour}$$

$$\rho = \frac{\lambda}{\mu} = \frac{3}{4} = 0.75$$

The maximum number of calling units in the system $N = 2$

The probability that there is no calling system in the queuing system,

$$P_0 = \frac{1-\rho}{1-\rho^{N+1}} = \frac{1-0.75}{1-0.75^{2+1}} = 0.432$$

$$P_n = P_0 \cdot \rho^n$$

Therefore,

$$P_1 = 0.432(0.75) = 0.324$$

$$P_2 = 0.432(0.75)^2 = 0.243$$

Average number of calling units in the system

$$X_S = 1(0.324) + 2(0.243)$$

$$= 0.81$$

Each calling system requires service time $\frac{1}{\mu} = 0.25$ hours

Average waiting time in the queue

$$W_q = 0.81(0.25) = 0.2025 \text{ hours}$$

$$= 0.2025 \times 60$$

$$W_q = 12.15 \text{ minutes}$$

Example 9.9

Assume that the goods trains are coming in a yard at the rate of 30 trains per day and suppose that the inter-arrival times follow an exponential distribution. The service time for each train is assumed to be exponential with an average of 36 minutes. If the yard can admit 9 trains at a time (there being 10 lines, one of which is reserved for shunting purposes), calculate the probability that the yard is empty and find the average queue length arrival.

Solution

$$\text{Arrival rate } \lambda = 30 \text{ trains per day}$$

$$= \frac{5}{4} \text{ trains per hour}$$

$$\lambda = \frac{5}{60.4} = \frac{1}{48} \text{ trains/minute}$$

$$\text{Service rate} = \mu = \frac{1}{36} \text{ trains/minute}$$

$$\rho = \frac{\lambda}{\mu} = \frac{1}{48} \times 36$$

$$\rho = 0.75$$

Probability that the yard is empty

$$P_0 = \frac{1-\rho}{1-\rho^{N+1}} = \frac{1-0.75}{1-(0.75)^{10}}$$

$$P_0 = 0.28$$

Average queue length is given by

$$X_q = \frac{\rho^2[1 - N\rho^{N+1} + (N-1)\rho^N]}{(1-\rho)(1-\rho^{N+1})}$$

$$= \frac{(0.75)^2\left[1 - 9(0.75)^8 + 8(0.75)^9\right]}{(1-0.75)\left[1-(0.75)^{10}\right]}$$

$$X_q = 1.55 \text{ trains}$$

Example 9.10

A supermarket has two girls serving at the counters. The customers arrive in a Poisson fashion at the rate of 12 per hour. The service time for each customer is exponential with *mean* 6 minutes. Find

i. The probability that an arriving customer has to wait for service
ii. The average number of customers in the system and
iii. The average time spent by a customer in the supermarket.

Solution

$$\text{Arrival rate } \lambda = 12 \text{ per hour}$$

$$\text{Service time} = \frac{1}{\mu} = 6 \text{ minutes}$$

$$\text{Service rate } \mu = \frac{1}{6} \text{ per minute}$$

$$\mu = 10 \text{ customers/hour}$$

$$C = 2$$

This is a problem on $m|m|C: \infty|FCFS$

Average number of customers in the system $= X_S$

$$\rho = \frac{\lambda}{\mu} = \frac{12}{10} = 1.2$$

$$\frac{\rho}{C} = \frac{12}{2} = 0.6 < 1$$

$$P_0 = \left[\sum_{n=0}^{C-1} \frac{\rho^n}{n!} + \frac{\rho^C}{C!}\left(\frac{1}{1-\frac{\rho}{C}}\right)\right]^{-1}$$

$$= \left[\sum_{n=0}^{2-1} \frac{1}{n!}(1.2)^n + \frac{1}{2!}(1.2)^2\left(\frac{1}{1-\frac{1.2}{2}}\right)\right]^{-1}$$

$$P_0 = \frac{1}{4}$$

Average queue length,

$$X_q = \frac{\rho^{C+1}}{(C-1)!(C-\rho)^2} \cdot P_0$$

$$= \frac{(12)^{2+1}}{(2-1)!(2-1.2)^2} \times \frac{1}{4}$$

$$= 0.675$$

Average number of customers in the system,

$$X_S = X_q + \rho$$

$$X_S = 0.675 - 1.2$$

$$X_S = 1.875 \text{ customers } (\approx 2 \text{ customers})$$

Average time spent by a customer in the supermarket,

$$W_S = \frac{X_S}{\lambda} = \frac{1.875}{12} = 0.156 \text{ hours}$$

$$W_S = 9.3 \text{ minutes}$$

Probability that an arriving customer has to wait for service,

$$= \frac{1}{C!} \cdot \rho^C \cdot \left(\frac{C}{C-\rho}\right) \cdot P_0 = \frac{1}{C!} \cdot \rho^C \cdot \left(\frac{1}{1-\frac{\rho}{C}}\right) \cdot P_0$$

$$= \frac{1}{2!} \cdot (1.2)^2 \left(\frac{1}{1-\frac{1.2}{2}}\right) \cdot \frac{1}{4}$$

$$= 0.45$$

Example 9.11

A petrol pump station has two pumps. The service time follows the exponential distribution with a mean of 4 minutes and the cars arrive for service in a Poisson process at the rate of 10 cars per hour. Find the probability that a customer has to wait for service and what proportion of time the pumps remain idle?

Solution

$$C = 2 \text{ pumps}$$

$$\text{Service time} = \frac{1}{\mu} = 4 \text{ minutes}$$

$$\text{Service rate} = \mu = \frac{1}{4} \text{ per minute}$$

$$\mu = 15 \text{ customers/hour}$$

$$\text{Arrival rate } \lambda = 10 \text{ cars/hour}$$

$$\rho = \frac{\lambda}{\mu} = \frac{10}{15} = 0.66$$

This is a problem on $m|m|c : \infty|FCFS$ model. $\frac{\rho}{C} < 1$.

$$P_0 = \left[\sum_{n=0}^{C-1} \frac{\rho^n}{n!} + \frac{\rho^C}{C!}\left(\frac{1}{1-\frac{\rho}{C}}\right)\right]^{-1}$$

$$P_0 = \left[\sum_{n=0}^{2-1} \frac{(0.66)^n}{n!} + \frac{(0.66)^2}{2!}\left(\frac{1}{1-\frac{0.66}{2}}\right)\right]^{-1} = \frac{1}{2}$$

Probability that a customer has to wait for service

$$= \frac{1}{C!} \cdot \rho^C \cdot \left(\frac{1}{1-\frac{\rho}{C}}\right) P_0$$

$$= \frac{1}{2!}(0.66)^2 \left(\frac{1}{1-\frac{0.66}{2}}\right) \cdot \frac{1}{2}$$

$$= 0.217$$

Probability that the pump is idle is

$$= P_0 = \frac{1}{2} = 50\%$$

Example 9.12

A tax consulting firm has four service stations (counters) in its office to receive people who have problems and complaints about their income, wealth and sales taxes. Arrivals average 100 persons in a 10-hour service day. Each tax advisor spends an irregular amount of time servicing the customer which has been found to have an exponential distribution. The average service time is 20 minutes. Calculate

i. The average number of customers in the system.
ii. Average number of customers waiting to be served.
iii. Average time a customer spends in the system.
iv. The average waiting time for a customer.
v. The probability that a customer has to wait before he gets the service.

Solution

$$C = 4$$

Arrival rate = 100 persons/10 hour day

$$\lambda = 10 \text{ persons/hour}$$

Service time = $\frac{1}{\mu} = 20$ minutes

$$\text{Service rate} = \mu = \frac{1}{20} \text{ per minute}$$

$$\mu = 3 \text{ per minute}$$

$$\rho = \frac{\lambda}{\mu} = \frac{10}{3} = 3.33$$

$$\frac{\rho}{C} < 1 \qquad P_0 = \left[\sum_{n=0}^{C-1} \frac{\rho^n}{n!} + \frac{\rho^C}{C!} \left(\frac{1}{1 - \frac{\rho}{C}} \right) \right]^{-1}$$

$$= \left[\sum_{n=0}^{4-1} \frac{\rho^n}{n!} + \frac{(3.33)^4}{4!} \left(\frac{1}{1 - \frac{0.33}{4}} \right) \right]^{-1}$$

$$P_0 = 0.0219$$

Number of customers in the queue (waiting to be serviced),

$$X_q = \frac{\rho^{C+1}}{(C-1)!(C-\rho)^2} \cdot P_0$$

$$= \frac{(3.33)^5}{3!(4-3.33)^2} \cdot 0.0219$$

$$X_q = 0.3 \text{ customers}$$

The average number of customers in the system,

$$X_S = X_q + \rho$$

$$= 0.3 + 3.33$$

$$X_S = 3.63 \text{ customers}$$

Average time a customer spends in the system

$$W_S = \frac{X_S}{\lambda}$$

$$= \frac{3.33}{10}$$

$$W_S = 0.333 \text{ hours}$$

$$W_S = 19.98 \text{ minutes}$$

The probability that a customer has to wait before he gets service

$$= \frac{1}{C!} \cdot \rho^C \cdot \left(\frac{1}{1 - \frac{\rho}{C}} \right) \cdot P_0$$

$$= \frac{1}{4!} \cdot (3.33)^4 \left(\frac{1}{1 - \frac{3.33}{4}} \right) \cdot 0.0219$$

$$= 0.6698 = 66.98\%$$

***Example* 9.13**

An oil company is constructing a service station on a highway. Traffic analysis indicates that customer's arrivals over most of the day would approximate a Poisson distribution with a mean of 30 automobiles per hour. Previous studies show that one pump could service a mean of 10 automobiles per hour with the service time distribution approximating the negative exponential. If 4 pumps are installed, find

i. The probability that an arrival would have to wait in line.

ii. The average waiting time, average time spent in the system and the average number of automobiles in the system.

Solution

$$\text{Arrival rate} = \lambda = 30/\text{hour}$$

$$\text{Service rate} = \mu = 10 \text{ automobiles/hour}$$

$$C = 4$$

This is a problem on $m|m|C{:}\ \infty|FCFS$ model.

$$\rho = \frac{\lambda}{\mu} = \frac{30}{10} = 3$$

$\frac{\rho}{C} = \frac{3}{4} < 1...,$

$$P_0 = \left[\sum_{n=0}^{C-1} \frac{\rho^n}{n!} + \frac{\rho^C}{C!} \left(\frac{1}{1 - \frac{\rho}{C}} \right) \right]^{-1}$$

$$P_0 = \left[\sum_{n=1}^{4-1} \frac{(3)^n}{n!} + \frac{(3)^4}{4!} \left(\frac{1}{1 - \frac{3}{4}} \right) \right]^{-1}$$

$$P_0 = 0.037$$

Number of automobiles in the queue,

$$X_q = \frac{\rho^{C+1}}{(C-1)!(C-\rho)^2} \cdot P_0$$

$$= \frac{(3)^5}{(4-1)!(4-3)^2} \cdot 0.037$$

$$X_q = 3.05$$

Number of automobiles in the system,

$$X_s = X_q + \rho$$
$$= 3.05 + 3$$
$$X_s = 6.05$$

Waiting time in the system,

$$W_S = \frac{X_s}{\lambda}$$
$$= \frac{6.05}{30} \text{ hours}$$
$$= \frac{1}{5} \text{ hours}$$
$$W_S = 12 \text{ minutes}$$

Waiting time in the queue,

$$W_q = \frac{X_q}{\lambda}$$
$$= \frac{3.05}{30} \text{ hours}$$
$$= \frac{1}{10} \text{ hours}$$
$$W_q = 6 \text{ minutes}$$

Probability that an arrival has to wait

$$= \frac{1}{C!} \cdot \rho^C \cdot \left(\frac{1}{1 - \frac{\rho}{C}} \right) P_0$$

$$= \frac{1}{4!} \cdot (3)^4 \cdot \left(\frac{1}{1 - \frac{3}{4}} \right) \cdot 0.037$$

$$= 0.4995$$
$$= 49.95\%$$

EXERCISES

1. A TV repairman finds that the time spent on his jobs has an exponential distribution with mean 30 minutes. If the repair starts in the order in which they come in and if the arrival of TV sets is approximately Poisson with an average rate of 10 per 8 hour day, what is the repairman's expected idle time each day? How many jobs are ahead of the average set just brought in?

2. The rate of arrival of customers at a public telephone booth follows Poisson distribution, with an average time of 10 minutes between one customer and the next. The duration of a phone call is assumed to follow exponential distribution with a mean time of 3 minutes.

   i. What is the probability that a person arriving at the booth will have to wait?
   ii. What is the average length of the non-empty queues that form from time to time?
   iii. The Mahanagar Telephone Nigam Limited will install a second booth when it is convinced that the customers would expect waiting for at least 3 minutes for their turn to make a call. By how much time should the flow of customers increase in order to justify a second booth?
   iv. Estimate the fraction of a day that the phone will be in use.
   v. What is the probability that it will take the customer more than 10 minutes altogether to wait for phone and complete his/her call?

3. A road transport company has one reservation clerk on duty at a time. He handles information of bus schedules and makes reservations. Customers arrive at a rate of 8 per hour and the clerk can service 12 customers on an average per hour. After stating your assumption, answer the following

   i. What is the average number of customers waiting for the service of the clerk?
   ii. What is the average time a customer has to wait before getting service?
   iii. The management is contemplating to install a computer system to handle the information and reservation. This is expected to reduce the service time from 5 to 3 minutes. The additional cost of having the new system works out to Rs. 50 per day. If the cost of goodwill of having to wait is estimated to be 12 paisa per minute spent waiting before being served. Should the company install the computer system? Assume 8 hours working day.

4. A bank plans to open a single-server drive in a banking facility at a particular centre. It is estimated that 20 customers will arrive each hour on an average. If on an average, it requires 2 minutes to process a customer's transaction, determine.

   i. The proportion of time that the system will be idle.
   ii. On the average, how long will a customer have to wait before reaching the server?
   iii. The fraction of customer who will have to wait?

5. In a maintenance shop, the inter-arrival times at tool crib are exponential with an average time of 10 minutes. The length of the service time (i.e., the amount of time taken by the tool crib operator to meet the needs of the maintenance man) is assumed to be exponentially distributed, with a mean of 6 minutes. Find

   i. The probability that a person arriving at the booth will have to wait.
   ii. The average length of the queue that forms and the average time that an operator spends in the *Q*. system.
   iii. The probability that an arrival will have to wait for more than 12 minutes for service and to obtain his tools.
   iv. Estimate the fraction of the day that the tool crib operator will be idle.
   v. The probability that there will be six or more operators waiting for the service

vi. The manager of the shop will install a second booth when an arrival would expect to have to wait 10 minutes or more for the service. By how much must the rate of arrival be increased in order to justify a second booth?

6. A company distributes its products by trucks, loaded at its only loading station. Both the company's trucks and the contractor's trucks are used for this purpose. It was found that on an average of every five minutes, one truck arrived and the average loading time was three minutes. 50% of the trucks belong to the contractor. Find

   i. The probability that a truck has to wait.

   ii. The waiting time of the truck that waits.

   iii. The expected waiting time of the contractor's trucks per day, assuming a 24-hour shift.

7. At a railway station, only one train is handled at a time. The railway yard is sufficient for only two trains to wait, while the other is given the signal to leave the station. Trains arrive at the station at an average rate of 6 per hour and the railway station can handle them on an average of 12 per hour. Assuming Poisson arrivals and exponential service distribution, find the steady-state probabilities for the various number of trains in the system. Also find the average waiting time of a new train coming into the yard.

8. In a car wash service facility, cars arrive for service according to a Poisson distribution with mean 5 per hour and the time for washing and cleaning each car varies, but is found to follow an exponential distribution with mean 10 minutes per car. The facility cannot handle more than one car at a time and has a total of 5 parking spaces.

   i. Find the effective arrival rate.

   ii. What is the probability that an arriving car will get service immediately upon arrival?

   iii. Find the expected number of parking spaces occupied?

9. A bank has two tellers working on savings accounts. The first teller handles withdrawals only. The second teller handles deposits only. It has been found that the service time distribution for both deposits and withdrawals is exponential with mean service time 3 minutes per customer. Depositors are found to arrive in Poisson fashion throughout the day with mean arrival rate of 16 per hour. Withdrawers are found to arrive in Poisson fashion throughout the day with mean arrival rate of 14 per hour. What would be the effect on the average waiting time for depositors and withdrawers if each teller could handle both withdrawals and deposits? What could be the effect if this could be accomplished by increasing the mean service time to 3.5 minutes?

10. A telephone exchange has two long-distance operators. The telephone company finds that during the peak load, long-distance calls arrive in a Poisson fashion at an average rate of 15 per hour. The length of service on their calls is approximately exponentially distributed with mean length of 5 minutes.

    i. What is the probability that a subscriber will have to wait for his long-distance call during the peak hour of the day?

    ii. If the subscribers wait and are serviced in turn, what is the expected waiting time?

11. Assume that customers arrive according to Poisson fashion and service is exponentially distributed at a small post office. Determine the minimum number of parallel service channels needed in each of the following situations to guarantee that the operation of the queuing situation will be stable (i.e., the queue length will not grow indefinitely).

    i. Customers arrive every 5 minutes and are served at the rate of 10 customers per hour.

    ii. The average inter-arrival time is 2 minutes and the average service time is 6 minutes.

    iii. The arrival rate is 30 customers per hour and the service rate per server is 40 customers per hour.

Chapter 10

REPLACEMENT PROBLEM

10.1 INTRODUCTION

In any system, the efficiency of an item deteriorates with time or sometimes the item fails completely. This breakdown affects the whole system adversely. In some situations, either the item is to be replaced by a new item or proper maintenance is required to restore the efficiency of the system at some level. When the maintenance cost is so large, it may be profitable to replace the item. Thus, a most economic replacement policy is necessary. This chapter deals with the mathematical tool to solve replacement problems.

10.2 REPLACEMENT OF ITEMS WHOSE MAINTENANCE COST INCREASES WITH TIME AND VALUE OF MONEY REMAINS SAME DURING THE PERIOD

Let C be the capital or purchase cost of the item.

S be the resale value of the item.

$U(t)$ be the maintenance or running cost at time t. If the item is used in the system for a period y, the total maintenance cost or cummulative running cost incurred during period Y will be

$$M(Y) = \int_0^y u(t)dt \tag{10.1}$$

The total cost incurred on the item during period

$$Y = \text{Capital cost of the item} + \text{Total maintenance cost in period } Y - \text{Resale value}$$

$$\text{Total cost} = C + M(Y) - S \tag{10.2}$$

Hence, average cost per unit incurred during period Y on the item is given by $G(Y)$

$$G(Y) = \frac{C + M(Y) - S}{Y} \tag{10.3}$$

Our aim is to find that value of Y for which $G(Y)$ is minimum

$$\frac{dG}{dY} = \frac{Y(0 + M'(Y) - 0) - (C + M(Y) - S)1}{Y^2}$$

$$\frac{dG}{dY} = \frac{M'(Y)Y}{Y^2} - \frac{(C + M(Y) - S)}{Y^2}$$

We know that

$$M(Y) = \int_0^y u(t)dt$$

Therefore,

$$M'(Y) = u(Y)$$

$$\frac{dG}{dY} = \frac{u(Y)}{Y} - \frac{(C-S)}{Y^2} - \frac{M(Y)}{Y^2}$$

$$= -\frac{(C-S)}{Y^2} + \frac{u(Y)}{Y} - \frac{1}{Y^2}\int_0^y u(t)dt \quad (10.4)$$

For G to be minimum, equate Equation (10.4) to zero.

$$-\frac{(C-S)}{Y^2} + \frac{u(Y)}{Y} - \frac{1}{Y^2}\int_0^y u(t)dt = 0 \quad (10.5)$$

We know that

$$u(Y) = \frac{C-S+M(Y)}{Y}$$

Comparing Equations (10.3) and (10.5), we get,

$$u(Y) = G(Y) \quad (10.6)$$

From Equation (10.6), we conclude that the item has to be replaced, where the average running cost is minimum. Minimum cost will occur at a time when the average cost becomes equal to the current maintenance cost.

Example 10.1

The cost of a machine is Rs. 6100 and its set-up value is Rs. 100. The maintenance cost found from experience are as follows

Year	1	2	3	4	5	6	7	8
Cost in Rs.	100	250	400	600	900	1200	1600	2000

When should the machine be replaced?

Solution

This problem can be solved by minimising $G(Y)$. Here C = Rs. 6100 and S = Rs. 100.

The value of $G(Y)$ can be calculated for different years by constructing the following table.

Year	$u(t)$ Rs.	$M(Y) = \sum_{t=1}^{Y} u(t)$	$C-S+M(Y)$ Rs.	$G(Y) = \frac{C-S+M(Y)}{Y}$ Rs.
1	100	100	6100 − 100 + 100 = 6100	6100.00
2	250	350	6350	3175.00
3	400	750	6750	2250.00
4	600	1350	7350	1837.50
5	900	2250	8250	1650.00
6	1200	3450	9450	1575.00
7	1600	5050	11,050	1578.57
8	2000	7050	13,050	1631.25

From the table, we find that $G(Y)$ is minimum in the 6th year, being equal to Rs. 1575. Also, the maintenance cost in the 7th year is more than Rs. 1575. Hence the machine should be replaced at the end of the 6th year; otherwise the average annual cost would increase.

Example 10.2

A fleet owner finds from his past record that the running cost per year and resale value of a truck whose purchase price is Rs. 6000 are as follows At what stage is the replacement due?

Year	1	2	3	4	5	6	7	8
Running cost	1000	1200	1400	1800	2300	2800	3400	4000
Resale value	3000	1500	750	375	200	200	200	200

Solution

Minimising $G(Y)$, we can determine the year of replacement. Here C = Rs. 6000.

Year	Running cost $u(t)$ Rs.	Cumulative running cost $M(Y) = \sum_{t=1}^{Y} u(t)$ Rs.	$C - S + M(Y)$ Rs.	$G(Y)$ Rs.
1	1000	1000	6000 – 3000 + 1000 = 4000	4000.00
2	1200	2200	6000 – 1500 + 2200 = 6700	3350.00
3	1400	3600	6000 – 750 + 3600 = 8850	2950.00
4	1800	5400	6000 – 375 + 5400 = 11,025	2756.25
5	2300	7700	6000 – 200 + 7700 = 13,500	2700.00
6	2800	10,500	6000 – 200 + 10,500 = 16,300	2716.67
7	3400	13,900	6000 – 200 + 13900 = 19,700	2814.28
8	4000	17900	6000 – 200 + 17900 = 23,700	2962.50

From the table, it is found that the value of $G(Y)$ is minimum in the 5th year. Hence the truck should be replaced by the end of the 5th year. If not, the average annual cost would increase.

Example 10.3

Machine A costs Rs. 9000. Annual operating costs are Rs. 200 for the first year and increase by Rs. 2000 every year. Determine the best age at which to replace the machine, if the optimum replacement policy is followed. What will be the average yearly cost of owning and operating the machine? Assume that the machine has no resale value when replaced and the future costs are not discounted.

Machine B costs Rs. 10,000. The annual operating cost is Rs. 400 for the first year and increases by Rs. 800 every year you have a machine of type A which is one year old. Should we replace it with B and if so when.

Suppose you are ready to replace machine A with another machine of the same type. When you hear that machine B will become available in a year, what should you do?

Solution

Machine A				
Year	$u(t)$	$M(Y)=\sum_{t=1}^{Y} u(t)$	$C-S+M(Y)$	$G(Y)$
1	200	200	9200	9200
2	2200	2400	11,400	5700
3	4200	6600	15,600	5200
4	6200	12,800	21,800	5450

Age of replacement is 3 years for machine A. It is clear from the table that machine A should be replaced at the end of the third year. The average yearly cost in this situation is Rs. 5200.

Machine B				
Year	$u(y)$	$M(Y)=\sum_{t=1}^{Y} u(t)$	$C-S+M(Y)$	$G(Y)$
1	400	400	10,400	10,400.00
2	1200	1600	11,600	5800.00
3	2000	3600	13,600	4533.33
4	2800	6400	16,400	4100.00
5	3600	10,000	20,000	4000.00
6	4400	14,400	24,400	4066.67

It is clear from the table above that if machine B is replaced after 5 years, the average cost per year is Rs. 4000 which is minimum.

Average yearly cost $G(Y)$ of machine A is less than that of machine B in the first and the second years but from the third year onwards machine B is economical. Therefore, we have to replace the machine A by machine B, since it is economical. Therefore, we have to replace the machine A by machine B after 2 years. Since the machine A is one year old now it should be replaced after one year from now.

As shown above, the machine A should be replaced after one year from now and machine B is also available at that time. Therefore the machine A should be replaced by machine B after one year from now.

Example 10.4

A firm is considering when to replace a machine whose price is Rs. 12,200; the scrap value of the machine is Rs. 200 only. From past experience, the maintenance of the machine is as given below.

Year	1	2	3	4	5	6	7	8
Maintenance cost	200	500	800	1200	1800	2500	3200	4000

Find when the new machine should be installed?

Solution

Year	$u(t)$	$M(Y) = \sum_{t=1}^{Y} u(t)$	$C - S + M(Y)$	$G(Y)$
1	200	200	12,200	12,200.00
2	500	700	12,700	6350.00
3	800	1500	13,500	4500.00
4	1200	2700	14,700	3675.00
5	1800	4500	16,500	3300.00
6	2500	7000	19,000	3166.67
7	3200	10,200	22,200	3171.40
8	4000	14,200	26,200	3275.00

From the table we find that $G(Y)$ is minimum in the 6th year being equal to Rs. 3166.67. Also, the maintenance cost in the 7th year is more than Rs. 3166.67.

***Example* 10.5**

A truck owner finds from his past records that the maintenance cost per year and resale value of a truck whose purchase price is Rs. 8000 are as follows

Year	1	2	3	4	5	6	7	8
Maintenance cost	1000	1300	1700	2200	2900	3800	4800	6000
Resale value	4000	2000	1200	600	500	400	400	400

When should the truck be replaced?

Solution

Year	$u(t)$ Rs.	$m(Y)$ Rs.	S Rs.	$C - S$ Rs.	$C - S + m(Y)$ Rs.	$G(Y)$ Rs.
1	1000	1000	4000	4000	5000	5000
2	1300	2300	2000	6000	8300	4150
3	1700	4000	1200	6800	10,800	3600
4	2200	6200	600	7400	13,600	3400
5	2900	90,100	500	7500	16,600	3320
6	3800	12,900	400	7600	20,500	3416
7	4800	17,700	400	7600	25,300	3614
8	6000	23,700	400	7600	31,300	3912

From the table we find that the average yearly cost $G(Y)$ is minimum in the 5th year being equal to Rs. 3320. Also, the maintenance cost in the 6th year is more than Rs. 3320. Therefore, the machine should be replaced at the end of the 5th year.

Example 10.6

The following are the costs of running a particular car to date and the forecast into the future. Assuming that the car will be replaced by a similar car, determine the best time to replace it. What will be the average yearly running cost, assuming C = Rs. 1000.

Year	Resale value	Petrol and tax during the year	All other running costs	$u(t)$
0	700	–	–	–
1	625	90	10	100
2	575	90	30	120
3	550	90	50	140
4	500	90	70	160
5	450	90	90	180
6	400	90	110	200
7	350	90	130	220
8	300	90	150	240

Solution

Year	$u(t)$	$M(Y)$	$C-S$	$C-S+M(Y)$	$G(Y)$
0	–	–	300	300	0
1	100	100	375	475	475.00
2	120	220	425	645	322.10
3	140	360	450	810	270.00
4	160	520	500	1020	255.40
5	180	700	550	1250	250.00
6	200	900	600	1500	250.00
7	220	1260	700	1960	2575.50

At the end of the 6th year, the car has to be replaced as the average annual cost increases for the 7th year.

Example 10.7

A machine owner finds from his past records that the cost per year of maintaining machine A whose purchase price is Rs. 6000 is as given below. At what age is replacement due?

Year	1	2	3	4	5	6	7	8
Maintenance cost	1000	1200	1400	1800	2300	2800	3400	4000
Resale value	3000	1500	720	375	200	200	200	200

The cost of machine B whose capacity is more than machine A is worth Rs. 8000. Estimates of the maintenance cost and the resale price for the machine are as follows. Calculate the replacement age of machine B.

Year	1	2	3	4	5	6	7	8
Maintenance cost	1200	1500	1800	2400	3100	4000	5000	6800
Resale value	4000	2000	1000	500	300	300	300	300

A machine owner has three machines of type *A*, 2 of these machines are 2 years old and the third is one year old. He is considering to purchase a new machine of type *B*. Assuming that the laws of flexibility due to fewer machines is of no importance, he continues to have sufficient work for 3 of the old machines. What should the policy be?

Solution

Year	$u(t)$ Rs.	$M(Y)$ Rs.	$C-S$ Rs.	$C-S+M(Y)$ Rs.	$G(Y)$ Rs.
1	1000	1000	3000	4000	4000.00
2	1200	2200	4500	6700	3350.00
3	1400	3600	5280	8880	2960.00
4	1800	5400	5625	11,025	2756.25
5	2300	7700	5800	13,500	2700.00
6	2800	10,500	5800	16,300	2716.66
7	3400	13,900	5800	19,700	2814.28
8	4000	17,900	5800	23,700	2962.50

Machine A has to be replaced at the end of the 5th year.

Year	$u(t)$ Rs.	$M(Y)$ Rs.	$C-S$ Rs.	$C-S+M(Y)$ Rs.	$G(Y)$ Rs.
1	1200	1200	4000	5200	5200.00
2	1500	2700	6000	8700	4350.00
3	1800	4500	7000	11,500	3833.35
4	2400	6900	7500	14,400	3600.00
5	3100	10,000	7700	17,700	3540.00
6	4000	14,000	7700	21,700	3616.66
7	5000	19,000	7700	26,700	3814.20
8	6800	25,800	7700	33,500	4187.50

Machine B should be replaced at the end of the 5th year.

The machine owner has 3 machines of type *A*, two of these are 2 years old and the third, one year old. The machine owner has to decide whether to retain the same 3 machines of type *A* in the 3rd year or replace them by two machines of type *B*.

In the 3rd year, the average annual cost for machine *A* is Rs. 2960. For 3 machines, the average cost = Rs. 2960 × 3 = Rs. 8880.

Type *B* machines have the capacity of one and a half times that of type *A* machines. Therefore, the average annual cost for machine *B* in the 3rd year is Rs. 3833.33. For two machines, the average cost = 3833.33 × 2 = Rs. 7666.66.

Two machines of type *B* can replace 3 machines of type *A*. The average annual cost is less for machine *B* than for machine *A*. Therefore, the machine owner can replace the 2 machines that are 2 years old and one machine that is one year old by two new machines of type *B*.

10.3 REPLACEMENT OF ITEMS WHOSE MAINTENANCE COST INCREASES WITH TIME AND VALUE OF MONEY ALSO CHANGES WITH TIME

Under this type of replacement problems, we find two different situations.

Situation 1 The maintenance cost varies with time and we have to find the optimum value of time at which the item should be replaced. The value of many items decreases at a constant rate which is known as its depreciation ratio or discounted factor denoted by d.

Situation 2 A manufacturer takes loan for a certain period at a given interest and agrees to repay it in a certain number of installments. Here, we find the most suitable period during which the loan should be repaid.

Let us assume that the expenditure incurred to purchase the equipment is at the beginning of the year. Let the equipment cost be C and the resale value of the item be zero. The maintenance costs are Rs. $S_1, S_2, S_3, \ldots, S_n$ during the 1st, 2nd, 3rd, ..., nth year respectively. Let us assume that it is desirable to replace the item after Y years of service. Then, we calculate the expenditure made in different years and their present values as follows

Year	Capital cost	Maintenance cost	Total cost in that year	Present value of total cost
1	C	S_1	$C + S_1$	$C + S_1$
2	·	S_2	S_2	$d \cdot S_2$
3	·	·	S_3	$d^2 S_3$
·	·	·	·	·
·	·	·	·	·
·	·	·	·	·
·	·	·	·	·
Y	·	S_Y	S_Y	$d^{Y-1} \cdot C_Y$
$Y + 1$	C	S_1	$C + S_1$	$d^Y (C + S_1)$
·	·	S_2	S_2	$d^{Y+1} \cdot S_2$
·	·	·	S_3	·
·	·	·	·	·
·	·	·	·	·
$2Y$	·	S_Y	S_Y	$d^{2Y-1} \cdot S_Y$

Hence from the table, the present value of the total future expenditure incurred in a Y yearly replacement policy.

$$\text{Total expenditure} = C + S_1 + dS_2 + d^2 S_3 + \cdots + d^{Y-1} S_Y + (C + S_1) d^Y + \cdots + d^{Y+1} + S_2 + \cdots + d^{2Y-1} \cdot S_Y$$

$$\text{Total expenditure} = \left(C + \sum_{i=0}^{Y} S_1 d^{i-0} \right) + d^Y \left(C + \sum_{i=0}^{Y} S_1 d^{i-1} \right) + d^{2Y} \left(C + \sum_{i=0}^{Y} S_1 d^{i-1} \right) + \ldots$$

$$G(Y) = \frac{C + \sum_{i=1}^{Y} S_i d^{i-1}}{\sum_{i=1}^{Y} d^{Y-1}}$$

Example 10.8

A machine costs Rs. 10,000 and the operating costs are Rs. 500 per year for the first five years. In the 6th and succeeding years the operating cost increases by Rs. 100 per year. Assuming a 10% discount rate of money per year, find the optimum length of time to hold the machine before we replace it.

Solution

The discount rate per rupee will be 0.1. Therefore, the depreciation ratio will be given by

$$d = \frac{1}{1+i} = \frac{1}{1+0.1} = 0.9091$$

$$C = \text{Rs. } 10{,}000$$

Following is the table constructed to calculate $G(Y)$.

The expression for average annual cost $G(Y) = \dfrac{C + \sum_{i=1}^{Y} S_i d^{i-1}}{\sum_{i=1}^{Y} d^{Y-1}}$

Year	S_i Rs.	d^{Y-1}	$S_i d^{Y-1}$ Rs.	$C + \sum_{i=1}^{Y} S_i d^{i-1}$ Rs.	$\sum_{i=1}^{Y} d^{Y-1}$	$G(Y)$ Rs.
1	500	1.0	500.00	10,500.00	1.0000	10,500.00
2	500	0.9091	454.55	10,954.55	1.9091	5738.00
3	500	0.8264	413.20	11,367.70	2.7350	4786.40
4	500	0.7513	375.65	11,743.40	3.4863	3368.40
5	500	0.6830	341.30	12,084.70	4.1693	2898.40
6	600	0.6209	372.54	12,457.24	4.7902	2600.00
7	700	0.5645	395.15	12,852.39	5.3547	2400.00
8	800	0.5131	410.48	13,262.87	5.8678	2260.27
9	900	0.4665	419.85	13,682.72	6.3343	2160.09
10	1000	0.4241	424.10	14,106.82	6.7584	2087.30
11	1100	0.3855	424.05	14,530.87	7.1439	2034.02
12	1200	0.3505	420.60	14,951.47	7.4944	1995.01
13	1300	0.3186	414.18	15,365.65	7.8130	1966.67
14	1400	0.2897	405.58	15,771.23	8.1027	1946.40
15	1500	0.2633	394.95	16,166.18	8.3660	1932.36
16	1600	0.2394	383.04	16,549.22	8.6054	1923.12
17	1700	0.2176	369.92	16,919.14	8.8230	1917.61
18	1800	0.1978	356.04	17,275.18	9.0208	1915.03
19	1900	0.1799	341.79	17,616.97	9.2047	1914.74
20	2000	0.1635	327.07	17944.04	9.3642	1916.23

At the end of the 19th year, the machine should be replaced; $G(Y)$ is minimum in this year.

Example 10.9

A firm pays Rs. 10,000 for its automobiles. The operating and maintenance costs are about Rs. 2500 per year for the first two years and then go up by Rs. 1500 per year approximately. Find when the automobile should be replaced if the discount rate is 0.1 per rupee?

Solution

$$\text{Discount rate} = 0.1 \text{ per rupee}$$

$$\text{Depreciation ratio} = \frac{1}{1+0.1} = 0.9091 = d$$

Year	S_i	d^{i-1}	$S_i d^{i-1}$	$c + \sum_{i=0}^{Y} S_1 d^{i-1}$	$\sum_{i=0}^{Y} d^{i-1}$	$G(Y)$
1	2500	1.00	2500	12,500	1.0000	12,500.00
2	2500	0.9091	2273	14,773	1.9091	7738.20
3	4000	0.8264	3306	18,079	2.7350	6610.23
4	5500	0.7513	4132	22,211	3.4868	6370.00
5	7000	0.6830	4781	26,992	4.1698	6473.20
6	8500	0.6209	5278	32,470	4.7907	6735.90
7	10,000	0.3645	5645	37,905	5.3553	7080.00

From the above table it is found that $G(5) > G(4)$. Hence the optimum length of time to hold the machine before replacement is 4 years.

10.4 PRESENT VALUE CONCEPT

Many replacement problems involve huge expenditure both for purchasing as well as for maintenance of the respective units. There can be a number of alternatives and one may like to choose the one which requires minimum expenditure. The selection can be made by an approach based on the concept of present value. The present value of a number of expenditure incurred over different periods of time represents their value at the current time. It is based upon the fact that the amount would be invested at interest rate i to produce the same amount of money at the end of the time, i.e., if an amount is to be spent in different years, what is its present worth?

$$\text{Present worth} = s \times \frac{(1+i)^n - 1}{i(1+i)^n}$$

where S = expenditure at different periods.

When the interest, as soon as it is due is added to the principal of the preceding period to get a new principal and the interest of the next period is calculated over this new principal, the interest is known as compound interest.

If P = Principal and i = Interest, the amount at the end of n years with the compound interest is given by

$$A = P(1 + i)^n$$

or
$$P = \frac{A}{(1+i)^n} = A \times \text{Present worth factor}$$

Therefore, present worth factor $= \frac{1}{(1+i)^n}$.

Example 10.10

A person is considering a machine for his factory. The related data about the alternative machines are as follows.

	Machine A	Machine B	Machine C
Present investment	Rs. 10,000	Rs. 12,000	Rs. 15,000
Annual cost	Rs. 2000	Rs. 1500	Rs. 1200
Life	10 years	10 years	10 years
Salvage value	Rs. 500	Rs. 1000	Rs. 1200

As an advisor of the company, you have been asked to select the best machine considering 12% normal rate of returns per year.

Solution

For annual costs, the present worth factor at the rate of return of 12%

$$= \frac{(1+i)^n - 1}{i(1+i)^n}$$

$$= \frac{(1+0.12)^{10} - 1}{0.12(1+0.12)^{10}}$$

Present worth factor $= 5.65$

For salvage value, the present worth factor at the rate of return of 12%

$$= \frac{1}{(1+i)^n} = \frac{1}{(1+0.12)^{10}}$$

Present worth factor $= 0.322$

The following table gives the present value of the total cost of each machine in a period of 10 years.

	Machine A	Machine B	Machine C
Present investment	Rs. 10,000	Rs. 12,000	Rs. 15,000
Annual cost	Rs. 2000 × 5.65 = Rs. 11,300	Rs. 1500 × 5.65 = Rs. 8475	Rs. 1200 × 5.65 = Rs. 6780
Salvage value	Rs. 500 × 0.322 = Rs. 161	Rs. 1000 × 0.322 = Rs. 322	Rs. 1200 × 0.322 = 386.4
Present value of total cost = $[A + B - C]$	Rs. 21,139	Rs. 20,153	Rs. 21,393.6

Present value of the total cost is minimum for machine B. The machine B having the least present value of the total cost should be purchased.

Example 10.11

A company is considering the purchase of a new grinder which will cost Rs. 10,000. The economic life of the machine is expected to be 6 years. The salvage value of the machine will be Rs. 2000. The average operating and maintenance costs are estimated to be Rs. 5000 per annum.

i. Assuming an interest rate of 10%, determine the present value of the future cost of the proposed grinder.

ii. Compare this grinder with the presently owned grinder that has an annual operating cost of Rs. 4000 and expected maintenance cost of Rs. 2000 in the second year with an annual increase of Rs. 1000 thereafter.

Solution

For operating and maintenance costs the present worth factor at 10% for six years

$$= \frac{(1+i)^n - 1}{i(1+i)^n}$$

$$= \frac{(1+0.1)^6 - 1}{0.1(1+0.1)^6}$$

Present worth factor = 4.355

For salvage value, present worth factor at 10% for 6 years

$$= \frac{1}{(1+i)^n}$$

$$= \frac{1}{(1+0.1)^6}$$

Present worth factor = 0.5645

(a) Present value of annual operating cost

= Rs. 5000 × 4.355

= Rs. 21,775

Present value of the salvage value

= 2000 × 0.564

= Rs. 1129

Hence the present value of total future costs

= Rs. 21,775 – Rs. 1129

= Rs. 20,646

(b) Here the annual operating and maintenance costs vary with time. The calculation for present value of total operating and maintenance costs are given below.

Year	Operating cost (Rs.)	Maintenance cost (Rs.)	Total operating and maintenance cost (Rs.)	P.w.f for single payment (Rs.)	Present value (Rs.)
1	4000	–	4000	0.909	3636
2	4000	2000	6000	0.826	4956
3	4000	3000	7000	0.751	5257
4	4000	4000	8000	0.683	5464
5	4000	5000	9000	0.621	5589
6	4000	6000	10,000	0.564	5640
Total present value = Rs. 30,542					

Now from part (a), it is observed that the present value of a new grinder is Rs. 20,646. Hence the cost saved if a new grinder is purchased = Rs. 30,542 – Rs. 20,646 = Rs. 9896.

The cost of new grinder is Rs. 10,000 and its cost savings are only Rs. 9896. The management is not advised to purchase a new grinder.

10.5 REPLACEMENT OF ITEMS THAT FAIL COMPLETELY

We know that the possibility of failure of any item increases with time. The nature of the system may be such that if the item fails, it may result in complete breakdown of the system. The breakdown implies the loss in production, idle inventory, idle labour and other losses. In such circumstances, it is advisable to formulate a suitable replacement policy.

1. Individual replacement policy.
2. Group replacement policy.

10.5.1 Individual Replacement Policy

Whenever an item fails, it should be replaced immediately. Such a replacement policy is called individual replacement policy.

10.5.2 Group Replacement Policy

In this policy, all the items in the system should be replaced after a certain period, irrespective of the fact that if any item fails before this time, it can be replaced individually. A group replacement policy consists of two steps

Step 1 Individual replacement of an item that fails after time t.

Step 2 Group replacement of existing items at a suitable time.

Here the individual replacement at the time of failure ensures the running of the system, whereas the group replacement after some time interval will reduce the probability of failure. The period for which the total cost incurred is minimum will be the optimum for replacement.

10.6 RATE OF REPLACEMENT AND TOTAL COST IN GROUP REPLACEMENT

Let all units in a system be replaced after a definite time t with the provision that individual replacement can be made if and when any unit in the system fails during this time.

Let

$F(t)$ be the number of units failing at time (t)

$P(x)$ be the probability of failure of any unit at age (x)

$P(t-x)$ be the probability of failure of any unit at age $(t-x)$

N be the total number of units in the system

$N.P(x)$ is the expected number of units to be replaced at time (x)

$P_{(x)} \cdot P(t-x)$ is the probability of failure of any unit at age x and again at age $(t-x)$

$P_{(x)} \cdot P(b-x).P(t-b)$ is the probability that any item fails at ages x, $(b-x)$ and $(t-b)$

Hence the total number of replacement in time (t) will be given by

$$F(t) = N\left[P(t) + \sum_{x=1}^{t-1} P_x \cdot P(t-x) + \sum_{b=2}^{t-1}\sum_{x=1}^{t-1} P(x) \cdot P(b-x)P(t-b)\right] \tag{10.7}$$

The number of units failing at the end of the nth period in any system can be calculated from Equation (10.7).

Cost of group replacement in time *t*

Let $C(t)$ be the cost of group replacement after the time period (t).

$$\text{The average cost per period} = \frac{C(t)}{(t)} \tag{10.8}$$

Let $C(1)$ be the unit cost of replacement in a group and $C(2)$ be the individual replacement cost on failure. Then,

$$\begin{aligned} C(t) &= NC_1 + C_2\,[\text{Number of failures at the end of period 1} \\ &\quad + \text{Number of failures at the end of period 2} + \cdots \\ &\quad + \text{Number of failures at the end of period } (t-1)] \\ C(t) &= NC_1 + C_2[F(1) + F(2) + F(3) + \cdots + F(t-1)] \end{aligned}$$

where $F(1)$, $F(2)$, $F(3)$ ... etc. are the number of failures.

Example 10.12

A computer contains 10,000 resistors. When any resistor fails, it is replaced. The cost of replacing a resistor individually is Re. 1. If all the resistors are replaced at the same time, the cost per resistor would be reduced to 35 paise. The percent saving $S(t)$ at the end of month t and $P(t)$ the probability of failure during the month t are as follows

t	0	1	2	3	4	5	6
$S(t)$	100	97	90	70	30	15	0
$P(t)$	–	0.03	0.07	0.2	0.4	0.15	0.15

What is the optimum replacement plan?

Solution

This problem consists of two parts

1. Individual replacement policy
2. Group replacement policy

It is known that no resistor will survive for more than six months. Thus a resistor which has survived for 5 months is sure to fail during the 6th month. We assume that the resistors failing during a month are replaced just at the end of that month.

Let $N(t)$ denote the number of resistors replaced at the end of the tth month. The different values of $N(t)$ are calculated as follows.

N_0: Number of resistors in the beginning

N_1: The number of resistors being replaced by the end of the first month = (Number of resistors during the first month) × (Probability that a resistor fails during the first month of its installation)

$$N_1 = N_0 \times P_0$$
$$N_1 = 10{,}000 \times 0.03 = 300 \text{ resistors}$$

N_2: Number of resistors replaced by the end of second month = (Number of resistors in the beginning × Probability that these fail in second month) + (Number of resistors replaced in the first month × Probability that these fail in second month)

$$\begin{aligned} N_2 &= N_0P_2 + N_1P_1 \\ &= (10{,}000 \times 0.07) + (300 \times 0.03) \\ N_2 &= 709 \text{ resistors} \end{aligned}$$

Similarly,

$$\begin{aligned} N_3 &= N_0P_3 + N_1P_2 + N_2P_1 \\ &= (10{,}000 \times 0.2) + (300 \times 0.07) + (709 \times 0.03) \\ N_3 &= 2042 \text{ resistors} \\ N_4 &= N_0\,P_4 + N_1P_3 + N_2P_2 + N_3P_1 \\ &= (10{,}000 \times 0.4) + (300 \times 0.2) + (709 \times 0.07) + \\ &\quad (2042 \times 0.03) \\ N_4 &= 4171 \text{ resistors} \\ N_5 &= N_0P_5 + N_1P_4 + N_2P_3 + N_3P_2 + N_4P_1 \\ &= (10{,}000 \times 0.15) + (300 \times 0.4) + (709 \times 0.2) + \\ &\quad (2042 \times 0.07) + (4171 \times 0.03) \\ N_5 &= 2030 \text{ resistors} \end{aligned}$$

$$N_6 = N_0P_6 + N_1P_5 + N_2P_4 + N_3P_3 + N_4P_2 + N_5P_1$$
$$= (10{,}000 \times 0.15) + (300 \times 0.15) + (709 \times 0.4) + (2042 \times 0.2) + (4171 \times 0.07) + (2030 \times 0.03)$$
$$N_6 = 2590 \text{ resistors}$$

It can be seen from the above calculation that $N(t)$ increases up to the fourth month and then decreases. It can also be seen that $N(t)$ will tend to increase and the value of $N(t)$ will oscillate till the system acquires a steady state.

$$\text{The expected life of each resistor} = \Sigma t \times P(t)$$

where

t is the month

$P(t)$ is the corresponding probability of failure.

$$\text{Life} = 1 \times (0.03) + 2 \times (0.07) + 3 \times (0.2) + 4 \times (0.4) + 5 \times (0.15) + 6 \times (0.15)$$
$$= 4.02 \text{ months}$$

Therefore, average number of resistors replaced every month

$$\frac{N}{\text{Mean life}} = \frac{10{,}000}{4.02} = 2488 \text{ resistors}$$

Average cost of monthly individual replacement policy

= Rs. 2488 [Cost per resistor = Re. 1]

Let us consider the policy of group replacement.

End of month	Total cost of group replacement	Cost/month in Rs.
1	300(1) + 10,000(0.35)	3800.00
2	(300 + 709)1 + 10,000(0.35)	2254.50
3	(300 + 709 + 2042)1 + 10,000(0.35)	2183.66
4	(300 + 709 + 2042 + 4171)1 + 10,000(0.35)	2680.50

Hence the minimum cost per month is obtained by group replacement of all resistors after three months with an average cost of Rs. 2183.66 per month.

10.7 STAFF PROBLEM

So far, we have discussed about the replacement policy of items in an organisation. The problem of staff replacement arises due to inefficiency, resignation, retirement or death of staff members from time to time. Therefore, to maintain suitable strength of staff members in a system, there is a need to formulate some recruitment policy. We assume that life distribution for the service of staff in a system is known.

Example 10.13

A research team plans to rise to a strength of 50 chemists and then remain at that level. The wastage of recruits depends on their length of service which is as follows

Year	1	2	3	4	5	6	7	8	9	10
Total percentage who have left up to the end of the year	5	36	59	63	68	73	79	87	97	100

What is the replacement per year necessary to maintain the strength; there are eight senior posts for which the length of service is the main criterion. What is the average length of service after which a new entrant can expect his promotion to one of the posts?

Solution

Here we are given the percentage of chemists who have left up to the end of a certain year. With the help of this distribution, we can find out the probability of the persons leaving during a certain year. It is certain that none of the recruited continue in service for more than 10 years, using the relation.

$$\text{Probability of failure} = \frac{M(t-1) - M(t)}{N}$$

where,

$M(t)$ is the number in service at any time t.

N is the number of items

The following table illustrates the probability of being in service.

Year	No. of persons who leave at the end of the year	No. of persons in service at the end of the year	Probability of leaving service at the end of the year	Probability of being in service at the end of the year
0	0	100	0.00	1.00
1	5	95	0.05	0.95
2	36	64	0.36	0.64
3	59	41	0.59	0.41
4	63	37	0.63	0.37
5	68	32	0.68	0.32
6	73	27	0.73	0.27
7	79	21	0.79	0.21
8	87	13	0.87	0.13
9	97	3	0.97	0.03
10	100	0	1.00	0.00
Total number who are in service at the end of 10th year = 436				

From the table, we find that with a recruitment policy of 100 every year, the total number of chemists serving in the organisation would have been 436. Hence, if we want to maintain a strength of 50 chemists, we should recruit

$$\frac{100}{436} \times 50 = 11.5 \approx 12 \text{ chemists per year}$$

Also we find the number of survivals after each year by multiplying the various values of probability of survival at the end of the year by 12.

We get

Year	0	1	2	3	4	5	6	7	8	9	10
Number of chemists in service	12	11	7	5	4	4	3	2	2	0	0

Now there are eight senior posts. From the table above, we find that there are 3 persons in service during the 6th year, 2 in the 7th year and 2 in the 8th year, i.e., in all, 7 persons are there in service from the 6th year to the 8th year. Hence promotion of new recruits will start by the end of the 5th year and will continue up to the 6th year.

Example 10.14

An airline requires 200 assistant hostesses, 300 hostesses and 50 supervisors. Girls are recruited at the age of 21. If still in service they retire at the age of 60. Given the following life table, determine the following.

1. How many girls should be recruited each year?
2. At what age should promotion take place?

Age	21	22	23	24	25	26	27	28	29	30	31	32	33	34
Number in service	1000	600	400	384	307	261	228	206	190	181	173	157	161	155

Age	35	36	37	38	39	40	41	42	43	44	45	46	47	48
Number in service	150	146	144	136	131	125	115	113	106	99	93	87	80	73

Age	49	50	51	52	53	54	55	56	57	58	59	60
Number in service	66	59	53	46	39	33	27	22	18	14	11	0

Solution

If a policy of recruiting 1000 girls every year is followed, the total number of girls in service between 21 and 59 years will be equal to 6480. But we want 200 + 300 + 50 = 550 girls in the organisation. Hence to maintain the strength of 550 hostesses, we should recruit

$$\frac{1000}{6480} \times 550 = 84.87 \approx 85 \text{ girls every year}$$

Now, if assistant hostesses are promoted at the age Y, up to age $(Y - 1)$ we require 200 assistant hostesses. Out of 550, there are 200 assistant hostesses. Hence out of a strength of 1000, there will be $\frac{200}{550} \times 1000 = 364$ assistant hostesses.

From the given table, this number is available up to 24 years. Hence the promotion of assistant hostesses will take place in the 25th year.

Again, out of 550 recruitments, we require 300 hostesses. So, if we require 1000, we will require $\frac{300}{550} \times 1000 = 545$ hostesses. Hence the number of hostesses and assistant hostesses in a recruitment of 1000 will be $364 + 545 = 909$, i.e., we require 91 supervisors whereas at the age of 46 only 87 will survive. Hence promotion of hostesses will take place in the 46th year.

Example 10.15

An airline whose staff are subject to the same survival rate as in Example 10.14 presently has staff whose ages are distributed as in the following table. It is estimated that for the next two years, the staff requirements will increase by 10% per year. If girls are to be recruited at the age of 21, how many should be recruited for the next year and at what age will promotion take place? How many should be recruited for the following year and at what age will the promotion take place.

Age	Assistant Hostesses	Age	Hostesses	Age	Supervisor
21	90	26	40	42	5
22	50	27	35	43	4
23	30	28	35	44	5
24	20	29	30	45	3
25	10	30	28	46	3
		31	26	47	3
		32	20	48	6
		33	18	49	2
		34	16	50	–
		35	12	51	–
		36	10	52	4
		37	8	53	3
		38	–	54	5
		39	6	55	–
		40	8	56	3
		41	8	57	2
				58	–
				59	2
Total	200	Total	300	Total	50

Solution

If a policy of recruiting 90 girls every year is followed, the number of girls in service between 21 and 59 years will be equal to 550. But we need 10% extra in the coming years.

Hence, to maintain a strength of $(1.1) \times 550 = 605$ hostesses, we should recruit $\frac{90}{550} \times 605 = 99$ girls every year at the age of 21.

Now, if assistant hostesses are promoted at the age Y, then up to $(Y - 1)$ years, we require 220 assistant hostesses out of 605.

Hence, out of a strength of 90, there will be $\frac{220}{605} \times 90 = 33$ assistant hostesses.

From the table given, this number is available up to 23 years. Hence the promotions of the assistant hostesses will take place in the 24th year. Again, out of 605 recruitments, we require 330 hostesses.

So if we recruit 90, we will require $\frac{330}{605} \times 90 = 49$ hostesses.

Hence the number of assistant hostesses and hostesses in a recruitment of 90 will be 33 + 49 = 82.

We require 17 supervisors; whereas at the age of 33 only 18 will survive. Hence promotion of hostesses will takes place in the 33rd year.

EXERCISES

1. A firm is using a machine whose purchase price is Rs. 12,000. The installation charges amount to Rs. 3000 and the machine has a scrap value of only Rs. 1500, because the firm has a monopoly of this type of work. The maintenance cost in various years is given in the following table.

Year	1	2	3	4	5	6	7	8	9
Cost (Rs.)	250	750	1000	1500	2100	2900	4000	4800	6000

Determine the optimum replacement period.

2. A simple engineering company has a machine whose purchase price is Rs. 80,000. The expected maintenance costs and resale price in different years are as given here.

Year	1	2	3	4	5	6	7
Maintenance cost (Rs.)	1000	1200	1600	2400	3000	3900	5000
Resale value ('000 Rs.)	75	72	70	65	58	50	45

After what time interval should the machine be replaced?

3. An electro-mechanical equipment has a purchase price of Rs. 7000. Its running costs per year and resale values are given here.

Year	1	2	3	4	5	6	7	8
Running cost (Rs.)	2000	2100	2300	2600	3000	3500	4100	4600
Resale value (Rs.)	4000	3000	2200	1600	1400	700	700	700

At what year is the replacement due?

4. Determine the optimum replacement interval of an equipment which costs Rs. 5200 and whose resale values and running costs are as follows

Year	1	2	3	4	5	6	7
Running cost (Rs.)	3500	2700	1800	1000	850	600	425
Resale value (Rs.)	600	850	1000	1250	1400	1450	2000

5. Determine the age at which the following type of machine has to be replaced.

   Cost price = Rs. 8000

Operating costs = Rs. 1000 for the first year, increasing by Rs. 500 every year

Resale value = Rs. 4000 for the first year, decreasing by Rs. 500 every year

6. The cost of a machine is Rs. 6500 and its scrap value is only Rs. 500 whenever disposed. The maintenance costs are found to be as under, on the basis of experience.

Year	1	2	3	4	5	6	7	8
Maintenance cost (Rs.)	100	250	400	600	900	1250	1800	2500

When should the machine be replaced to get maximum advantage?

7. The following table gives the running costs per year and resale prices of a certain equipment whose purchase price is Rs. 5000.

Year	1	2	3	4	5	6	7	8
Resale value (Rs.)	1500	1600	1800	2100	2500	2900	3400	4000
Running cost (Rs.)	3500	2500	1700	1200	800	500	300	200

In which year is the replacement due?

8. A piece of equipment costs Rs. 7500 initially and requires Rs. 400 to be spent on its maintenance in the first year. The maintenance cost would increase by Rs. 500 per year in each of the subsequent years. Determine the optimal replacement for the machine when

   i. future costs are not discounted

   ii. future costs are discounted at the rate of 10% per annum.

   The cost of a new machine is Rs. 5000. The maintenance cost during the nth year is given by $R_n = 500(n - 1)\ n = 1, 2, \ldots$ Suppose that the discount rate per year is 0.05, after how many years will it be economical to replace the machine by a new one?

9. A manufacturer has to decide between two alternative machines M_1 and M_2 about which pertinent information is as follows

	M_1	M_2
Cost	Rs. 5000	Rs. 2500
Maintenance cost	Rs. 800 p.a for years 1, 2, ..., 5 increasing by Rs. 200 every year	Rs. 1200 p.a for years 1, 2, ..., 6 increasing by Rs. 200 every year
Scrap value	Nil	Nil
Cost of capital	(to be used as discount rate) 10% p.a	

   i. Determine the optimal replacement period of M_1 and M_2.

   ii. Which of the two is a better alternative?

10. An electronic machine has a component that is subject to failure and needs replacement immediately when this occurs. A total of 200 such machines are used in a company. The cost of each replacement is Rs. 80 per item and for group replacement, it is Rs. 5000 for all the 200 machines. Records of previous failure rates indicate the following pattern.

10 percent would fail during the first month

20 percent would fail during the second month

20 percent would fail during the third month

30 percent would fail during the fourth month

20 percent would fail during the fifth month

Assuming that group replacement can be carried out only at the month end, what group replacement policy would you suggest? Consider an 8-month period for cost evaluation.

11. The distribution of life of a special type of bulb used by a company is as follows

Life in months	1	2	3	4	5	6
Proportion of bulbs failing	0.08	0.12	0.20	0.30	0.20	0.10

The cost of replacing an individual bulb when it fails is Rs. 5 while the cost when replaced on a group basis is Rs. 2 per bulb. The company has one thousand bulbs of this type installed.

i. Calculate the average life of a bulb.

ii. Determine if the policy of group replacement is superior to the policy of individual replacement. Consider the periods of 1 month, 2 months and 3 months for replacement.

12. A computer has a large number of components all of which have to be fully working and failures have to be replaced to maintain operational efficiency. Past observations have shown the following mortality pattern. If the component is replaced individually, the cost of single replacement is Rs. 20, whereas if all components are replaced concurrently, the cost of a single component is Rs. 7. Decide whether it would be worthwhile to have a mortality statistics. The observed number of components is 500.

End of week	1	2	3	4	5	6
No. of components Servicing	455	375	250	75	15	0

13. A public sector bank requires 400 clerks, 250 officers and 50 managers for a certain state. The employees are recruited at the age of 21 as clerks. The promotions take place only from within, on the basis of the length of service. You are given the following life table.

Age (Years)	21	22	23	24	25	26	27	28	29	30
No. of employees in service	1000	640	560	450	400	355	300	205	190	182
Age (Years)	31	32	33	34	35	36	37	38	39	40
No. of employees in service	175	170	162	155	150	148	145	142	140	132
Age (Years)	41	42	43	44	45	46	47	48	49	50
No. of employees in service	125	118	112	105	101	100	95	90	88	85
Age (Years)	51	52	53	54	55	56	57	58	59	60
No. of employees in service	80	72	70	63	55	40	32	25	20	0

Determine

(a) How many employees should be recruited each year?

(b) At what age should the promotion take place?

Chapter 11

DYNAMIC PROGRAMMING

11.1 INTRODUCTION

The operations research techniques are used to find the optimal solution, under constraints. Dynamic programming is an OR technique used to break up the decisions into smaller components and then to recombine previous decisions in a certain form to obtain the desired solution. Dynamic programming is used to solve problems related to all fields of engineering.

In the late 1940s, Richard Bellman developed the concept of dynamic programming. Bellman has also authored a number of books on dynamic programming. Other authors who contributed to the concept of dynamic programming include Aris, Nem hauser, L.G. Mitten, Phillips and D.J.White.

11.2 DYNAMIC PROGRAMMING SYSTEM

Let us assume that we know the input variables and we are supposed to use these variables in an optimum quantity. An organisation system consists of multiple stages at which decisions are to be made. Each one of the decisions made at different stages has a positive or negative return. This return function depends on the input state variable and the decision.

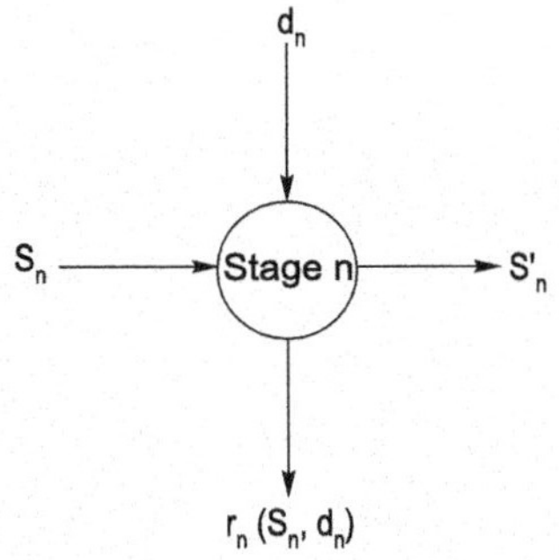

Figure 11.1 Single stage system

In Figure 11.1,

S_n denotes input state variable

S'_n denotes output state variable

n is the stage number

d_n is the decision

$r_n(S_n, d_n)$ is the return function

Consider the example where a carpenter has certain quantity of wood (input state variable). He has options to use the wood to make chairs, tables, cots, windows and doors (decision variables). Each type of furniture gives a return.

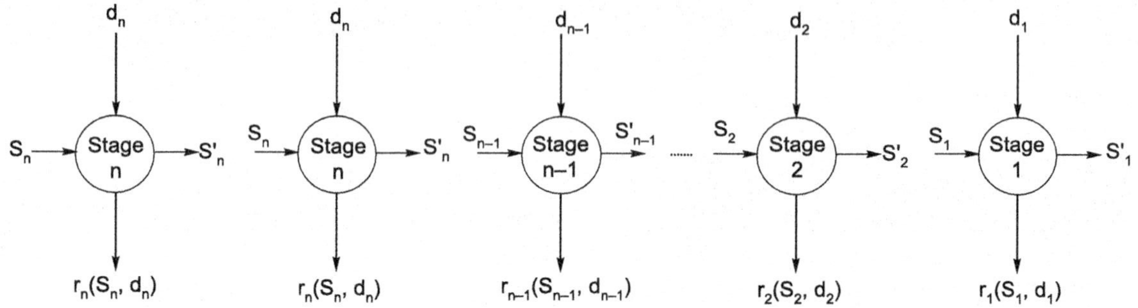

Figure 11.2 An N stage system

Figure 11.2 illustrates a multistage system. Here, there are N stages and all of them are linked by transition functions.

In Figure 11.2,

$S_{n-1} = S_n \times d_n$

S_{n-1} = input to $n-1$th stage

S_n = input to nth stage

d_n = decision at nth stage

The output state variable of stage (n) is used as the input variable of stage ($n - 1$).

i.e., $$S'_n = S_{n-1}$$

In the multistage system shown in Figure 11.2, the stages are numbered in a direction opposite to the flow of information.

Let $f_n(S_n, d_n)$ be the accumulated total return calculated over n-stages.

$f'_n(S_n, d_n)$ be the accumulated optimum return for a particular input state variable S_n.

For a particular input state variable (S_n) i.e., certain quantity of wood available, the carpenter can make several decisions (d_n), i.e., making chairs, tables, cots, windows and doors. Among these decision variables, decisions d'_n gives an optimum return for n stage system.

Therefore, the accumulated optimum return can be written as

$$f'_n(S_n) = \underset{d_n, d_{n-1}, \ldots, d_1}{\text{Optimum}} \left\{ r_n(S_n, d_n) * r_{n-1}(S_{n-1}, d_{n-1}) * \ldots r_1(S_1, \ldots, d_1) \right\}$$

*Transition function represents the mathematical operation such as addition, subtraction or division.

In a multistage system consisting of N stages, a decision variable at stage N affects the decision at stage $N-1$ of decision variable at stage 2 affects the decision variable at stage 1.

Optimum return for one-stage system with input state variable S_1 can be written as

$$f_1'(S_1) = \underset{d_1}{\text{Optimum}}\left\{r_1(S_1,\ d_1)\right\}$$

Similarly, optimum return for two-stage system is

$$f_2'(S_2) = \underset{d_2}{\text{Optimum}}\left\{r_2(S_2,\ d_2) + f_1'(S_1)\right\}$$

We know that

$$\begin{aligned} S_{n-1} &= S_n * d_n \\ S_1 &= S_2 * d_2 \end{aligned}$$

Then

$$f_2'(S_2) = \underset{d_2}{\text{Optimum}}\left\{r_2(S_2,\ d_2) + f_1'(S_2 * d_2)\right\}$$

For an n stage system, the optimum return is

$$f_n'(S_n) = \underset{d_n}{\text{Optimum}}\left\{r_n(S_n,\ d_n) + f_{-1}'(S_n * d_n)\right\}$$

where $n = 1, 2, 3, ..., N$

Example 11.1

An electric utility is building a large generating station and a regional power distribution network to serve a large number of customers scattered over a wide area. Because of the large area to be served, they decide to build 10 substations to handle the distribution of power to the specific location.

The problem is where to locate the substation along the transmission line so as to minimise the cost of building high-tension lines, assuming that local distribution costs are roughly the same no matter where the substation is built in a given local area.

The number of sites is possible for each substation but not every site in one area can be reached from every site in the next area for a number of reasons including geographic inaccessibility due to terrain features such as lakes and mountains and the inability to purchase or lease right a way for the lines.

The following diagram represents the problem with accessible sites connected by areas labelled with the cost of connecting them. Determine the minimum cost power distribution plan.

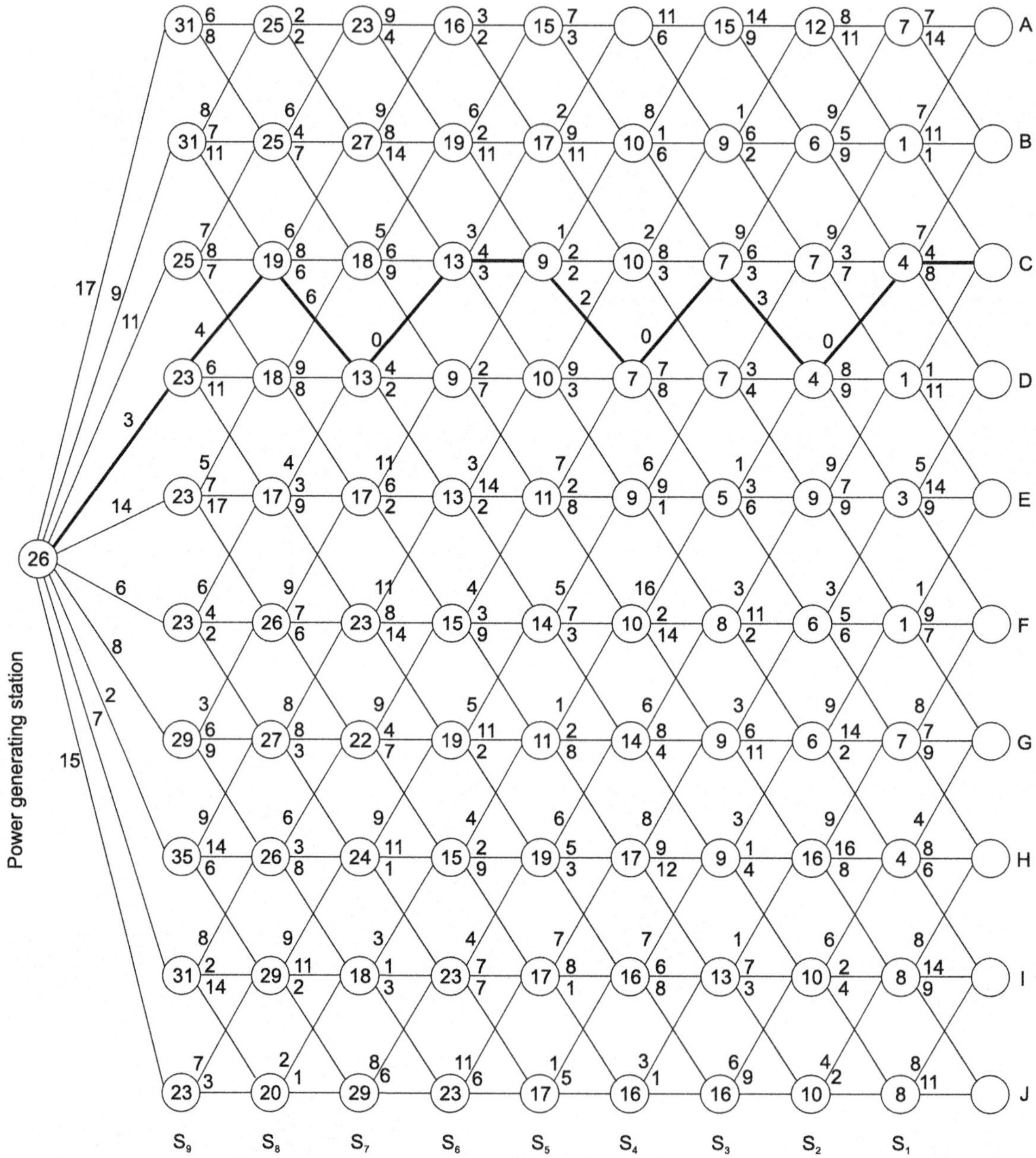

Figure 11.3 Transmission lines

Number the stages in a direction opposite to the flow of information.

Optimum return for stage 1

$$f_1'(S_1) = \underset{d_1}{\text{Optimum}} \left\{ r_1(S_1, d_1) * f_0'(S_0) \right\}$$

Table 11.1 Stage 1

S_1	R	L	M	Optimum d_1	$f_1'(S_1)$
A	14	–	7	M	7
B	1	7	11	R	1
C	8	7	4	M	4
D	11	–	1	M	1
E	3	5	11	R	3
F	7	1	9	L	1
G	9	8	7	M	7
H	6	4	8	L	4
I	9	8	14	L	8
J	–	8	11	L	8

Accumulated optimum return for stage 2

$$f_2'(S_2) = \underset{d_2}{\text{Optimum}} \left\{ r_2(S_2, d_2) * f_1'(S_1) \right\}$$

Accumulated optimum return for stage 3

$$f_3'(S_3) = \underset{d_3}{\text{Optimum}} \left\{ r_3(S_3, d_3) * f_2'(S_2) \right\}$$

Table 11.2 Stage 2

S_2	R	L	M	Optimum d_2	$f_2'(S_2)$
A	12	–	15	R	12
B	13	16	6	M	6
C	8	9	7	M	7
D	12	4	9	L	4
E	10	9	10	L	9
F	13	6	6	L, M	6
G	6	10	21	R	6
H	16	16	20	R, L	16
I	12	10	10	L, M	10
J	–	12	10	M	10

Table 11.3 Stage 3

S_3	R	L	M	Optimum d_3	$f_3'\ (S_3)$
A	15	–	26	R	15
B	9	13	12	R	9
C	7	15	13	R	7
D	13	7	7	L, M	7
E	12	5	12	L	5
F	8	12	17	R	8
G	27	9	12	L	9
H	14	9	17	L	9
I	13	17	17	R	13
J	–	16	19	L	16

Accumulated optimum return for stage 4

$$f_4'(S_4) = \underset{d_4}{\text{Optimum}}\left\{ r_4(S_4, d_4) * f_3'(S_3) \right\}$$

Accumulated optimum return for stage 5

$$f_4'(S_4) = \underset{d_4}{\text{Optimum}}\left\{ r_4(S_4, d_4) * f_3'(S_3) \right\}$$

Table 11.4 Stage 4

S_4	R	L	M	Optimum d_4	$f_4'\ (S_4)$
A	–	15	26	L	15
B	13	23	10	M	10
C	10	11	15	R	10
D	13	7	14	L	7
E	9	13	14	R	9
F	23	21	10	M	10
G	23	14	17	L	14
H	25	17	18	L	17
I	24	16	19	L	16
J	–	16	17	L	16

Table 11.5 Stage 5

S_5	R	L	M	Optimum d_5	$f'_5(S_5)$
A	13	–	22	R	13
B	21	17	19	L	17
C	9	11	12	R	9
D	12	10	16	L	10
E	18	14	11	M	11
F	17	14	17	L	14
G	25	11	16	L	11
H	19	20	22	R	19
I	17	24	24	R	17
J	–	17	21	L	17

Accumulated optimum return for stage 6

$$f'_6(S_6) = \underset{d_6}{\text{Optimum}}\left\{r_6(S_6, d_6) * f'_5(S_5)\right\}$$

Accumulated optimum return for stage 7

$$f'_7(S_7) = \underset{d_7}{\text{Optimum}}\left\{r_7(S_7, d_7) * f'_6(S_6)\right\}$$

Table 11.6 Stage 6

S_6	R	L	M	Optimum d_6	$f'_6(S_6)$
A	19	–	16	M	16
B	20	19	19	L, M	19
C	13	20	13	R, M	13
D	18	9	12	L	9
E	16	13	25	L	13
F	20	15	17	L	15
G	21	19	22	L	19
H	26	15	21	L	15
I	24	23	24	L	23
J	–	28	23	M	23

Table 11.7 Stage 7

S_7	R	L	M	Optimum d_7	$f_7'(S_7)$
A	23	–	25	R	23
B	27	25	27	R,M	27
C	18	24	19	R	18
D	15	13	13	L,M	13
E	17	20	19	R	17
F	33	24	23	M	23
G	22	24	23	R	22
H	24	28	26	R	24
I	26	18	24	L	18
J	–	31	29	M	29

Accumulated optimum return for stage 8

$$f_8'(S_8) = \underset{d_8}{\text{Optimum}} \left\{ r_8(S_8, d_8) * f_7'(S_7) \right\}$$

Accumulated optimum return for stage 9

$$f_9'(S_9) = \underset{d_9}{\text{Optimum}} \left\{ r_9(S_9, d_9) * f_8'(S_8) \right\}$$

Table 11.8 Stage 8

S_8	R	L	M	Optimum d_8	$f_8'(S_8)$
A	29	–	25	M	25
B	25	29	31	R	25
C	19	33	26	R	19
D	25	18	22	L	18
E	32	17	26	L	17
F	28	26	30	L	26
G	27	31	30	R	27
H	26	28	27	R	26
I	31	33	29	M	29
J	–	20	30	L	20

Table 11.9 Stage 9

S_9	R	L	M	Optimum d_9	$f_9'(S_9)$
A	33	–	31	M	31
B	31	33	32	R	31
C	25	32	27	R	25
D	28	23	24	L	23
E	43	23	24	L	23
F	29	23	30	L	23
G	35	29	33	L	29
H	35	36	40	R	35
I	34	35	31	M	31
J	–	36	23	M	23

Table 11.10 Stage 10

S_{10}	A	B	C	D	E	F	G	H	I	J	Optimum d_{10}	$f_{10}'(S_{10})$
Power status	48	40	36	26	37	29	37	37	38	38	D	26

Transmission line connecting the substation is given below.

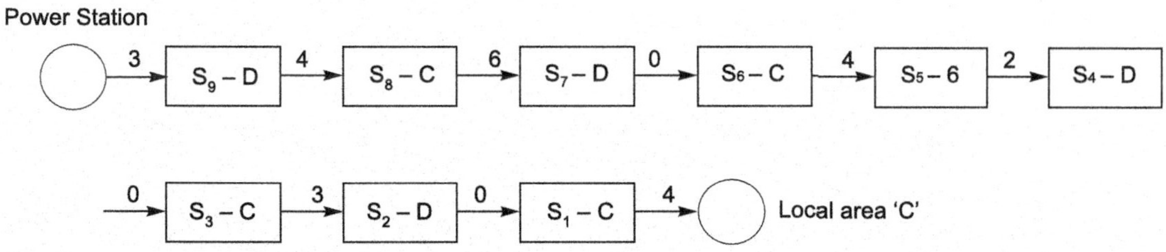

Figure 11.4 Transmission line

The minimum cost for power distribution is 26 units.

Example 11.2

A salesman located in city A decided to travel to city B. The arrow represents the route in cities. The distance in kilometers is indicated for each route. The salesman problem is to find the shortest route that covers all these selected cities from city A to city B.

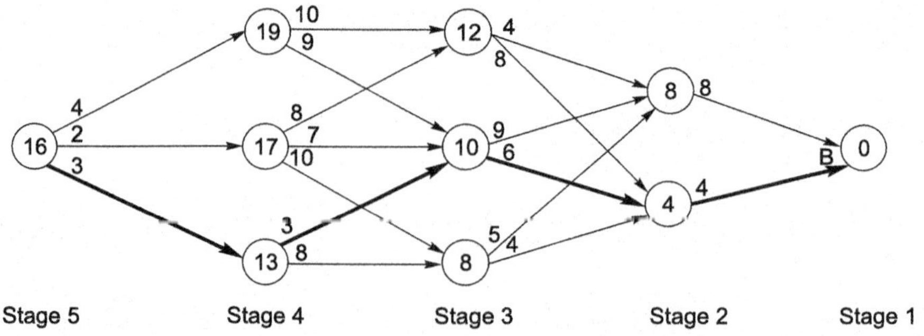

Figure 11.5 Route network

Number the stages in a direction opposite to the flow of information.

Optimum return for stage 1

Stage 1

$$f_1'(S_1) = \underset{d_1}{\text{Optimum}} \left\{ r_1(S_1, d_1) * f_0'(S_0) \right\}$$

$$f_1'(S_1) = 0$$

Optimum return for stage 2

$$f_2'(S_2) = \underset{d_2}{\text{Optimum}} \left\{ r_2(S_2, d_2) * f_1'(S_1) \right\}$$

Table 11.11 Stage 2

S_2	1	2	3	Optimum d_2	$f_2'(S_2)$
1	8	–	–	8	4
2	4	–	–	4	–

Optimum return for stage 3

$$f_3'(S_3) = \underset{d_3}{\text{Optimum}} \left\{ r_3(S_3, d_3) * f_2'(S_2) \right\}$$

Table 11.12 Stage 3

S_3	1	2	3	Optimum d_3	$f_3'(S_3)$
1	12	12	–	12	–
2	17	10	–	10	8
3	13	8	–	8	–

Optimum return for stage 4

$$f_4'(S_4) = \underset{d_4}{\text{Optimum}}\left\{r_4(S_4, d_4) * f_3'(S_3)\right\}$$

Table 11.13 Stage 4

S_4	1	2	3	Optimum d_4	$f_4'(S_4)$
1	22	19	–	19	–
2	20	17	18	17	13
3	13	16	–	13	–

Optimum return for stage 5

$$f_5'(S_5) = \underset{d_5}{\text{Optimum}}\left\{r_5(S_5, d_5) * f_4'(S_4)\right\}$$

Table 11.14 Stage 5

S_5	1	2	3	Optimum d_5	$f_5'(S_5)$
1	23	19	16	16	16

Shortest route with distance 16 kms

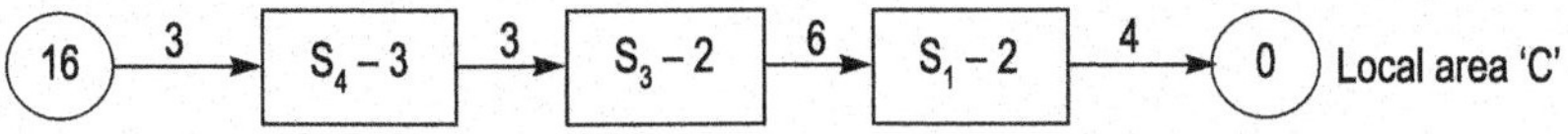

Figure 11.6 Shortest route

11.3 DYNAMIC PROGRAMMING FOR PROFIT OPTIMISATION

A textile marketingcompany has received order from a banner-making agency. The requirements of the banner-maker are as follows

3 ft cloth at Rs. 210 per roll

4 ft cloth at Rs. 350 per roll

5 ft cloth at Rs. 450 per roll

6 ft cloth at Rs. 580 per roll

The textile marketing company has only 17 ft of cloth to meet these orders. How many of each cloth type rolls should be marketed to maximise total profit?

This problem can be represented as a linear programming problem as follows

$$\underset{(d,\,S)}{\text{Maximise}}\{d_1P_1 + d_2P_2 + \cdots + d_nP_n\}$$

subject to

$$d_1C_1 + d_2C_2 + \ldots + d_nC_n \leq M$$

where,

$d_1, d_2 + \ldots + d_n$ are decision variables

$d_i \geq 0$ for $i = 1, 2, 3, \ldots, n$

P_i = Profit per roll

C_i = Consumption per roll

M = Maximum limit consumption

To solve this problem, we can make use of dynamic programming technique. As we discussed in the previous section, here also we have to define the problem stages, decision variables, input state variables, return function and transition function.

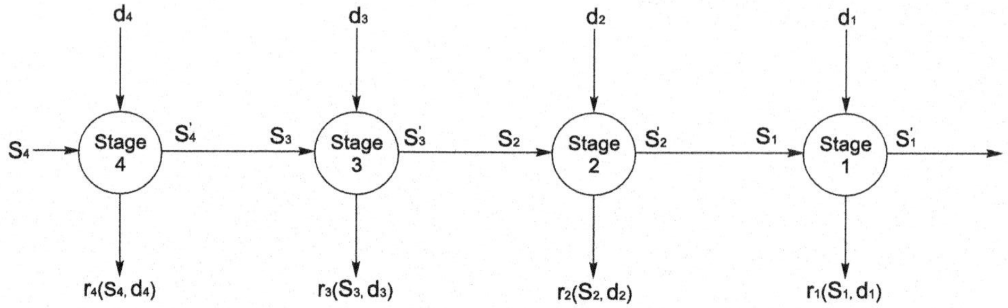

Figure 11.7 Multistage dynamic programming

In this problem the orders for different types of cloth rolls are the stage. The figure above illustrates the four-stage dynamic programming.

Input state variable (S_n) is the remaining length of cloth to meet the requirement of the order at stage n and all remaining stages.

The decision variable (d_n) is the number of rolls of cloth to be sold at each stage.

The return function r_n (S_n, d_n) is the revenue gained by selling 'd_n' rolls of the nth order.

The transaction function f_n' (S_n, d_n) is the amount of cloth available at the nth stage.

At the nth stage, the dynamic programming problem is given by

$$f_n'(S_n) = \underset{0 \leq d_n \leq}{\text{Maximum}} \left\{ r_n(S_n, d_n) * f_{n-1}'(S_{n-1}) \right\}$$

where

$S_{n-1} = S_n - d_n L_n$

L_n is the length of the cloth at stage 'n'

$$f_0'(S_0) = 0$$

As we mentioned earlier, Let us take different types of cloth rolls as stages.

i.e.,

Stage 1 ⇒ 3 ft cloth

Stage 2 ⇒ 4 ft cloth

Stage 3 ⇒ 5 ft cloth

Stage 4 ⇒ 6 ft cloth

Stage 1

$$f_1'(S_1) = \underset{0 \le S_1 d_1 \le \frac{S_1}{L_1}}{\text{Maximum}} \left\{ r_1(S_1, d_1) * f_0'(S_0) \right\}$$

$$\frac{S_1}{L_1} = \frac{17 \text{ ft}}{3 \text{ ft}} = 5 \text{ rolls}$$

$$\therefore d_1 \le 5 \text{ rolls}$$

For $d_1 = 0$ rolls, we get zero in the first column. First order requires a minimum of 3 ft cloth. Therefore with $S_1 = 0$ ft, 1 ft and 2 ft we cannot meet requirements. Put dash for $S_1 = 0,1$ and 2 in the second Column. For $d_1 = 1$ and $S_1 = 3$ ft and 5 ft we can make one roll of 3 ft cloth. It costs Rs. 210 per roll.

Similarly for $d_1 = 2$ rolls and $S_1 = 6,7$ and 8 ft we can make two rolls of 3 ft cloth. It costs Rs. 420 and so on.

Table 11.15 illustrates stage 1.

Table 11.15

S_1 ft	d_1 rolls						d_1'	$f_1'(S_1)$
	0	1	2	3	4	5		
0	0	–	–	–	–	–	0	0
1	0	–	–	–	–	–	0	0
2	0	–	–	–	–	–	0	0
3	0	Rs. 210	–	–	–	–	1	210
4	0	210	–	–	–	–	1	210
5	0	210	–	–	–	–	1	210
6	0	–	420 (3 ft × 2)	–	–	–	2	420
7	0	–	420	–	–	–	2	420
8	0	–	420	–	–	–	2	420
9	0	–	–	630	–	–	3	630
10	0	–	–	630	–	–	3	630
11	0	–	–	630	–	–	3	630
12	0	–	–	–	840	–	4	840
13	0	–	–	–	840	–	4	840
14	0	–	–	–	840	–	4	840
15	0	–	–	–		1050	5	1050
16	0	–	–	–		1050	5	1050
17		–	–	–		1050	5	1050

$f_1'(S_1)$ and d_1' are tabulated above.

Stage 2

$$f_2'(S_2) = \underset{0 \le d_2 \le \frac{S_2}{L_2}}{\text{Maximum}} \left\{ r_2(S_2, d_2) * f_1'(S_1) \right\}$$

$$\frac{S_2}{L_2} = \frac{17\text{ ft}}{4\text{ ft}} = 4 \text{ rolls}$$

$$d_2 \le 4 \text{ rolls}$$

For $d_2 = 0$ rolls, $f_2'(S_2) = f_1'(S_1)$. It is shown in the first column of Table 11.16. For $d_2 = 1$ roll and $S_2 = 0$, 1, 2 and 3, we cannot meet the requirements. Put dash for $S_2 = 0$, 1, 2 and 3, we cannot meet the requirements. Put dash for $S_2 = 0$, 1, 2 and 3 in the second column in Table 11.16. For $S_2 = 4$, 5 and 6 ft, we can make one roll of 4 ft cloth. It costs Rs. 350. For $S_2 = 7$, 8 and 9 ft, we can make one roll of 4 ft cloth and one roll of 3 ft cloth.

Similarly, for $S_2 = 10$, 11 and 12 ft, we can make two rolls of 3 ft cloth in addition to one roll of 4 ft. cloth. For $d_2 = 2$ rolls, we need minimum of $S_2 = 8$ ft, we can make one roll of 3 ft cloth in addition to two rolls of 4 ft cloth. The total cost is Rs. 910. For $d_3 = 3$ rolls, we need minimum of $S_2 = 12$ ft and it costs Rs. 1050 and so on.

Table 11.16

S_2 ft	d_2 rolls					d_2'	$f_2'(S_2)$
	0	1	2	3	4		
0	0	–	–	–	–	0	0
1	0	–	–	–	–	0	0
2	0	–	–	–	–	0	0
3	210	–	–	–	–	0	210
4	210	Rs. 350 (4 ft × 1)	–	–	–	1	350
5	210	350	–	–	–	1	350
6	420	350	–	–	–	0	420
7	420	560 (4 ft + 3 ft)	–	–	–	1	560
8	420	560 (4 ft × 2)	700	–	–	2	700
9	630	560	700	–	–	2	700
10	630	770 (4 ft + 3ft × 2)	700	–	–	1	770
11	630	770 (8ft + 3ft)	910	–	–	2	910
12	840	770 (4 ft × 3)	910	1050	–	3	1050
13	840	980 (4ft + 3ft × 3)	910	1050	–	3	1050
14	840	980 (8ft + 3ft × 2)	1120	1050	–	2	1120
15	1050	980 (12 ft + 3 ft)	1120	1260	–	3	1260
16	1050	1190 (4 ft × 3 ft × 4)	1120	1260 (4 ft × 4)	–	4	1400
17	1050	1190 (8 ft + 3 ft × 3)	1330	1260	1400	4	1400

Stage 3

$$f_3'(S_3) = \underset{0 \le d_3 \le \frac{S_3}{L_3}}{\text{Maximum}} \left\{ r_3(S_3, d_3) * f_2'(S_2) \right\}$$

$$\frac{S_3}{L_3} = \frac{17\text{ ft}}{5\text{ ft}} = 3 \text{ rolls}$$

$$d_3 \le 3 \text{ rolls}$$

Table 11.17

S_3 ft	d_3 rolls				d_3'	$f_3'(S_2)$
	0	1	2	3		
0	0	–	–	–	0	0
1	0	–	–	–	0	0
2	0	–	–	–	0	0
3	210	–	–	–	0	210
4	350	–	–	–	0	350
5	350	450 (5 ft × 1)	–	–	1	450
6	420	450	–	–	1	450
7	560	450	–	–	0	560
8	700	560 (5 ft + 3 ft)	–	–	0	700
9	700	800 (5 ft + 4 ft)	–	–	1	800
10	770	800	900 (5 ft × 2)	–	2	900
11	910	870 (5 ft + 3ft × 2)	900	–	0	910
12	1050	1010 (5 ft + 4 ft + 3 ft)	900	–	0	1050
13	1050	1150 (5 ft + 4ft × 2)	1110 (10 ft + 3ft)	–	1	1150
14	1120	1150	1250 (10 ft + 4ft)	–	2	1250
15	1260	1220 (5 ft + 4 ft + 3ft × 2)	1250	1350 (5 ft × 3)	3	1350
16	1400	1220	1320 (10 ft + 3ft × 2)	1350	0	1400
17	1400	1550 (5 ft + 4 ft × 3)	1460 (10 ft + 4 ft + 3ft)	1350	1	1500

Stage 4

$$f_4'(S_4) = \underset{0 \le d_4 \le \frac{S_4}{L_4}}{\text{Maximum}} \left\{ r_4(S_4, d_4) * f_3'(S_3) \right\}$$

$$\frac{S_4}{L_4} = \frac{17\text{ ft.}}{6\text{ ft.}} = 2 \text{ rolls}$$

$$d_4 \le 2 \text{ rolls}$$

Table 11.18

S_4 ft	d_4 rolls			d_4'	$f_4'(S_2)$
	0	1	2		
0	0	–	–	0	0
1	0	–	–	0	0
2	0	–	–	0	0
3	210	–	–	0	210
4	350	–	–	0	350
5	450	–	–	0	450
6	450	–	–	1	580
7	560	580(6 ft × 1)	–	1	580
8	700	580	–	0	700
9	800	580	–	0	800
10	900	790 (6 ft + 3 ft)	–	1	930
11	910	930(6 ft + 4 ft)	–	1	1030
12	1050	1030	1160(6 ft × 2)	2	1160
13	1150	1140(6 ft + 4 ft + 3 ft)	1160	2	1160
14	1250	1280(6 ft + 5 ft + 3 ft)	1160	1	1280
15	1350	1380(6 ft + 5 ft + 4 ft)	1370(12 ft + 3 ft)	1	1380
16	1400	1480(6 ft + 5 ft × 2)	1510(12 ft + 4 ft)	2	1510
17	1500	1480	1610(12 ft. + 5 ft.)	2	1610

$$f_4'(S_4) = \text{Rs. } 1610$$

The 17 mm roll can be cut into 2 rolls of 6 ft and 1 roll of 5 ft.

Chapter 12

SEQUENCING

12.1 INTRODUCTION

Sequencing is the process of selecting the parts of the product in an order in the transformation process from raw material to finished goods. The problem of sequencing has been encountered in production shops where different products are to be processed over various combination of machines.

Take the case of n number of jobs to be processed on m number of machines. Theoretically $(n!)^m$ number of sequences can be used to process n jobs on m machines. It is very difficult to solve such problems.

A general sequencing problem can be defined as follows. In the above problem of processing n jobs on m machines, the actual or expected time required by the jobs on each machine and the order of the machines for each job are known. Our problem is to find that sequence out of $(n!)^m$ possible sequences which minimises the total time, i.e., the time from start of the first job up to the completion of the last job should be minimum.

Let T be the total elapse time.

A_i is the time estimated for the ith job on the machine A.

Then we determine the sequence of jobs which minimises T.

Following are the different types of sequencing problems

1. n jobs are to be processed on 2 machines A and B in the order AB.
2. n jobs are to be processed on 3 machines A, B and C in the order ABC.
3. Two jobs are to be processed on m machines.

12.2 ASSUMPTIONS

1. Processing time of the jobs are independent of the order in which they are to be processed.
2. Time taken by the jobs to move from one machine to another is negligible.
3. Each job once started on the machine is to be processed up to the completion of that job.
4. A job starts on the machine as soon as the job and the machine are empty.

12.3 *n* JOBS AND 2 MACHINES

12.3.1 Algorithm

1. Select the smallest processing time, out of A_i's and B_i's where A and B are machines and $i = 1, 2, 3, \ldots, n$.

If it is A_r, do the rth job first. If it is B_s, do the sth job last.

2. Now $(n - 1)$ jobs are left over. Repeat step 1 with the remaining number of jobs.

Example 12.1

We have 5 jobs each of which has to go through the machines A and B in the order A, B. Processing times are given below.

Job no.	Processing time in hours	
	Machine A	Machine B
1	5	2
2	1	6
3	9	7
4	3	8
5	10	4

Determine a sequence of these jobs that will minimise the total elapse time.

Solution

Step 1 Select the smallest processing time.

Here it is job 2, having smallest processing time on machine A. Therefore, do it first and delete this job.

Step 2 Select the smallest processing from among the remaining jobs.

Here it is job 1, having smallest processing time on machine B. Therefore, do it last and delete this job.

Step 3 Repeat the above steps giving highest priority to jobs on machine A and least priority to jobs on machine B.

Sequence of the jobs

2	4	3	5	1

Table below illustrates the total elapse time and idle time of machine A and machine B.

Job no.	Machine A		Machine B		Idle time	
	Time in	Time out	Time in	Time out	A	B
2	0	1	1	7	–	1
4	1	4	7	15	–	–
3	4	13	15	22	–	–
5	13	23	23	27	–	1
1	23	28	28	30	2	1
Total idle time					2	3

Total elapse time = 30 hours

Idle time on Machine A = 2 hours

Idle time on Machine B = 3 hours

From the aforementioned table, it is observed that the job 3 is finished on machine B in 22 hours. Here machine is empty. But the next job, i.e., job 5 is under process on machine A. Machine B can process the job 5 after 23 hours only. Therefore, machine is idle for one hour.

Similarly, even though machine B is empty after 27 hours, it has to wait till job 1 is completed on machine A. Here also, the idle time for machine B is one hour.

The machine A is empty after 28 hours. But the total elapse time is 30 hours. Hence, machine A is idle for 2 hours.

Processing time of job 2 on machine A is one hour. Machine B can take up job 2 in sequence after one hour. Therefore, machine B is idle for one hour in the beginning.

Example 12.2

Find the sequence that minimises the total elapse time required to process the following jobs through the machines A and B in the order A, B.

Job no.	Processing time in hours	
	Machine A	Machine B
1	2	6
2	5	8
3	4	1
4	3	2
5	2	3
6	1	5

Solution

Step 1 Select the smallest processing time.

Here, it is job 6 having the smallest processing time of one hour on machine A and job 3 having the same processing time of one hour on machine B. Tie occurs. As the jobs are to be processed on machine A first, job 6 is processed first on machine A. Delete this job.

Step 2 Select the smallest processing time from among the remaining jobs. Job 3 is having the smallest processing time on machine B. Do it last and delete this job.

Step 3 Repeat the above steps giving the highest priority to jobs on machine A and least priority to jobs on machine B.

Sequence of jobs

6	1	5	2	4	3

or

6	5	1	2	4	3

The table below illustrates the total elapse time and idle time of machines A and B.

Job no.	Machine A		Machine B		Idle time	
	Time in	Time out	Time in	Time out	A	B
6	0	1	1	6	–	1
1	1	3	6	12	–	–
5	3	5	12	15	–	–
2	5	10	15	23	–	–
4	10	13	23	25	–	–
3	13	17	25	26	9	–
	Total idle time				9	1

Total elapse time = 26 hours

Idle time on Machine A = 9 hours

Idle time on Machine B = 1 hour

From the above table, it is found that the processing time of job 6 on machine A is one hour. Machine B can take up job 6 in sequence after one hour. Therefore, machine B is idle for one hour in the beginning. Machine A is empty after 17 hours. But the total elapse time is 26 hours. Hence, machine A is idle for 9 hours.

Example 12.3

A book binder has one printing press, one binding machine and the manuscripts of a number of different books. The time required to perform the printing and binding operations for each book are known. Determine the order in which books should be processed on the machines in order to minimise the total elapse time.

Book no.	Printing time	Binding time
1	30	80
2	120	100
3	50	90
4	20	60
5	90	30
6	110	10

Solution

Step 1 Select the book having the shortest processing time. If the printing time required is less, do it first. If the binding time is less do it last.

Step 2 Select the book 4 having the smallest printing time. Do it first and delete this job. Book 6 requires the shortest binding time. Do it last and delete this job.

Step 3 Repeat steps 1 and 2.

Sequence of Jobs

4	1	3	2	5	6

The table below illustrates the total elapse time and idle time of the printing machine and binding machine.

Book no.	Printing machine		Binding machine		Idle time	
	Time in	Time out	Time in	Time out	*P*	*B*
4	0	20	20	80	–	20
1	20	50	80	160	–	–
3	50	100	160	250	–	–
2	100	220	250	350	–	–
5	220	310	350	380	–	–
6	310	420	420	430	10	40
	Total idle time				10	60

Total elapse time = 430 hours

Idle time on printing machine = 10 hours

Idle time on Binding Machine = 60 hours

Here, printing time of book no. 4 is 20 hours, binding machine can take up book no. 4 in sequence after 20 hours. Therefore, binding machine is idle for 20 hours in the beginning. Binding of book no. 5 is finished in 380 hours. But the next book i.e., book no. 6 is ready by 420 hours. The binding machine is idle for 40 hours.

The printing machine is empty after 420 hours. But the total elapse time is 430 hours. Hence, printing machine is idle for 10 hours.

12.4 *n* JOBS AND 3 MACHINES IN THE ORDER ABC

If the problem is to determine the optimal sequence of n jobs to be performed on three machines A, B and C in order, such a problem can be solved by the method of n jobs and two machines under the following limitations.

1. The smallest processing time for machine A is greater than or equal to the greatest processing time for machine B.

or

2. The smallest processing time for machine C is greater than or equal to the greatest processing time for machine B.

To solve such a problem, we replace the three machines by two fictitious machines G and H with the corresponding process time given by

$$G_i = A_i + B_i$$
$$H_i = B_i + C_i$$

Example 12.4

Determine the optimal sequence of jobs which minimises the total elapse time based on the following information.

Job no.	Processing time in hours		
	Machine A	Machine B	Machine C
1	3	3	5
2	8	4	8
3	7	2	10
4	5	1	7
5	2	5	6

Solution

Step 1

$$\text{Check Min } A_i \geq \text{ Max } B_i$$

$$2 \ngeq 5$$

or

$$\text{Min } C_i \geq \text{ Max } B$$

$$5 \geq 5$$

The condition is satisfied.

Step 2 Replace three machines by two machines.

i.e.,

$$G_i = A_i + B_i$$

$$H_i = B_i + C_i$$

Job no.	Processing time in hours	
	Machine G	Machine H
1	6	8
2	12	12
3	9	12
4	6	8
5	7	11

Step 3 Determine the sequence of jobs on fictitious machines G and H using two machines and n jobs method.

1	4	5	3	2

Step 4 Determine the total elapse time and idle time on machines A, B and C using the sequence obtained.

The following table illustrates the total elapse time and idle time of machines A, B and C.

Job no.	Machine A		Machine B		Machine C		Idle time		
	Time in	Time out	Time in	Time out	Time in	Time out	A	B	C
1	0	3	3	6	6	11	–	3	6
4	3	8	8	9	11	18	–	2	–
5	8	10	10	15	18	24	–	1	–
3	10	17	17	19	24	34	–	2	–
2	17	25	25	29	34	42	17	19	–
	Elapse time						17	27	6
Total elapse time = 42 hours									

Example 12.5

A book binder has to perform printing, binding and finishing operations on a number of different books in the order, printing, binding and finishing. The time required to perform these operations are known. Determine the sequence in which the books should be processed in order to minimise the total time required.

Books	Printing time in hours	Binding time in hours	Finishing time in hours
1	4	5	8
2	9	6	10
3	8	2	6
4	6	3	7
5	5	4	11

Solution

Step 1 Name printing, binding and finishing processes as A, B and C respectively and check

$$\text{Min } A_i \geq \text{Max } B_i$$

$$\text{i.e., } 412 \ngeq 6$$

Or

$$\text{Min } C \geq \text{Max } B_i$$

i.e.,

$$6 \geq 6$$

Condition is satisfied.

Step 2 Replace three machines by two machines.

i.e.,

$$G_i = A_i + B_i$$

$$H_i = B_i + C_i$$

Job no.	Processing time in hours	
	Machine G	Machine H
1	9	13
2	15	16
3	10	8
4	9	10
5	9	15

Step 3 Determine the sequence of jobs on fictitious machines G and H using the method of n jobs and two machines.

1	4	5	2	3

Step 4 Determine the total elapse time and idle time on the printing machine (A), binding machine (B) and finishing machine (C) using the sequence obtained. The table below illustrates the total elapse time and idle time of machines A, B and C.

Book	Printing–Machine A		Binding– Machine B		Finishing– Machine C		Idle time		
	Time in	Time out	Time in	Time out	Time in	Time out	A	B	C
1	0	4	4	9	9	17	–	4	9
4	4	10	10	13	14	24	–	1	–
5	10	15	15	19	24	35	–	2	–
2	15	24	24	30	30	45	–	5	–
3	24	32	32	34	45	51	19	19	–
	Elapse time						19	31	9
Total elapse time = 51 hours									

12.5 TWO JOBS AND n MACHINES

When two jobs are to be processed on n machines, the graphical method is used to determine the optimum sequence and optimum time required to complete each job.

Step 1 The bar chart is plotted for n machines on a graph with Job 1 along X-axis and Job-2 along Y-axis with reference to the sequence and processing time on different machines is given in the problem.

Step 2 Paths are drawn to determine the sequence of machines to complete the jobs with the minimum time.

Step 3 Priority of Job-1 and Job-2 on n machines is determined from the graph.

Example 12.6

Use graphical method to minimise the time needed to process the following jobs on the machines shown. For each machine, find the job which should be done first. Also, calculate the time needed to complete the jobs.

Job-I	Sequence	A	B	C	D
	Time	2	4	5	1
Job-II	Sequence	D	B	A	C
	Time	2	5	3	6

Solution

Step 1 Plot the graph of Job-I vs Job-II and draw the bars for machines with reference to sequence and time given.

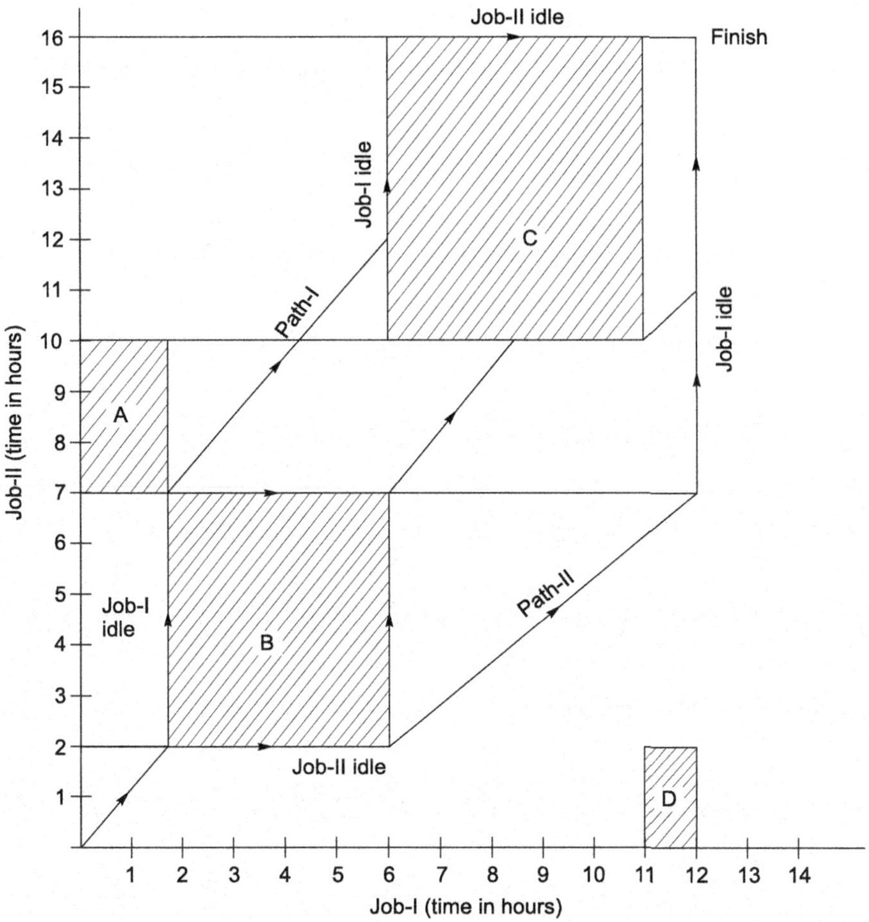

Figure 12.1 Sequence of jobs and processing time on different machines

Step 2 Identify the paths to perform Job-1 and Job-2 on machines and determine the idle time and the total elapse time for each job.

Path-I from the graph

Idle time for Job-I	=	5 + 5 = 10 hours
Total elapse time for Job-I	=	22 hours
Idle time for Job-II	=	6 hours
Working hours for Job-II	=	16 hours
Total elapse time for job-II	=	22 hours

***Note* 1** Total length of vertical lines along path-I represents the idle time for Job-I and the total length of horizontal lines along path-I represents the idle time for Job-II.

***Note* 2** Idle time for Job-I is the working time for Job-II and vice versa.

***Note* 3** Inclined line along path-I represents working time for both Job-I and Job-II.

Similarly

Path-II from the graph

Idle time for Job-I = 8 hours

Working time for Job-I = 12 hours

Total elapse time for Job-I = 20 hours

Idle time for Job-II = 4 hours

Working time for Job-II = 16 hours

Total elapse time for Job-II = 20 hours

Therefore, path-II is better minimum time route and the total elapse time is 20 hours.

Step 3 Determine the priority of jobs on each machine with reference to path-II.

On machine A, Job-I is done first.

On machine B, Job-I is done first.

On machine C, Job-I is done first.

On machine D, Job-II is done first.

Example 12.7

The time spent in processing two jobs on six machines A, B, C, D, E and F and necessary technological order of machines are as follows

Job-I	$A-20$	$C-10$	$D-10$	$B-30$	$E-25$	$F-15$
Job-II	$A-10$	$C-30$	$B-15$	$D-10$	$F-15$	$E-20$

Use graphical method to determine an optimal sequence of jobs which minimises the elapse time.

Solution

Step 1 Plot the graph of Job-I vs Job-II and draw the bars for machines with reference to the sequence and the time given.

Step 2 Identify the paths to perform job one and job two on machines and determine idle time and total elapse time for each job.

Path-I from the graph

Idle time for Job-I = $10 + 10 + 15 = 35$ hours

Total elapse time for Job-I = 145 hours

Idle time for Job-II = $40 + 5 = 45$ hours

Total elapse time for Job-II = 145 hours

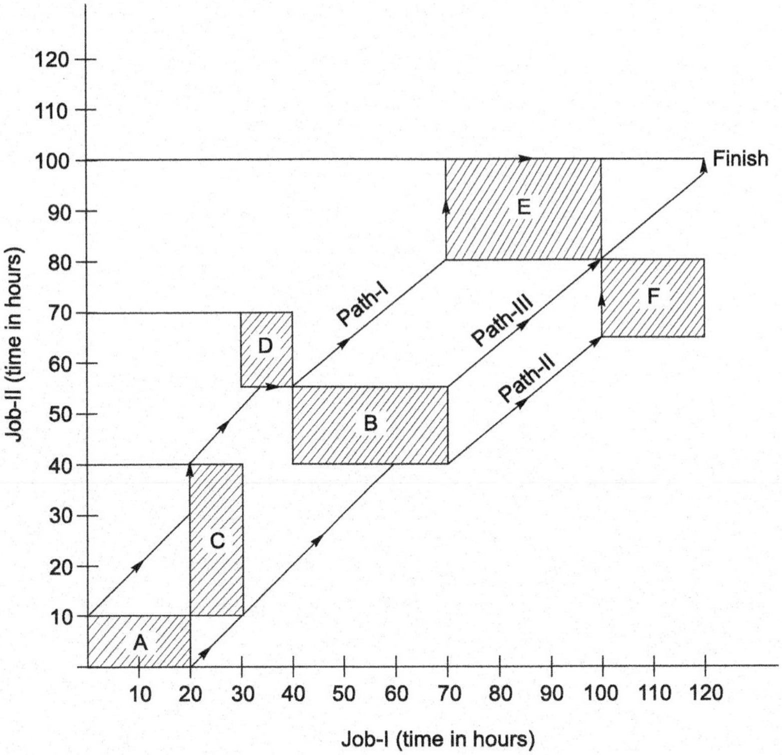

Figure 12.2 Sequence of jobs and processing time on different machines

Similarly,

Path-II from the graph

Idle time for Job-I = 15 + 5 = 20 hours

Total elapse time for Job-I = 130 hours

Idle time for Job-II = 20 + 10 = 30 hours

Total elapse time for Job-II = 130 hours

Therefore path-II is the shortest path and total elapse time is 130 hours.

Step 3 Determine the priority of jobs on each machine with respect to path-II.

On machine *A*, Job-I is done first.

On machine *B*, Job-I is done first.

On machine *C*, Job-I is done first.

On machine *D*, Job-I is done first.

On machine *E*, Job-I is done first.

On machine *F*, Job-II is done first.

EXERCISES

1. Six jobs go on machine-I and then machine-II. The order of completion of the job has no significance. The following table gives the machining time in hours for six jobs and two machines.

Job no.	Machine-I	Machine-II
1	5	7
2	9	4
3	4	8
4	7	3
5	8	9
6	6	5

Find the significance of the jobs that minimises the total elapse time to complete all the jobs.

2. Find the sequence that minimises the total elapse time required to complete the following jobs.

Job no.	1	2	3	4	5	6
Machine *A*	4	8	3	6	7	5
Machine *B*	6	3	7	2	8	4

3. We have five jobs each of which must go through the two machines in the order *A*, *B*. Processing time is given below. Find the sequence of jobs.

Job no.	1	2	3	4	5
Machine *A*	10	2	18	6	20
Machine *B*	4	12	14	16	8

4. Determine the optimal sequence of jobs, which minimises the total elapse time based on the following information.

Job no.	Machine *A*	Machine *B*	Machine *C*
1	2	5	4
2	6	9	8
3	7	4	3
4	8	7	6
5	4	8	5

5. Two jobs are to be processed on four machines *A*, *B*, *C*, and *D*. The technological order of these jobs on machines are as follows.

Job-I	*A*	*B*	*C*	*D*
Job-II	*D*	*B*	*A*	*C*

Processing times are as follows

Job-I	$A = 4$	$B = 6$	$C = 7$	$D = 3$
Job-II	$A = 4$	$B = 7$	$C = 5$	$D = 8$

Find the optimal sequence of jobs on each of the machines and the optimal total time.

6. Use graphical method to minimise the time needed to process the following jobs on machines A, B, C, D and E. For each machine, find the job which should be done first. Also calculate the total time needed to complete the job.

Job-I	Sequence	A	B	C	D	E
	Time in minutes	3	4	2	6	2
Job-II	Sequence	B	C	A	D	E
	Time in minutes	5	4	3	2	6

7. Use graphical method to minimise the time needed to process the following jobs on machines shown. For each machine, find the job which is to be done first. Also, calculate the time needed to complete the jobs.

Job I	Sequence	A	B	C	D
	Time in minutes	2	4	4	1
Job-II	Sequence	D	B	A	C
	Time in minutes	2	6	3	5

8. The time spent in processing two jobs on five machines A, B, C, D and E, and the technological ordering of machines are as follows

Job-I	$A = 20$	$C = 10$	$D = 10$	$B = 10$	$E = 30$
Job-II	$B = 15$	$A = 10$	$C = 30$	$D = 10$	$E = 20$

Use graphical method to determine an optimal sequence of jobs which minimises the total elapse time.

Chapter 13

INTEGER PROGRAMMING PROBLEMS

13.1 INTRODUCTION

Linear programming is a method used to determine optimum quantity of products to be manufactured or procured considering the limited resources that are available to the managers during a specified period of time. If the items under consideration are two, graphical solution method is used to find the quantity of products to be manufactured. If the items are more than two, simplex method is used to find the quantity of products to be manufactured. In order to check the feasibility of the solution, we use dual simplex method.

In all these cases, we obtain the solution to either maximise or minimise the Z value. The quantity obtained will not be an integer normally. e.g., $X_1 = \frac{3}{2}, X_2 = \frac{15}{4}, X_3 = \frac{81}{17}$, etc.

If the products are measurable in kilograms, litres, and metres, the solution obtained by solving LPP is appropriate. If the products are unit quantities, the quantity of products obtained should be integers, e.g., pen, scooter, bus, lathe, dolls, etc. If the solution of an LPP is not an integer, accurate results cannot be obtained. In order to overcome this limitation, the integer programming problem technique has been developed.

Mathematically, integer programming problem is represented as follows

$$\text{Optimize } f(x) = C_1x_1 + C_2x_2 + \ldots + C_nx_n$$

subject to

$$a_{i1}x_1 + a_{i2}x_2 + \ldots + a_{ij}x_{ij} = b_i$$

where i varies from 1 to m constraints

$x_j \geq 0$ j varies from 1 to n variables

x_j are integer values.

13.2 GOMORY'S CONSTRAINT

Integer programming problem technique is a modified method of the simplex technique. The solution of a given problem is obtained initially by the simplex method. The equation with real number is split into integer part and fraction part. Then using fractional part, an equation is obtained.

For example,

$$0x_1 + 1x_2 + \frac{3}{10} S_1 - \frac{1}{10} S_2 = \frac{1}{10}$$

$$(0+0)x_1 + (1+0)x_2 + \left(0 + \frac{3}{10}\right) S_1 + \left(-1 + \frac{9}{10}\right) S_2 = 1 + \frac{4}{5}$$

i.e.,

$$\frac{3}{10} S_1 + \frac{9}{10} S_2 \geq \frac{4}{5}$$

Rewritten as

$$-\frac{3}{10} S_1 - \frac{9}{10} S_2 \leq -\frac{4}{5}$$

A slack variable g_1 is used to bring this inequality to equality.

The equation obtained is Gomorian constraint.

$$-\frac{3}{10} S_1 - \frac{9}{10} S_2 + g_1 = -\frac{4}{5}$$

13.3 PURE AND MIXED INTEGER PROGRAMMING PROBLEMS

If all the variables are restricted to take only integer value in an integer programming problem, it is called pure integer programming problem.

If only some variable is restricted to take integer value in an integer programming problem, it is called mixed integer programming problem.

13.4 ZERO–ONE PROGRAMMING PROBLEM

If all the variables are allowed to take values either one or zero in an integer programming problem, it is called zero–one programming problem.

13.5 GOMORY'S ALL-INTEGER CUTTING PLANE METHOD

Following are the steps involved in solving an all-integer programming problem by Gomory's method.

Step 1 Solve the given problem by using simplex method and obtain optimum solution.

Step 2 If all the variables in the solution have integer value, accept the solution of step 1. If at least one variable in the optimal solution of step 1 has fractional value, introduce a secondary constraint called Gomory's constraint in the basic table.

Step 3 Find the new optimum solution by dual simplex method.

Step 4 If the solution value of the variable is an integer, same is the solution of the given integer programming problem. If not, repeat step 2 and introduce Gomory's constraint in the basic table and repeat step 3.

This method of finding solution for an integer programming problem is also called cutting plane method as the Gomory's constraint introduced, cuts the area of the feasible region in the graphical solution of the problem which has no integer valued solution. The Gomory's constraints convert the non-integer solution to integer solution.

Example 13.1

Channapatna Toys is manufacturing two types of dolls on machine A and machine B. Doll X requires 2 hours on machine A and 6 hours on machine B. Doll Y takes 5 hours on machine A and 5 hours on machine B. The time available on machine A is 16 hours and that on machine B is 30 hours. Profit for each doll is Re. 1. Determine the number of dolls of both the types to be manufactured to maximise the profit. Use LPP to solve the problem.

Solution

Formulation

$$\text{Maximise } Z = X_1 + X_2$$

subject to

$$\begin{aligned} 2X_1 + 5X_2 &\le 16 \\ 6X_1 + 5X_2 &\le 30 \\ X_1 &\le 0 \\ X_2 &\le 0 \end{aligned}$$

Add slack variable and bring inequality to equality.

$$\begin{aligned} 2X_1 + 5X_2 + S_1 &= 16 \\ 5X_1 + 5X_2 + S_2 &= 30 \end{aligned}$$

Here, there are four variables and two constraints Therefore, keep 4 – 2 = 2 variables at zero.

Let us keep X_1 and X_2 at zero.

Thus $S_1 = 16$

$S_2 = 30$

Objective function $Z = X_1 + X_2$

$$Z - X_1 - X_2 = 0$$

Write the simplex problem in table form.

Basic	X_1	X_2 ↓PC	S_1	S_2	Solution	
Z	–1	–1	0	0	0	
S_1	2	5*	1	0	16	$\frac{16}{5} = 3.2$ →PR
S_2	6	5	0	1	30	$\frac{30}{5} = 6$

Look for the maximum value with a negative sign in the Z row. It is –1. Consider X_2 variable corresponding to –1. X_2 tries to enter the basic column. This X_2 column is called a pivot column.

Which leaves the basic column?

Ignore all zeroes and negative values in the pivot column. Divide the elements in the solution column by the corresponding elements in the pivot column.

i.e.,
$$\frac{16}{5} = 3.2$$

and
$$\frac{30}{5} = 6$$

Select the minimum of these two. It is 3.2. This row is called pivot row.

Now X_2 enters the basic column and S_1 leaves.

An element which is common to both pivot row and pivot column is called pivot element. In the previous table, 5 is the pivot element. It is common to both pivot row S_1 and pivot column X_2. (* represents pivot element). The solution is optimum when all the values in the 'Z' row are either positive or zero. Here there are two variables with negative sign. Therefore, the solution is not optimum.

Iteration 1

$$\text{NPE} = \frac{\text{OPE}}{\text{PE}}$$

$$\text{NPE}: \quad \frac{2}{5} \quad 1 \quad \frac{1}{5} \quad 0 \quad \frac{16}{5}$$

For all other equations,

$$\text{New equation} = \text{Old equation} - \begin{pmatrix}\text{Corresponding} \\ \text{pivot column element}\end{pmatrix} \times (\text{NPE})$$

Old Z	-1	-1	0	0	0
$-(-1)$ NPE	$\frac{2}{5}$	1	$\frac{1}{5}$	0	$\frac{16}{5}$
New Z	$-\frac{3}{5}$	0	$\frac{1}{5}$	0	$\frac{16}{5}$
Old S_2	6	5	0	1	30
-5 NPE	-2	-5	-1	0	-16
New S_2	4	0	-1	1	14

Formulate the new basic table.

Basic	X_1 ↓PC	X_2	S_1	S_2	Solution	
Z	$-\frac{3}{5}$	0	$\frac{1}{5}$	0	$\frac{16}{5}$	
S_1	$\frac{2}{5}$	1	$\frac{1}{5}$	0	$\frac{16}{5}$	8
S_2	4_*	0	-1	1	14	$\frac{14}{4} = 3.5$ →PR

Z row has a negative value $-\frac{3}{5}$. Therefore, the solution is not optimum.

Iteration 2

X_1 enters the basic column. X_1 is the pivot column. Ignore –ve value in the pivot column. Divide the elements in the solution column by the corresponding element in the pivot column.

i.e.,

$$\frac{\frac{16}{5}}{\frac{2}{5}} = 8$$

and

$$\frac{14}{4} = 3.5$$

Select the minimum of these two. It is 3.5. Row S_2 is called pivot row.

NPE	1	0	$-\frac{1}{4}$	$\frac{1}{4}$	$\frac{7}{2}$
Old Z	$-\frac{3}{5}$	0	$\frac{1}{5}$	0	$\frac{16}{5}$
$-\left(-\frac{3}{5}\right)$ NPE	$\frac{3}{5}$	0	$-\frac{3}{20}$	$\frac{3}{20}$	$\frac{21}{10}$
New Z	0	0	$\frac{1}{20}$	$\frac{3}{20}$	$\frac{53}{10}$
Old X_2	$\frac{2}{5}$	1	$\frac{1}{5}$	0	$\frac{16}{5}$
– NPE	$-\frac{2}{5}$	0	$\frac{2}{20}$	$-\frac{2}{20}$	$-\frac{28}{20}$
New X_2	0	1	$\frac{3}{10}$	$-\frac{1}{10}$	$\frac{9}{5}$

Formulate the new basic table.

Basic	X_1	X_2	S_1	S_2	Solution
Z	0	0	$\frac{1}{20}$	$\frac{3}{20}$	$\frac{53}{10}$
X_2	0	1	$\frac{3}{10}$	$-\frac{1}{10}$	$\frac{9}{5}$
X_1	1	0	$-\frac{1}{4}$	$\frac{1}{4}$	$\frac{7}{2}$

All the values in the Z row are either positive or zero; therefore, the solution is optimum.

$$\text{Max } Z = \frac{11}{10}$$

subject to

$$X_1 = \frac{7}{2}$$

$$X_2 = \frac{9}{5}$$

Here, both the variables are fractional. But their fractional parts are $\frac{4}{5}$ and $\frac{1}{2}$. Out of these, $\frac{4}{5}$ is the largest, which is of x_2 lying in the first row of the table. Taking the first row as the source row, the corresponding equation is

$$0x_1 + 1x_2 + \frac{3}{10}S_1 - \frac{1}{10}S_2 = \frac{9}{5}$$

Splitting the integer and fractional parts, we get

$$x_2 + \left(0 + \frac{3}{10}\right)S_1 + \left(-1 + \frac{9}{10}\right)S_2 = 1 + \frac{4}{5}$$

Considering the fractional part,

$$\frac{3}{10}S_1 + \frac{9}{10}S_2 \geq \frac{4}{5}$$

Multiplying both sides by minus sign,

$$-\frac{3}{10}S_1 - \frac{9}{10}S_2 \leq -\frac{4}{5}$$

Hence the Gomory's constraint is

$$-\frac{3}{10}S_1 - \frac{9}{10}S_2 + g_1 = \frac{-4}{5}$$

where G_1 is the slack variable.

Adding this secondary constraint equation in the optimum table, we get the modified new table as follows

↓ PC

Basic	X_1	X_2	S_1	S_2	g_1	Solution	
Z	0	0	$\frac{1}{20}$	$\frac{3}{20}$	0	$\frac{53}{10}$	
X_2	0	1	$\frac{3}{10}$	$-\frac{1}{10}$	0	$\frac{9}{5}$	
X_1	1	0	$-\frac{1}{4}$	$\frac{1}{4}$	0	$\frac{7}{2}$	
g_1	0	0	$-\frac{3}{10}$ *	$-\frac{9}{10}$	1	$-\frac{4}{5}$	→ PR

This table shows variable solution but $Z \geq 1$ for all values. Therefore dual simplex algorithm can be used to solve the problem. Compute $-\frac{Z}{g}$ (i.e., divide Z row by $-g_1$) and ignore zeroes.

$$\frac{-0}{0}, \frac{-0}{0}, \frac{\frac{1}{20}}{\left(\frac{-3}{10}\right)}, \frac{\frac{3}{20}}{\left(\frac{-9}{10}\right)}, \frac{-0}{1}$$

We get $\quad 0 \quad 0 \quad \frac{1}{6} \quad \frac{1}{6} \quad 0$

Here, G_1 row with negative solution value leaves. It is pivot row. Column with minimum positive value in $-\frac{Z}{g_1}$ equation enters. It is column S_1. S_1 is the pivot column. $-\frac{3}{10}$ is the pivot element.

NPE	0	0	1	3	$\frac{10}{3}$	$\frac{8}{3}$
Old x_2	0	1	$\frac{3}{10}$	$\frac{1}{10}$	0	$\frac{9}{5}$
$-\frac{3}{10}$ NPE	0	0	$-\frac{3}{10}$	$-\frac{9}{10}$	1	$\frac{4}{5}$
New x_2	0	1	0	-1	1	1
Old x	1	0	$-\frac{1}{4}$	$\frac{1}{4}$	0	$\frac{7}{2}$
$-\left(-\frac{1}{4}\right)$ NPE	0	0	$\frac{1}{4}$	$\frac{3}{4}$	$-\frac{5}{6}$	$\frac{2}{3}$
New x_1	1	0	0	1	$-\frac{5}{6}$	$\frac{25}{6}$
Old Z	0	0	$\frac{1}{20}$	$\frac{3}{20}$	0	$\frac{53}{10}$
$\frac{-1}{20}$ NPE	0	0	$-\frac{1}{20}$	$-\frac{3}{20}$	$\frac{1}{6}$	$-\frac{2}{15}$
New Z	0	0	0	0	$\frac{1}{6}$	$\frac{31}{6}$

Formulate the new basic table.

Basic	x_1	x_2	S_1	S_2	g_1	Solution
Z	0	0	0	0	$\frac{1}{6}$	$\frac{31}{6}$
X_2	0	1	0	–1	1	1
X_1	1	0	0	1	$-\frac{5}{6}$	$\frac{25}{6}$
S_1	0	0	1	3	$-\frac{10}{3}$	$\frac{8}{3}$

The optimum solution thus obtained is still non-integer. Now, again we have to construct a Gomory's constraint. This time, taking second row, having fractional variable $x_1 = \frac{25}{6}$ as a source row, we get the corresponding equation.

$$x_1 + S_2 - \frac{5}{6} g_1 = \frac{25}{6}$$

Splitting the integer and fractional part, we get

$$x_1 + S_2 + (-1 + \frac{1}{6}) g_1 = 4 + \frac{1}{6}$$

Considering the fractional part

$$\frac{1}{6} g_1 \geq \frac{1}{6}$$

Multiplying both sides by minus sign, we get

$$-\frac{1}{6} g_1 \leq -\frac{1}{6}$$

Hence the Gomory's constraint is

$$-g_1 + g_2 = -1$$

where, g_2 is the second slack variable.

Adding this constraint equation in the above optimum table, we get the modified new table as follows

Basic	x_1	x_2	S_1	S_2	g_1 ↓PC	g_2	Solution	
Z	0	0	0	0	$\frac{1}{6}$	0	$\frac{31}{6}$	
x_2	0	1	0	–1	1	0	1	
x_1	1	0	0	1	$-\frac{5}{6}$	0	$\frac{25}{6}$	
S_1	0	0	1	3	$-\frac{10}{3}$	0	$\frac{8}{3}$	
g_2	0	0	0	0	–1*	1	–1	→PR

Divide Z row by $-g_2$ and ignore zeroes.

$$-\frac{Z}{g_2}: \quad 0 \quad 0 \quad 0 \quad 0 \quad -\frac{1}{6} \quad 0 \quad \frac{-31}{6}$$

Row g_2 with negative solution value leaves and column g_1 with minimum $-\frac{Z}{g_2}$ value enters. The pivot element is -1.

New pivot equation

(NPE)	0	0	0	0	1	-1	1
Old Z	0	0	0	0	$\frac{1}{6}$	0	$\frac{31}{6}$
$-\frac{1}{6}$ (NPE)	0	0	0	0	$-\frac{1}{6}$	$\frac{1}{6}$	$-\frac{1}{6}$
New Z	0	0	0	0	0	$\frac{1}{6}$	5
Old x_2	0	1	0	-1	1	0	1
-1 (NPE)	0	0	0	0	-1	1	-1
New x_2	0	1	0	-1	0	1	0
Old x_1	1	0	0	1	$-\frac{5}{6}$	0	$\frac{25}{6}$
$-\frac{5}{6}$ (NPE)	0	0	0	0	$\frac{5}{6}$	$-\frac{5}{6}$	$\frac{5}{6}$
New x_1	1	0	0	1	0	$-\frac{5}{6}$	5
Old S_1	0	0	1	3	$-\frac{10}{3}$	0	$\frac{8}{3}$
$\frac{10}{3}$ (NPE)	0	0	0	0	$\frac{10}{3}$	$-\frac{10}{3}$	$\frac{10}{3}$
New S_1	0	0	1	3	0	$-\frac{10}{3}$	6

Formulate the basic table

Basic	x_1	x_2	S_1	S_2	g_1	g_2	Solution
Z	0	0	0	0	0	$\frac{1}{6}$	5
x_2	0	1	0	–1	0	1	0
x_1	1	0	0	1	0	$-\frac{5}{6}$	5
S_1	0	0	1	3	0	$-\frac{10}{3}$	6
g_1	0	0	0	0	1	–1	1

At this iteration, optimum solution with integer values is obtained. The solution is

$$\text{Max } Z = 5$$

subject to

$$x_2 = 0, x_1 = 5$$

Example 13.2

The owner of a readymade garments store makes two types of shirts known as zee shirts and button down shirts. He makes a profit of Re. 1 and Rs. 4/shirt on zee and button down shirts respectively. He has two tailors A and B at his disposal to stitch these shirts. Tailor A and tailor B can devote atmost 7 hours and 15 hours per day, respectively. Both these shirts are to be stitched by both the tailors. Tailor A and B spend two hours and five hours respectively in stitching the zee shirts and they take four hours and three hours respectively in stitching a button down shirt. How many shirts of both the types should be stitched in order to maximise daily profits.

(a) Set up and solve the LPP.

(b) If the optimal solution is not integer-valued, use the Gomory's technique to derive the optimal integer solution.

Solution

Formulation

$$\text{Maximise } Z = X_1 + X_2$$

subject to

$$2X_1 + 4X_2 \leq 7$$
$$5X_1 + 3X_2 \leq 15$$
$$X_1, X_1 \geq 0$$

Add slack variable and bring inequalities to equality.

$$2X_1 + 4X_2 + S_1 = 7$$
$$5X_1 + 3X_2 + S_2 = 15$$

Here, there are four variables and two constraints. Therefore, keep 4 – 2 = 2 variables at zero.

Let us keep X_1 and X_2 at zero.

Then

$$S_1 = 7$$
$$S_2 = 15$$
$$\text{Objective function } Z = X_1 + 4X_2$$
$$Z - X_1 - 4X_2 = 0$$

Write the simplex problem in table form

↓PC

Basic	x_1	x_2	S_1	S_2	Solution	
Z	–1	–4	0	0	0	
S_1	2	4*	1	0	7	$\frac{7}{4} = 1.75$ → PR
S_2	5	3	0	1	15	$\frac{15}{3} = 5$

Look for the maximum value with a negative sign in the Z row. It is –4. Consider X_2 variable corresponding to –1. X_2 tries to enter the basic column. This X_2 column is called a pivot column.

Which leaves the basic column?

Ignore all zeroes and negative values in the pivot column. Divide the elements in the solution column by the corresponding elements in the pivot column.

i.e.,

$$\frac{7}{4} = 1.75$$
$$\frac{15}{3} = 5$$

Select the minimum of these two. It is 1.75.

This row is called the pivot row. Now X_2 enters the basic column and S_1 leaves.

An element which is common to both pivot row and the pivot column is called pivot element. In the above table, 4 is the pivot element. It is common to both pivot row S_1 and pivot column X_2. The * symbol represents the pivot element.

The solution is optimum when all the values in the 'Z' row are either positive or zero. Here, there are two variables with negative sign. Therefore, the solution is not optimum.

Iteration 1

$$\text{NPE} = \frac{\text{OPE}}{\text{PE}}$$

NPE: $\frac{1}{2}$ 1 $\frac{1}{4}$ 0 $\frac{7}{4}$

For all other equations, new equations are determined as follows

Old Z	–1	–4	0	0	0
–(–4) NPE	2	4	1	0	7
New Z	1	0	1	0	7
Old S_2	5	3	0	1	15
–(3) NPE	$-\frac{3}{2}$	–3	$-\frac{3}{4}$	0	$-\frac{21}{4}$
New S_2	$\frac{7}{2}$	0	$-\frac{3}{4}$	1	$-\frac{39}{4}$

Formulate the new basic table.

Basic	x_1	x_2	S_1	S_2	Solution
Z	1	0	1	0	7
X_2	$\frac{1}{2}$	1	$\frac{1}{4}$	0	$\frac{7}{4}$
S_2	$\frac{7}{2}$	0	$-\frac{3}{4}$	1	$\frac{39}{4}$

Solution is optimum but the X_2 variable has fractional solution value $\frac{7}{4}$.

Taking X_2 row as the source row, the corresponding equation is.

$$\frac{1}{2}X_1 + X_2 + \frac{1}{4}S_1 = \frac{7}{4}$$

Splitting the integer and fractional parts, we get

$$\left(0+\frac{1}{2}\right)X_1 + X_2 + \left(0+\frac{1}{4}\right)S_1 = 1+\frac{3}{4}$$

Considering fractional part

$$\frac{1}{2}X_1 + \frac{1}{4}S_1 \geq \frac{3}{4}$$

Multiplying both sides by minus sign

$$-\frac{1}{2}X_1 - \frac{1}{4}S \leq -\frac{3}{4}$$

Hence the Gomory's constraints is

$$-\frac{1}{2}X_1 - \frac{1}{4}S + g_1 \leq -\frac{3}{4}$$

where, g_1 is the slack variable.

Adding this secondary constraint in the above optimum table, we get the modified new table as follows

↓PC

Basic	x_1	x_2	S_1	S_2	g_1	Solution
Z	1	0	1	0	0	7
x_2	$\frac{1}{2}$	1	$\frac{1}{4}$	0	0	$\frac{7}{4}$
S_2	$\frac{7}{2}$	0	$\frac{7}{2}$	1	0	$\frac{39}{4}$
g_1	$-\frac{1}{2}$	0	$-\frac{1}{4}$ *	0	1	$-\frac{3}{4}$ →PR

This table shows variable solution but $Z \geq 1$ for all values. Therefore, the dual simplex algorithm can be used to solve the problem.

Compute $\frac{-z}{g}$ (i.e., divide Z row by $-g_1$) and ignore zeroes.

We get $-\frac{Z}{g} =$ 2 0 4 0 0 $-\frac{28}{3}$

Here g_1 row with negative solution value leaves. It is the pivot row. Column with minimum (negative) value in $-\frac{Z}{g}$ equation enters. It is column S_1. S_1 is the pivot column. The pivot element is $-\frac{1}{4}$.

NPE	2	0	1	0	– 4	3
Old Z	1	0	1	0	0	7
–(1) NPE	–2	0	–1	0	4	–3
New Z	–1	0	0	0	4	4
Old X_2	$\frac{1}{2}$	1	$\frac{1}{4}$	0	0	$\frac{7}{4}$
$-\frac{1}{4}$ NPE	$-\frac{1}{2}$	0	$-\frac{1}{4}$	0	1	$-\frac{3}{4}$
New X_2	0	1	0	0	1	1
Old S_2	$\frac{7}{2}$	0	$\frac{3}{4}$	1	0	$\frac{39}{4}$
$-\left(\frac{3}{4}\right)$ NPE	$-\frac{3}{2}$	0	$-\frac{3}{4}$	0	3	$-\frac{9}{4}$
New S_2	2	0	0	1	3	$\frac{30}{4}$

Formulate the new basic table,

Basic	x_1	x_2	S_1	S_2	g_1	Solution
Z	–1	0	0	0	4	4
x_2	0	1	0	0	1	1
S_2	2	0	0	1	3	$\frac{30}{4}$
S_1	2	0	1	0	–4	3

At this iteration solution with integer values is obtained. The solution is

$$\text{Max } Z = 4$$

subject to

$$X_2 = 1$$
$$X_1 = 0$$

Example 13.3

A washing powder manufacturing company gives advertisement on radio and television to improve its share in the market. The company pays Rs.15 per minute for advertisement on radio and Rs. 300 per minute of advertisement on TV. It is estimated that the sales potential on TV is 25 times more than that on radio. The management is interested to advertise its product on radio at least twice that on TV. Determine the optimum time of advertisement to be given on radio and TV to maximise the washing powder sales.

Solution

Formulation

$$\text{Maximise } Z = 25X_1 + X_2$$

subject to

$$15X_1 + 300X_2 \leq 500$$
$$X_1 \geq 2X_2$$
$$X_1 - 2X_2 \leq 0$$
$$X_1 \geq 0$$
$$X_2 \geq 0$$

Bringing inequality to equality, we get

$$15X_1 + 300X_2 + S_1 = 500$$
$$X_1 - 2X_2 + S_2 = 0$$

where S_1 and S_2 are stack variables. Here there are four variables and two constraints. Therefore, keep $4 - 2 = 2$ variables at zero.

Let us keep X_1 and X_2 at zero. Then

$$S_1 = 500$$
$$S_2 = 0$$

Objective function $Z = 25X_1 + X_2$

$$Z - 25X_1 - X_2 = 0$$

Write the simplex problem in table form.

↓PC

Basic	x_1	x_2	S_1	S_2	Solution	
Z	–25	–1	0	0	0	
S_1	15*	300	1	0	500	→ PR
S_2	1	–2	0	1	0	

Look for the maximum value with a negative sign in the Z row. It is –25. Consider X_1 variable corresponding to –25. X_1 tries to enter the basic column. This X_1 column is called the pivot column.

Which leaves the basic column?

Ignore all zeroes and negative values in the pivot column. Divide the elements in the solution column by the corresponding elements in the pivot column.

i.e., $\frac{500}{15}$

S_1 leaves. S_1 row is pivot row.

An element which is common to both pivot row and pivot column is called the pivot element. In the above table, 15 is the pivot element. It is common to both pivot row S_1 and pivot column X_1. The * symbol represents the pivot element.

The solution is optimum when all the values in the Z row are either positive or zero. Here there are two variables with negative sign. Therefore the solution is not optimum.

Iteration I

NPE	1	20	$\frac{1}{15}$	0	$\frac{500}{15}$
Old Z	–25	–1	0	0	0
–(–25)NPE	25	500	$\frac{5}{3}$	0	$\frac{2500}{3}$
New Z	0	499	$\frac{5}{3}$	0	$\frac{2500}{3}$
Old S_2	1	–2	0	1	0
–(1)NPE	–1	–20	$-\frac{1}{15}$	0	$\frac{1500}{15}$
New S_2	0	–22	$-\frac{1}{15}$	1	$-\frac{500}{15}$

Formulate the new basic table.

Basic	x_1	x_2	S_1	S_2	Solution
Z	0	499	$\frac{5}{3}$	0	$\frac{2500}{3}$
X_1	1	20	$\frac{1}{15}$	0	$\frac{500}{15}$
S_2	0	–22	$-\frac{1}{15}$	1	$-\frac{500}{15}$

All the values in Z row are either positive or zero. Therefore, the solution is optimum.

$$\text{Max } Z = \frac{2500}{3}$$

subject to

$$X_1 = \frac{500}{15}$$

$$X_2 = 0$$

Here variable X_1 is fractional.

Taking X_1 row as the source row, the corresponding equation is

$$X_1 + 20X_2 + \frac{1}{15}S_1 + 0S_2 = \frac{500}{15}$$

Splitting the integer and fractional parts, we get

$$\left(0+\frac{1}{15}\right)S_1 \geq 33+\frac{1}{3}$$

Considering the fractional part,

$$\frac{1}{15}S_1 \geq \frac{1}{3}$$

Multiplying both sides by minus sign

$$-\frac{1}{15}S_1 \leq -\frac{1}{3}$$

Hence Gomory's constraint is

$$-\frac{1}{15}S_1 + g_1 = -\frac{1}{3}$$

where g_1 is the slack variable.

Adding this secondary constraint equation in the above optimum table, we get the modified new table as follows

		↓PC					
Basic	x_1	x_2	S_1	S_2	g_1	Solution	
Z	0	499	$\frac{5}{3}$	0	0	$\frac{2500}{3}$	
x_1	1	20	$\frac{1}{15}$	0	0	$\frac{500}{15}$	
S_2	0	–22	$-\frac{1}{15}$	1	0	$-\frac{500}{15}$	
g_1	0	0	$-\frac{1}{15}$ *	0	1	$\frac{-1}{3}$	→ PR

This table shows variable solution but $Z \geq 0$ for all values. Therefore, dual simplex algorithm can be used to solve the problem.

Compute $-\frac{Z}{g_1}$ (i.e., divide Z row by $-g_1$) and ignore zeroes.

$$-\frac{0}{0}, -\frac{499}{0}, \frac{5}{3}\left(\frac{15}{1}\right), -\frac{0}{0}, -\frac{0}{1}, \frac{2500}{\not{3}}(\not{3})$$

Column with minimum value (non-zero) in $\frac{-z}{g}$ equation enters. g_1 row with minimum solution value leaves. It is pivot row.

Here S_1 enters and g_1 leaves. $-\frac{1}{15}$ is the pivot element.

NPE	0	0	1	0	–15	5
Old Z	0	499	$\frac{5}{3}$	0	0	$\frac{2500}{3}$
$-\frac{5}{3}$ NPE	0	0	$-\frac{5}{3}$	0	25	$-\frac{25}{3}$
New Z	0	499	0	0	25	825
Old S_2	0	–22	$-\frac{1}{15}$	1	0	$-\frac{500}{15}$
$\frac{1}{15}$ NPE	0	0	$\frac{1}{15}$	0	–1	$\frac{5}{15}$
New S_2	0	–22	0	1	–1	–33
Old X_1	1	20	$\frac{1}{15}$	0	0	$\frac{500}{15}$
$-\frac{1}{15}$ NPE	0	0	$-\frac{1}{15}$	0	$\frac{1}{15}$	$-\frac{5}{15}$
New X_1	1	20	0	0	$\frac{1}{15}$	33

Formulate the new basic table.

Basic	x_1	x_2	S_1	S_2	g_1	Solution
Z	0	499	0	0	25	825
x_1	1	20	0	0	$\frac{1}{15}$	33
S_2	0	–22	0	1	–1	–33
S_1	0	0	1	0	–15	5

At this iteration, optimum solution with integer values is obtained. The solution is

$$\text{Max } Z = 825$$

subject to

$$X_1 = 33$$
$$X_2 = 0$$

13.6 GEOMETRICAL INTERPRETATION OF THE CUTTING PLANE METHOD

Consider a maximisation problem

$$\text{Max } Z = X_1 + X_2$$

subject to

$$2X_1 + 5X_2 \leq 16 \quad (1)$$
$$6X_1 + 5X_2 \leq 30. \quad (2)$$

To determine the graphical solution, we bring the inequalities to equality.

Consider equation (1), bringing inequality to equality.

We get $$2X_1 + 5X_2 = 16$$

When $X_1 = 0, X_2 = \frac{16}{5} = 3.2$ and when $X_2 = 0, X_1 = 8$. The values are (0, 3.2) and (8, 0).

Consider equation (2) bringing inequality to equality. We get

$$6X_1 + 5X_2 = 30$$

If $X_1 = 0, X_2 = 6$, and if $X_2 = 0, X_1 = 5$. The values are (0, 6) and (5, 0).

Plot the graph using X_1 and X_2 values.

Optimum solution occurs at the corners of the feasible region. Therefore, name the corners and determine the value of Z at all corners.

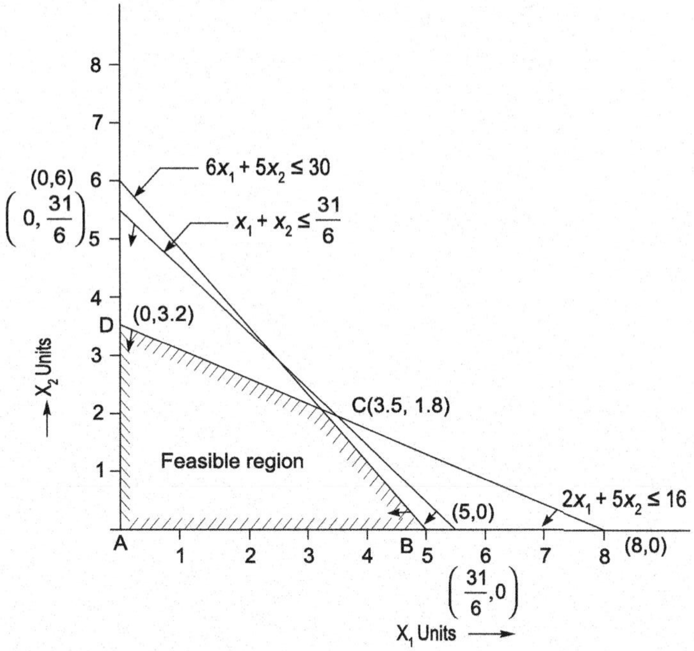

Figure 13.1 Graphical representation

$$\text{Max } Z = X_1 + X_2$$

At corner A, $X_1 = 0$ and $X_2 = 0$

$$Z = 0 + 0$$
$$Z = 0$$

At corner B, $X_1 = 5$ and $X_2 = 0$

$$Z = 5 + 0$$
$$Z = 5$$

At corner C, $X_1 = 3.5$ and $X_2 = 1.8$

$$Z = 3.5 + 1.8$$
$$Z = 5.3$$

At corner D, $X_1 = 0$ and $X_2 = 3.2$

$$Z = 3.2$$

Maximum Z occurs at corner C.

$$\text{Max } Z = 5.3$$

subject to

$$X_1 = 3.5$$
$$X_2 = 1.8$$

Figure 13.1 illustrates the feasible region.

The solution obtained is optimum but non-integer-valued.

We know that

$$2X_1 + 5X_2 + S_1 = 16$$

and

$$6X_1 + 5X_2 + S_2 = 30$$

where S_1 and S_2 are slack variables used to bring inequality to equality.

We rewrite the equations as

$$S_1 = 16 - 2X_1 - 5X_2$$
$$S_2 = 30 - 6X_1 - 5X_2$$

We introduce the Gomory's constraint

$$\frac{3}{10}S_1 + \frac{9}{10}S_2 \geq \frac{4}{5}$$

Substituting the value of S_1 and S_2 in this equation

$$\frac{3}{10}(16 - 2X_1 - 5X_2) + \frac{9}{10}(30 - 6X_1 - 5X_2) \geq \frac{4}{5}$$

We get

$$X_1 + X_2 \leq 5\frac{1}{6}$$

Bringing inequality to equality,

$$X_1 + X_2 = \frac{31}{6}$$

Put $X_1 = 0$ then $X_2 = \frac{31}{6}$

Put $X_2 = 0$ then $X_1 = \frac{31}{6}$

The values are $\left(0, \frac{31}{6}\right)$ and $\left(\frac{31}{6}, 0\right)$.

Plot the line using these values in the graph.

This constraint cuts off some region of the feasible region. Figure 13.1 illustrates the cut-off region. Similarly, another Gomory's constraint is $g_1 \geq 1$.

Here, we have

$$-\frac{3}{10}S_1 - \frac{9}{10}S_2 + g_1 = -\frac{4}{5}$$

$$g_1 = \left(\frac{3}{10}S_1 + \frac{9}{10}S_2\right) - \frac{4}{5}$$

$$= \frac{3}{10}(16 - 2 \times 1 - 5 \times 2) + \frac{9}{10}(30 - 6 \times 1 - 5 \times 2) - \frac{4}{5}$$

$$= 31.8 + 6 \times 1 - 6 \times 2$$

$$g_1 \geq 1$$

$$\therefore 31.8 - 6X_1 - 6X_2 \geq 1$$

$$6X_1 + 6X_2 \leq 30.8$$

$$X_1 + X_2 \leq 5.103$$

This constraint also cuts off some region of feasible region. Hence this method is called cutting plane method. The area cut off is very small. Therefore, it is not shown in the graph.

13.7 BRANCH AND BOUND METHOD

Branch and Bound method is used to solve pure and mixed integer programming problems. In this method, a given linear programming problem is solved graphically and solution is obtained. If the solution value of any variable is non-integer, the given linear programming problem is divided into two sub-problems.

For example, X_n is a non-integer optimum solution value. We take

$$X_i < X_n < X_i + 1$$

where X_i and $X_i + 1$ are integer values

Feasible integer value should satisfy one of the two conditions.

$$X_n \leq X_i$$

and $$X_n \geq X_i + 1$$

These two constraints cannot be used simultaneously in the given linear programming problem. These constraints can be added separately to the given LPP and two sub-problems can be formed. This is known as branching of original problem into two sub-problems. In this branching process, a portion of the feasible region of the graphical solution which has no integer solution is cut.

For example, optimal solution value of two variables is as follows

$$X_1 = \frac{7}{2} \text{ and } X_2 = \frac{9}{5}$$

$$X_1 = \frac{7}{2} \text{ is not an integer.}$$

Using branch and bound method, we can take $3 \leq X_1 \leq 4$ in an integer-valued solution.

i.e., $$X_1 \leq 3 \quad (1)$$

or $$X_1 \geq 4 \quad (2)$$

By adding the constraints (1) and (2) separately to the given linear programming problem, two sub-problems can be obtained after branching into two sub-problems, both the problems are solved. If any one of the two problems gives optimum integer solution, the problem is not further branched. If any sub-problem gives non-integer solution for variables, it is again branched until each sub-program gives either integer valued optimum solution or there is evidence that it will not give a better solution.

Optimum among all the solutions of sub-programs is selected as the overall optimum value of the given problem.

Example 13.4

Solve the following problem by Branch and Bound method.

$$\text{Max } Z = X_1 + X_2$$

subject to

$$3X_1 + 2X_2 \leq 12 \quad (1)$$

$$X_2 \leq 2 \quad (2)$$

$$X_1, X_2 \geq 0 \text{ and are integers.}$$

Solution

Solution by graphical method

Consider equation (1)

$$3X_1 + 2X_2 \leq 12$$

Bringing inequality to equality, we get

$$3X_1 + 2X_2 = 12$$

When $X_1 = 0$ $X_2 = 6$, and when $X_2 = 0$, $X_1 = 4$. The values are (0, 6) and (4, 0).

Consider equation (2)

$$X_2 \leq 2$$

Bringing inequality to equality, we get

$$X_2 = 2$$

Plot the graph using X_1, X_2 values and obtain the feasible region.

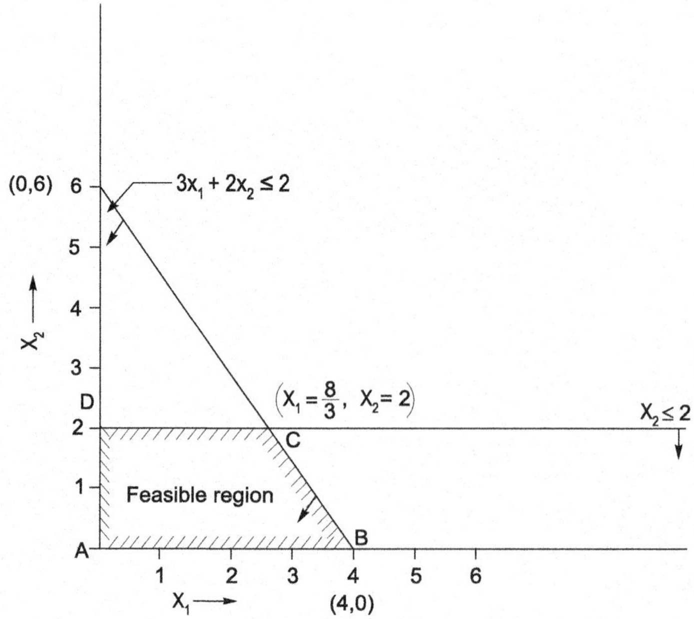

Figure 13.2 Graphical solution with feasible region

Optimum solution occurs at the corners of the feasible region. Name the corners and find the value of Z at all corners.

$$\text{Max } Z = X_1 + X_2$$

At corner A, $X_1 = 0$ and $X_2 = 0$

$$Z = 0 + 0$$
$$Z = 0$$

At corner B, $X_1 = 4$ and $X_2 = 0$

$$Z = 4 + 0$$
$$Z = 4$$

At corner C $X_1 = \frac{8}{3}$ and $X_2 = 2$

$$Z = \frac{8}{3} + 2$$
$$Z = \frac{14}{3}$$

At corner D, $X_1 = 0$ and $X_2 = 6$

$$Z = 0 + 6$$
$$Z = 6$$

Optimum solution occurs at corner C.

$$Z = \frac{14}{3}$$

subject to

$$X_1 = \frac{8}{3} \text{ and } X_2 = 2$$

Here the solution is optimum but not integer-valued. Using Branch and Bound method,

$$X_1 = \frac{8}{3} \text{ a non-integer can be written as}$$
$$2 \le X_1 \le 3$$

Then we form the two sub-problems as given below.

Problem 2		Problem 3
Max $Z = X_1 + X_2$		Max $Z = X_1 + X_2$
subject to $3X_1 + 2X_2 \le 12$		subject to $3X_1 + 2X_2 \le 12$
$X_2 \le 2$	and	$X_2 \le 2$
$X_1 \le 2$		$X_1 \le 3$
$X_1, X_2 \ge 0$		$X_1, X_2 \ge 0$

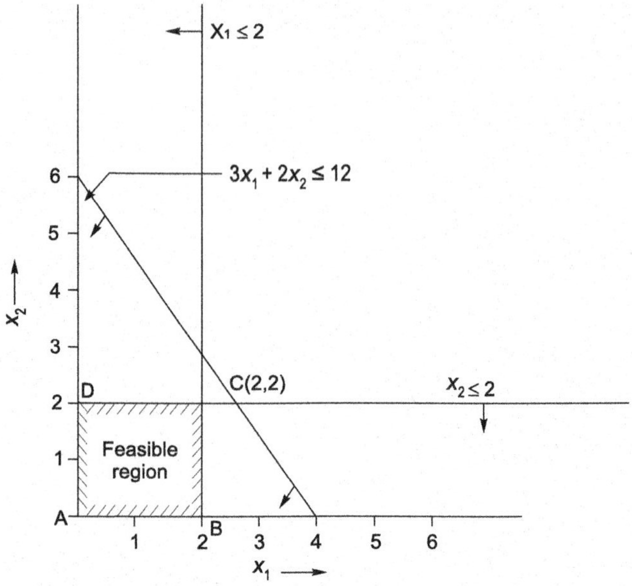

Figure 13.3 Graphical solution for Problem 2

For the solution of Problem 2 we plot the Figure 13.3. Optimum solution occurs at corners of the feasible region. Name the corners and find the value of Z at all these corners.

Optimum solution for Problem 2 is $Z = 4$, subject to $X_1 = 2$ and $X_2 = 2$. X_1 and X_2 are integers for Problem 2. Hence there is no need to branch this problem further.

For the solution of Problem 3, we plot the Figure13.4 as follows

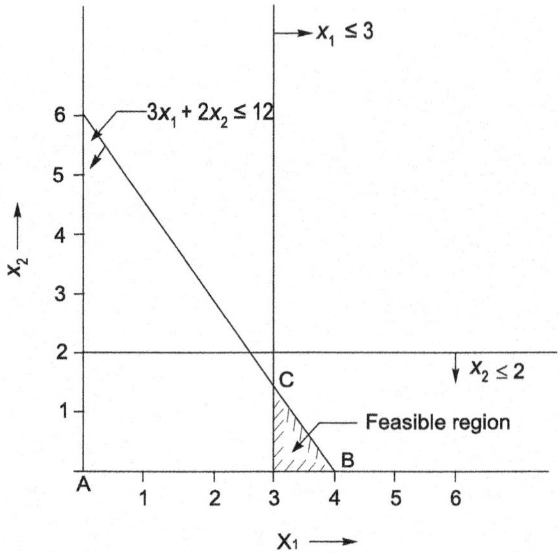

Figure 13.4 Graphical solution for problem 3

Optimum solution occurs at corners of the feasible region. Name the corners and find the value of Z at all these corners.

Optimum solution to Problem 3 is

$$\text{Max } Z = \frac{9}{2}$$

subject to

$$X_1 = 3$$

$$X_2 = \frac{3}{2}$$

X_2 is non-integer, therefore it needs further branching.

i.e., $$1 < X_2 \leq 2$$

We form two sub-problems by adding two constraints

$$X_2 \leq 1 \text{ and } X_2 \geq 2 \text{ respectively}$$

Problem 4		Problem 5
Max $Z = X_1 + X_2$		Max $Z = X_1 + X_2$
subject to $3X_1 + 2X_2 \leq 12$		subject to $3X_1 + 2X_2 \leq 12$
$X_2 \leq 2$	and	$X_2 \leq 2$
$X_1 \geq 3$		$X_1 \geq 3$
$X_2 \leq 1$		$X_2 \geq 2$
$X_1, X_2 \geq 0$		$X_1, X_2 \geq 0$

Graphical solution for Problem 4 is as follows

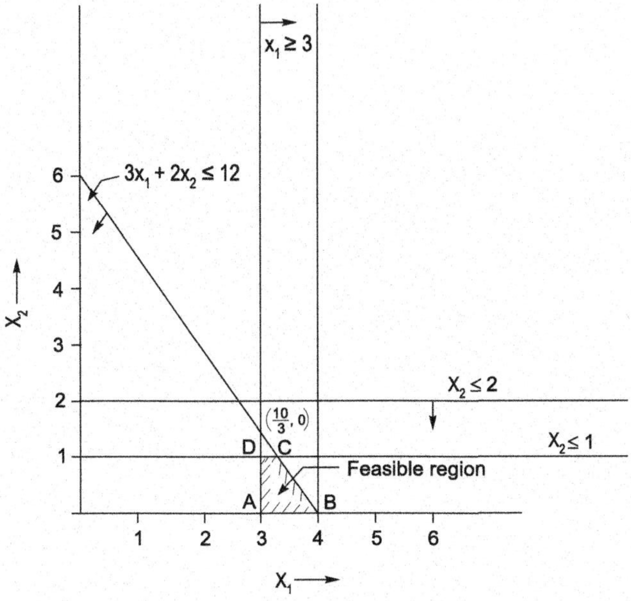

Figure 13.5 Graphical solution for problem 4

From the Figure 13.5 it is observed that $X_2 \leq 2$ does not form the feasible region. Hence, it is redundant. The optimum solution to Problem 4 is

$$\text{Max } Z = \frac{13}{3}$$

subject to

$$X_1 = \frac{10}{3}$$
$$X_2 = 1$$

From the figure, it is obvious that further branching of the problem will not improve the restriction

$$X_1 = \frac{10}{3} \text{ is not an integer}$$
$$X_1 \leq 3 \text{ and } X_1 \geq 4$$

The optimum solution is $X_1 = 3$ and $X_2 = 1$. And $X_1 = 4$ and $X_2 = 0$ respectively. These X_1 and X_2 values also give $Z = 4$. Hence, overall maximum value of objective function is

$$\text{Max } Z = 4$$

subject to

$$X_1 = 2$$
$$X_2 = 2$$

or

$$X_1 = 3 \text{ and } X_2 = 1$$

or

$$X_1 = 4 \text{ and } X_2 = 0$$

It is observed from the problem that no feasible solution can be obtained.

Example 13.5

Solve the following problem using Branch and Bound method of Integer Programming.

$$\text{Max } Z = 3X_1 + 4X_2$$

subject to

$$3X_1 - X_2 \leq 12 \quad (1)$$
$$3X_1 + 11X_2 \leq 0 \quad (2)$$
$$X_1, X_2 \geq 0 \text{ and are integers}$$

Graphical solution

Consider equation (1). Bringing inequality to equality, we get

$$3X_1 - X_2 = 12$$

Put

$$X_1 = 0, X_2 = -12$$
$$X_2 = 0\ X_1 = 4$$

The values are (0, –12) and (4, 0).

Consider equation (2). Bringing inequality to equality, we get

$$3X_1 + 1X_2 = 0$$

When $X_1 = 0, X_2 = 0$. The values are (0, 0).

Plot the graph using (X_1, X_2) values.

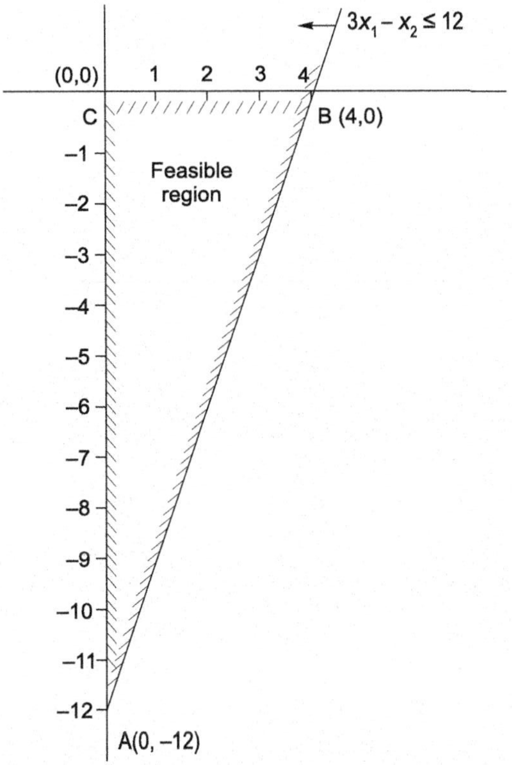

Figure 13.6 Graphical solution with feasible region

From the graph, optimum solution occurs at corner *B*.

$$\text{Max } Z = 12$$
$$X_1 = 4$$
$$X_2 = 0$$

Here, X_1 and X_2 are integers; there is no need to branch this problem further.

EXERCISES

1. Solve the LPP by Gomory's cutting plane method.

$$\text{Max } Z = 4X_1 + 6X_2 + 2X_3$$

subject to

$$4X_1 - 4X_2 \leq 5$$
$$-X_1 + 6X_2 \leq 5$$
$$-X_1 + X_2 + X_3 \leq 5$$
$$X_1, X_2, X_3 \geq 0 \text{ and are integers}$$

2. Solve by Gomory's method..

$$\text{Max } Z = 2X_1 + 2X_2$$

subject to

$$5X_1 + 3X_2 \leq 8$$
$$2X_1 + 4X_2 \leq 8$$
$$X_1, X_2 \geq 0 \text{ and are integers}$$

3. Solve by Gomory's method.

$$\text{Max } Z = X_1 + 2X_2$$

subject to

$$X_1 + 2X_2 \leq 12$$
$$4X_1 + 3X_2 \leq 12$$
$$X_1, X_2 \geq 0 \text{ and are integers}$$

4. Solve the following LPP by Branch and Bound method.

$$\text{Max } Z = 2X_1 + 2X_2$$

subject to

$$5X_1 + 3X_2 \leq 8$$
$$X_1 + 2X_2 \leq 4$$
$$X_1, X_2 \geq 0 \text{ and are integers}$$

5 Solve the following LPP by Branch and Bound method.

$$\text{Max } Z = 3X_1 + 2X_2$$

subject to

$$2X_1 + 2X_2 \leq 7$$
$$X_1 \leq 2$$
$$X_2 \leq 2$$
$$X_1, X_2 \geq 0 \text{ and are integers.}$$

Chapter 14

DECISION ANALYSIS

14.1 INTRODUCTION

Decision analysis is the discipline comprising procedures, methods and tools for identifying and assessing important aspects of a decision. Decision analysis is used to recommend the course of action to good decision that results in maximum expected utility.

14.2 HISTORY OF DECISION ANALYSIS

The term decision analysis was coined in 1964 by Ronald A Howard. Decision trees and influence diagrams are used to represent decision analysis problems graphically. Both these tools represent the alternatives available to describe the decision maker's uncertainty, measure of evaluation to achieve their objectives. Probability distribution tools are used to determine uncertainty. The decision-making involves risk as future is always uncertain and the objectives are conflicting one another. Decision analysis helps to choose, the decision whose consequences result in maximum expected utility. Multi-attribute utility functions are used to handle the conflicting objectives. Decision analysis with the help of mathematical optimization methods result in quantitative approach for decision-making.

14.3 A PROTOTYPE EXAMPLE 1

The Reliance Petrochemicals India Limited has a piece of land. The experts have reported to the company that there is a 25% chance of getting oil in that land. The Mangalore Petrochemical Refineries Limited has offered to purchase that land for Rs. 9 lakhs. The cost of drilling is Rs. 10 lakhs. If the oil is found, the expected revenue is Rs. 80 lakhs. If the land is found dry, the loss will be Rs. 10 lakhs. The profit, if the oil is found, is Rs. 70 lakhs.

The Reliance Company working with limited resources considers the loss of Rs.10 lakhs as a serious matter. If the oil is not found, the company has to take a decision whether to drill for oil or to sell the land under uncertainty.

14.4 DECISION-MAKING WITHOUT EXPERIMENTATION

In the problem discussed in section 14.3, the decision alternatives available are, whether to drill for oil or to sell the land. The outcome of possible decision alternatives are called state of nature. The decision-maker should know the pay-off (gain or loss) for each combination of decision alternative and state of nature. The pay-off is the quantitative measure of the value of gain or loss for a pair of course of action. A pay-off matrix is used for representing the pay-off, i.e., gain or loss for each combination of decision alternative and state of nature. This situation can be compared to a two-person game discussed under the chapter game theory.

Here, the decision maker has to select one of the decision alternatives; then, nature selects the possible state of nature. Each combination of decision alternative and state of nature gives a pay-off. The pay-off matrix should be prepared to find the optimum decision alternative.

In two-person zero-sum game of game theory, both the players know the pay-off for the combination of their own strategy and that of the opponent. In the present analysis, nature acts as a passive player as it chooses the strategy randomly; hence, the game theory will not give optimum decision strategy.

		State of nature	
		Oil	Dry
Decision alternative	Drill for oil	70	10
	Sell the land	9	9
	Probability	0.25	0.75

Figure 14.1 Pay-off matrix for decision analysis and pay-off in lakhs of Rupees

Note: Player 1 is the decision maker; Player 2 is Nature.

14.4.1 Maximin Principle

As we discussed in the game theory, we have to determine the minimum assured gain for each decision alternative for Player I. We have to choose the maximum among the minimum assured pay-offs for decision maker player II.

		Player II		
		Oil	Dry	Minimum
Player I	Drill	70	10	10
	Sell	9	9	9

Figure 14.2 Maximum pay-off criterion

From Figure 14.2 minimum pay-off for drilling is Rs. 10 lakhs and that for sales is Rs. 9 lakhs. Hence, it is advisable to sell the land. The decision-maker using this principle should be cautious.

14.4.2 Most Likely Principle

		State of Nature	
		Oil	Dry
Decision maker	Drill	70	10
	Sell	9	9
	Probability	0.25	0.75 Maximum

Figure 14.3 Most likely principles for state of nature

Figure 14.3 illustrates the most likely principle for state of nature. The state of nature looks for the decision alternative that gives maximum pay-off. We have to choose this decision alternative. The largest probability is 0.75. This principle prefers to sell the land. It ignores the profit of Rs. 70 lakhs if the company drills the land and finds oil.

14.4.3 Bayes' Decision Rule

In this rule, the best available estimates of probabilities for respective state of nature is considered for each decision alternative and pay-off is calculated for each decision alternative. The decision alternative with maximum pay-off is selected.

$$\begin{aligned}\text{Drilling land pay-off } \vartheta &= P_1D_{11} + P_2D_{12}\\ &= 0.25(70) + 0.75\,(-10)\\ &= \vartheta = 10\end{aligned}$$

where,

D_{11} - Profit for drilling and oil

D_{12} - Profit for drilling and no oil

P_1 - Probability of getting oil

P_2 - Probability of no oil

$$\begin{aligned}\text{Selling land pay off } \vartheta &= P_1D_{21} + P_2D_{22}\\ \vartheta &= 0.25(9) + 0.75\,(9)\\ \vartheta &= 9\end{aligned}$$

where,

D_{21} - Profit for sell and oil

D_{22} - Profit for sell and no oil

Pay-off for drilling land is larger. Therefore, this alternative is selected.

14.4.4 Sensitivity Analysis

We know that the sum of the two probabilities must be equal to one. Increase in probability for one state of nature automatically causes decrease in probability in other state of nature. If the probability of getting oil varies from 20% to 40% proportionately the probability of getting no oil varies from 80% to 60%.

$$\begin{aligned}\text{Pay-off drill} &= \text{Pay-off sell}\\ 800P &= 100 + 90 \text{ (From Bayes' rule)}\\ \therefore P &= \frac{190}{800}\\ &= 0.2375\end{aligned}$$

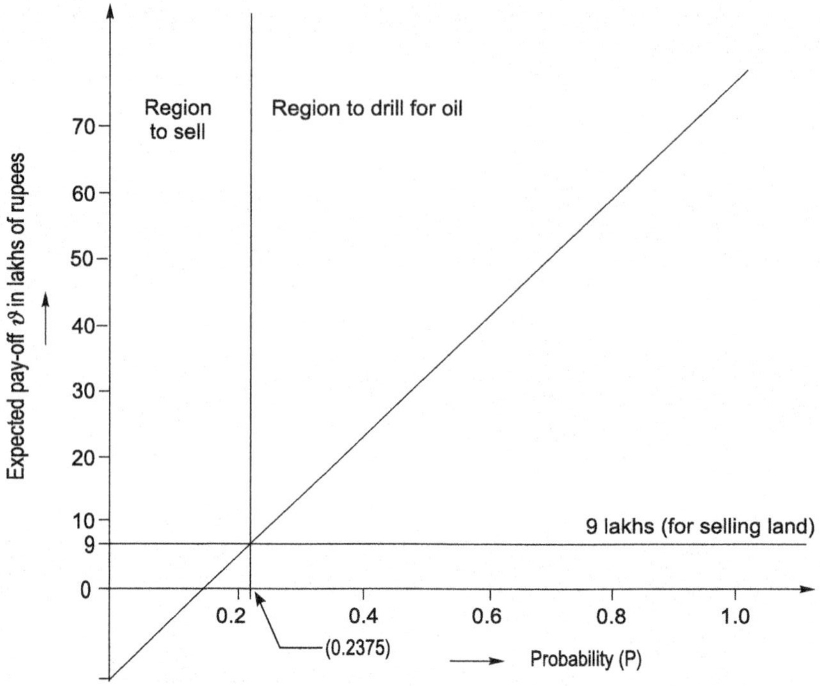

Figure 14.4 Probability of getting oil '*P*'

We can suggest the Reliance Petrochemicals Limited to sell the land, if the probability of getting oil is less than 23.75%. If the probability of getting oil is more than 23.75% we can suggest the company to drill the land.

14.5 DECISION-MAKING WITH EXPERIMENTATION

14.5.1 A Prototype Example 1

Experimentation may be carried out to improve the estimation of the probability of getting oil in the case study discussed in the previous section. Seismic testing of the land analyses the geological structure of the land through sound waves. It determines sounding patterns that indicate whether the pattern is favourable or not favourable.

Let the cost of seismic sounding test be Rs. 3 lakhs.

Let,

SSN – Seismic Sounding not favourable for State of Nature of oil

SSF – Seismic Sounding favourable for State of Nature of oil

and P – SSN (State of Nature – oil) = Say 0.4 (Probability).

P – SSF (State of Nature – oil) = 1 – 0.4 = 0.6 (Probability).

Because the sum of the probability of favourable and not favourable condition is one.

Similarly

SSN Seismic Sounding not favourable for State of Nature – No oil (Dry)

SSF Seismic Sounding favourable for State of Nature – No oil (Dry)

and P – SSN (State of Nature – no oil) = Say 0.8 (Probability)

P – SSF (State of Nature – no oil) = 1 – 0.8 = 0.2.

Then Seismic Sounding Favourable (SSF) for state of nature – no oil.

Posterior Probability

n – Number of possible states of nature, i.e., oil, no oil, etc.

P(State = State i) Prior probability that true State of Nature

State i for $i = 1, 2, 3, ..., n$

P[State = State i / (Finding = Finding j)] posterior probability that State of Nature is State i and Finding = Finding j, for $j = 1, 2, 3, ..., n$.

j number of findings of experimentation.

Combining the probabilities

$$P(\text{State} = \text{State } i)$$

and

$$P(\text{State} = \text{State } i / \text{Finding} = \text{Finding } j)$$

$$P(\text{State} = \text{State } i/\text{Finding} = \text{Finding } j) = \frac{P(\text{State} = \text{State } i, \text{Finding} = \text{Finding } j)}{P(\text{Finding} = \text{Finding } j)}$$

$$P(\text{Finding} = \text{Finding } j) = \sum_{k=1}^{n} P(\text{State} = \text{State } k, \text{Finding} = \text{Finding } k)$$

$$P(\text{State} = \text{State } i, \text{Finding} = \text{Finding } j) = \frac{P(\text{Finding} = \text{Finding } j, \text{State} = \text{State } i)}{P(\text{State} = \text{State } i)}$$

Formula for corresponding posterior probability is

$$P(\text{State} = \text{State } i, \text{Finding} = \text{Finding } j) = \frac{P(\text{Finding} = \text{Finding } j / \text{State} = \text{State } i)\ P(\text{State} = \text{State } i)}{\sum_{k=1}^{n} P(\text{Finding} = \text{Finding } j / \text{State} = \text{State } k)\ P(\text{State} = \text{State } k)}$$

Posterior probabilities are

$$P[\text{State} = \text{Oil/Finding}]\ \text{SSN} = \frac{0.4(0.25)}{0.4(0.25) + 0.8(0.75)} = \frac{1}{7} = 0.1428$$

$$P[\text{State} = \text{Dry/Finding}]\ \text{SSN} = 1 - 0.1428 = 0.8571$$

$$P[\text{State} = \text{Oil/Finding}]\ \text{SSF} = \frac{0.6(0.25)}{0.6(0.25) + 0.2(0.75)} = \frac{1}{2} = 0.5$$

$$\text{P[State} = \text{Dry/Finding] SSF} = 1 - 0.5 = 0.5$$

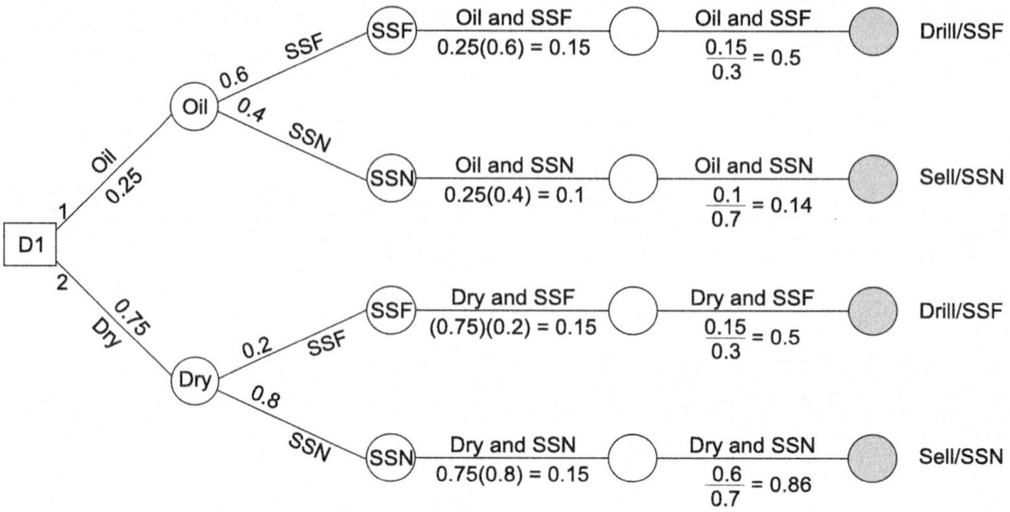

Figure 14.5 Probability tree for Reliance Petrochemicals company

Probability $P(\text{SSF}) = 0.15 + 0.15 = 0.3$

$P(\text{SSN}) = 0.1 + 0.6 = 0.7$

Figure 14.5 illustrates the decision tree. First column shows the prior probabilities. The second column shows the conditional probabilities. Multiplying the probabilities in the first and the second column, we get the joint probability in the third column. Fourth column shows the posterior probability.

Expected pay-offs if the finding of experiments is Seismic Sounding Not Favourable (SSN)

$$\begin{aligned}
\text{Pay-off [Drill/Finding] SSN} &= P_1D_{11} + P_2D_{12} - \text{Testing (expected)} \\
&= 0.1428(70) + 0.8571(-10) - 3 \\
&= -15.7 \\
\text{Pay-off [Sell/Finding] SSN} &= P_1D_{21} + P_2\,D_{22} - \text{Testing (expected)} \\
&= 0.1428(9) + 0.8571(9) - 3 \\
&= 6
\end{aligned}$$

Expected pay-offs if the findings of experiments is Seismic Sounding Favourable (SSF)

$$\begin{aligned}
\text{Pay-off [Drill/Finding] SSF} &= P_1D_{11} + P_2D_{12} - \text{Testing (expected)} \\
&= 0.5(70) + 0.5(10) - 3 \\
&= 27 \\
\text{Pay-off [sell/finding] SSF} &= P_1D_{21} + P_2D_{22} - \text{Testing (expected)} \\
&= 0.5(9) + 0.5(9) - 3 \\
&= 6
\end{aligned}$$

Our objective is to maximise the pay-off from the decision tree.

$$\begin{aligned}
P(\text{SSN}) &= 0.7 \\
P(\text{SSF}) &= 0.3
\end{aligned}$$

	Expected pay-off excluding seismic testing	Expected pay-off including cost of testing
SSN	9	6
SSF	30	27

Expected pay-off with experimentation

$$= 0.7(9) + 0.3(30)$$
$$= 15.3$$

Expected value of experimentation

$$= 15.3 - 10$$
$$= 5.3 \text{ lakhs}$$

This value exceeds the cost of conducting experiments. Therefore, experimentation should be done.

14.5.2 A Prototype Example 2

An economically strapped student has brought his failed computer to a local repair shop. The technician, being a friend of the student, would like to fix it with the least expenditure of money. He guesses that one of the four parts has probably failed. They are *A, B, C* and *D* with the cost of replacement Rs. 1000, Rs. 2000, Rs. 300 and Rs. 800 respectively. We use the letters *A, B, C* and *D* to represent the failures caused by the corresponding parts and letter *E* to be the failure caused by the part other than the four parts.

Three tests *X, Y* and *Z* are available to help locate the failed part and can be performed at a cost of Rs. 500, Rs. 700 and Rs. 800 respectively. If *E* is the result or if the technician decides to abandon testing, the mother board can be replaced at a cost of Rs. 5000. This is guaranteed to fix the problem. Before any testing is done, we estimate from past experience the probability that each of the five events is causing the failure. These are called prior probabilities. The events are mutually exclusive and their probabilities sum to 1.

In this problem, the technician has two options to decide.

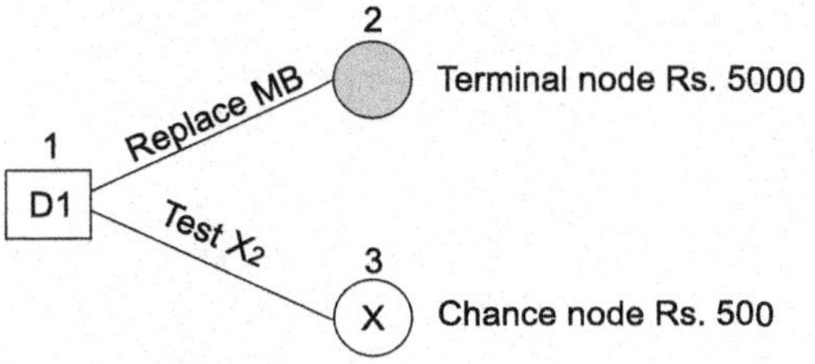

Figure 14.6 First decision D_1

Figure 14.6 illustrates the two decisions available to the technician, whether to replace the mother board or to perform the '*X*' test.

The decision to replace the mother board is represented by a shaded circle. It represents the terminal node at which the process stops. The cost associated with the replacement of mother board is Rs. 5000. Another decision to perform test *X* leads to node 3. This is a chance node because the result of this node is uncertain.

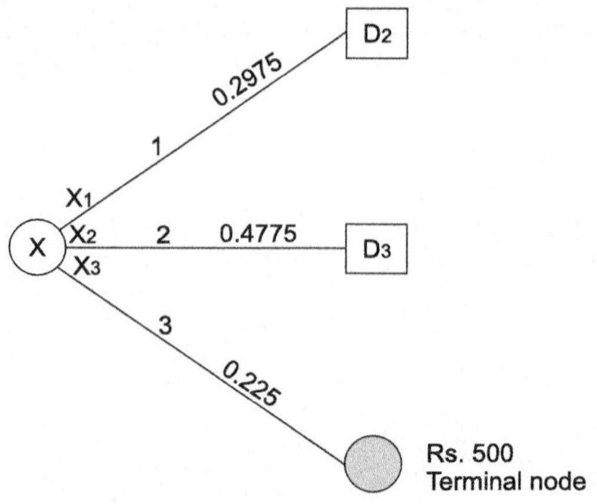

Figure 14.7 Possible results of test 'X'

Probabilities of outcome of test X are not given in the problem. We can approximate the probabilities from the experiments. X_1, X_2 and X_3 are the outcome of test X. X_1 may be the failure due to A or B. X_2 failure due to C or D. X_3 is to replace the motherboard.

Approximated probabilities of outcomes X_1, X_2 and X_3 are given below.

		A	B	C	D	E
	X_1	0.6	0.7	0.15	0.05	0.1
Test results	X_2	0.2	0.2	0.55	0.85	0.3
	X_3	0.2	0.1	0.3	0.1	0.6

Figure 14.8 Probabilities $P[x_j/Y]_j = 1, 2, ..., n$

Probability

$$P(X_i, A) = P(x_i \mid y)\, P(y)$$

$$P(X_1) = P(X_1, A) + P(X_1, B) + P(X_1, C) + P(X_1, D) + P(X_1, E)$$

$$= (0.6)(0.15) + 0.7(0.2) + (0.15)(0.3) + (0.05)(0.25) + (0.1)(0.1)$$

$$PX_1 = 0.2975$$

$$P(X_2) = P(X_2, A) + P(X_2, B) + P(X_2, C) + P(X_2, D) + P(X_2, E)$$

$$= (0.2)(0.15) + (0.2)(0.2) + (0.55)(0.30) + (0.85)(0.2) + 0.3(0.1)$$

$$P(X_2) = 0.4775$$

$$P(X_3) = P(X_3, A) + P(X_3, B) + P(X_3, C) + P(X_3, D) + P(X_3, E)$$

$$= (0.2)(0.15) + (0.1)(0.2) + (0.3)(0.3) + (0.1)(0.2) + 0.6(0.1)$$

$$P(X_3) = 0.225$$

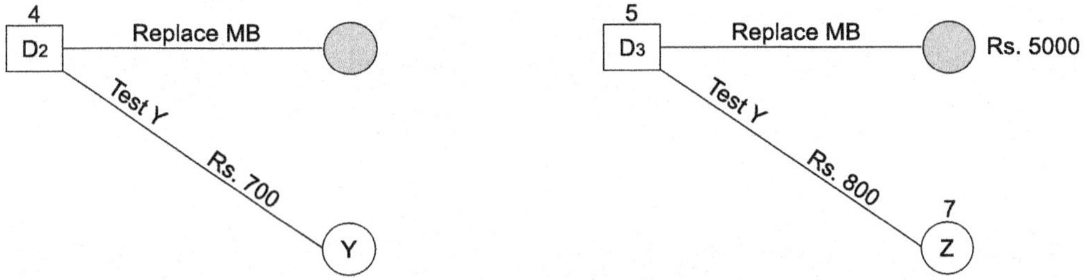

Figure 14.9 Second round decision

If test X indicates X_1, the technician has two options. He can either use test Y or replace the motherboard.

Similarly, if test X indicates X_2, the technician has two options. He can use either test Z or replace the motherboard. Decisions D_2 for X_1, and D_2 for X_2 are illustrated in Figure 14.9.

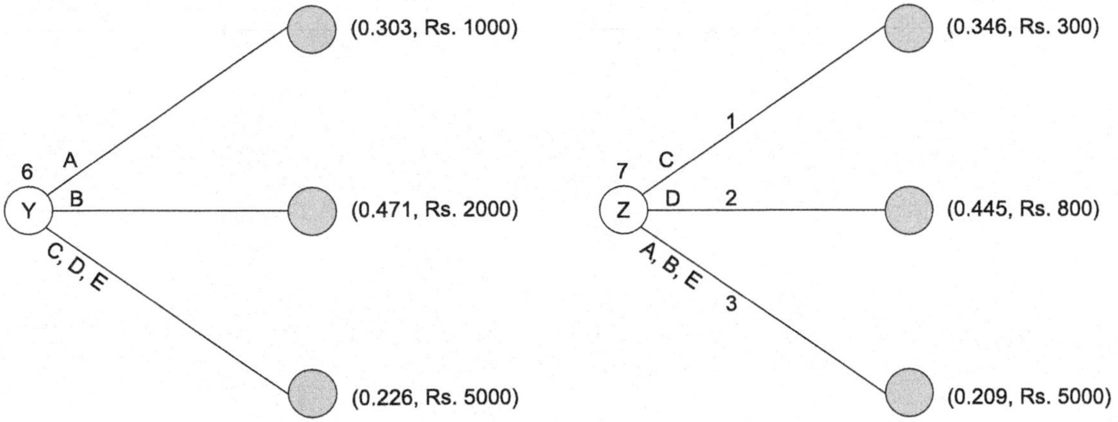

Figure 14.10 Test results with posterior probabilities

Test Y identifies the causes of failure due to components A or B or C, D, E. If the failure is not due to problems A or B, the mother board will be replaced.

Similarly, test Z identifies the causes of failure due to components C or D or A, B, E. If the failure is not due to problems C or D the mother board is replaced. This decision pattern is illustrated in Figure 14.10.

The probability of the event Y yielding result for problem A

$$P(Y|X_1) = \frac{P(X_i, Y)}{P(X_i)}$$

Therefore,

$$P(A|X_1) = \frac{P(X_1, A)}{P(X_1)} = \frac{(0.6)(0.15)}{(0.2975)} = 0.3025$$

Similarly probability of Y yielding results for problem B.

$$P(B|X_1) = \frac{P(X_1, B)}{P(X_1)} = \frac{(0.7)(0.12)}{(0.2975)} = 0.471$$

Similarly probability for problem C, D and E are

$$P(C|X_1) = \frac{P(X_1, C)}{P(X_1)} = \frac{(0.15)(0.3)}{(0.2975)} = 0.151$$

$$P(D|X_1) = \frac{P(X_1, D)}{P(X_1)} = \frac{(0.05)(0.25)}{(0.2975)} = 0.042$$

$$P(E|X_1) = \frac{P(X_1, E)}{P(X_1)} = \frac{(0.1)(0.1)}{(0.2975)} = 0.034$$

$$P(C, D, E|X_1) = 0.151 + 0.042 + 0.034 = 0.226$$

The probability of the events Z yielding result for problem A

$$P\ Z(X_i) = \frac{P(X_i, Z)}{P(X_i)}$$

Therefore,

$$P(A|X_2) = \frac{P(X_2, A)}{P(X_2)} = \frac{(0.2)(0.15)}{(0.4775)} = 0.063$$

The probability of the events Z yielding result for problem B

$$P(B|X_2) = \frac{P(X_2, B)}{P(X_2)} = \frac{(0.2)(0.2)}{(0.4775)} = 0.084$$

Similarly, probability for problem C, D and E are

$$P(C|X_2) = \frac{P(X_2, C)}{P(X_2)} = \frac{(0.55)(0.3)}{(0.4775)} = 0.346$$

$$P(D|X_2) = \frac{P(X_2, D)}{P(X_2)} = \frac{(0.85)(0.25)}{(0.4775)} = 0.445$$

$$P(E|X_2) = \frac{P(X_2, E)}{P(X_2)} = \frac{(0.3)(0.1)}{(0.4775)} = 0.063$$

$$P(A, B, E|X_2) = 0.063 + 0.084 + 0.063$$

$$= 0.209$$

Decision Tree

The complete decision tree for the prototype example 2 discussed above is shown in Figure 14.11 below.

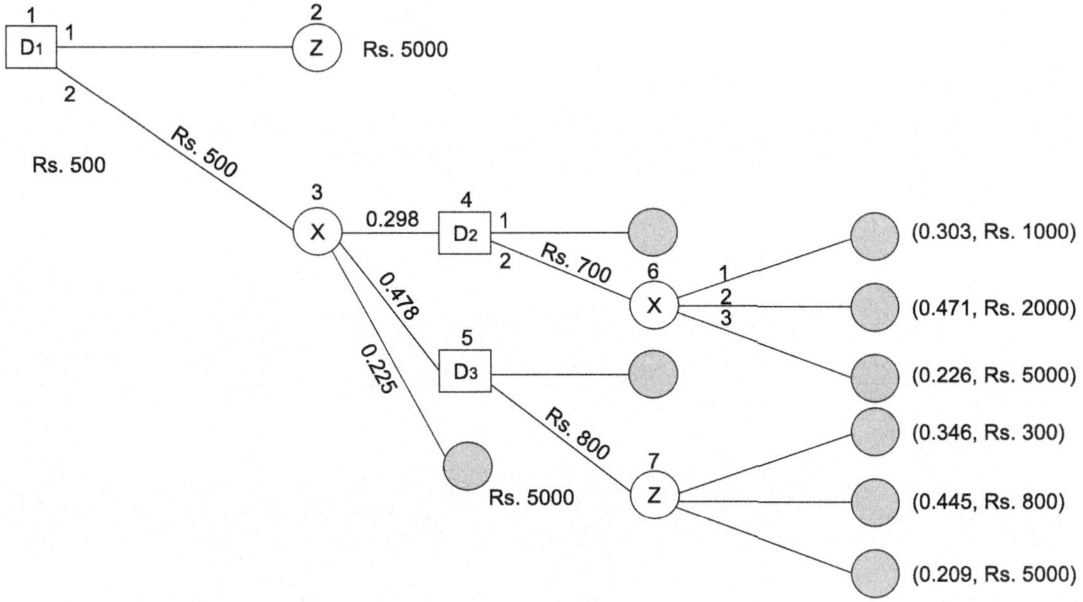

Figure 14.11 Decision tree for prototype example 2

14.6 THE GENERAL DECISION TREE MODEL

Decision tree is the graphical description of a sequential process. The decision tree consists of nodes and branches. The decision nodes are shown as square and chance nodes are shown as white circles. Shaded circles represent terminal nodes. The lines are the branches connecting the nodes. The branches represent specific decisions and outcomes of chance events. Figure 14.12 illustrates the general decision tree model.

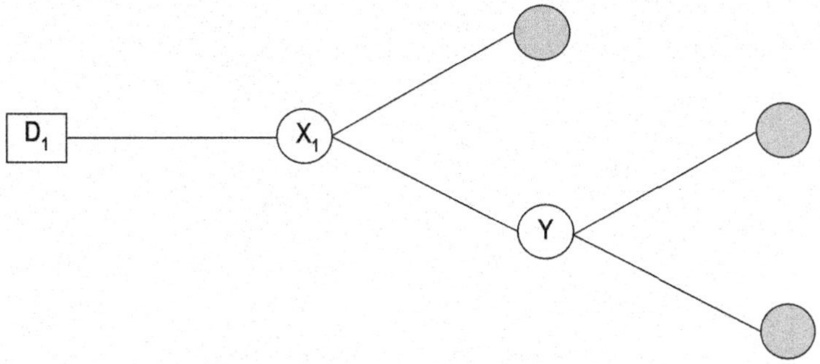

Figure 14.12 Decision tree model

The decision tree is drawn from left to right. Branches leave nodes going to the right and enter nodes coming from the left Node 1, precedes all other nodes. It is called the root node.

The decision node represents a choice that must be made in the sequential process. The capital letter in bold within the square node is the name of the set of possible decisions. The circular nodes represent a chance event that may occur during the sequential process. The uppercase letters in bold within the circle is the name of the set of possible outcomes of the event. The branches that leave the node describe the individual outcomes in the set.

Every branch leaving a chance node must be assigned a number which is the conditional probability of that particular outcome. This probability is written as $P(i/\text{Path})$ probability of outcome. The decision function specifies the optimal course of action to be taken at each decision node.

A solution to the problem is a specification of the optimal decision at each decision node. This is called the optimal policy.

EXERCISES

1. A bank manager wants to develop a decision policy to assist the bank's credit officers. Consider a situation involving Rs. 20 lakh loan. The following information has been gathered from historical records regarding the percentage of customers in various income categories who seek such loans.

Customer type	Income (lakhs)	Percentage
A	>10	10
B	8 to 10	20
C	6 to 8	40
D	<6	30

There are several steps in the decision process for each customer. The officer decides whether to reject the loan application, accept the application or call for a credit check. Whenever a customer is rejected, there is a loss of goodwill, which can be reflected as a cost that depends on the class of customer. When the application is accepted, there is a probability that the customer will default and the loan principal is lost. If the customer does not default, the gain is the interest income from the loan. All this information is shown in the following table.

Customer type	Loss from rejection Rs.	Probability of default	Loss on default Rs. in lakhs	Income from loan Rs.
A	1,00,000	0.05	20	2,00,000
B	5,00,000	0.10	20	2,20,000
C	2,00,000	0.10	20	2,50,000
D	0	0.15	20	3,00,000

The cost of a credit check is Rs.3000. The results of the check are "good risk" or "bad risk". The good risk customer has a probability of default 0.05 less than the entries in the table above. The bad risk customer has a probability of default 0.05 more than the entries in the table above. After the credit check, the officer must accept or reject the loan application. When a customer is rejected, assume that the alternative investment for the 20 lakhs has a return of Rs. 1.5 lakhs. Show the decision tree for this situation and find the optimal decision policy.

2. You own a stretch of land in Bombay high, where there is the possibility of discovering oil. Your options are as follows

a_1 Drill for oil yourself.

a_2 Lease the site to someone else to drill.

a_3 Lease the site but maintain an interest in the results.

There are four possible outcomes regarding the success of the oil well

q_1 6,00,000 barrel well

q_2 4,00,000 barrel well

q_3 1,00,000 barrel well

q_4 dry hole

Your profit (in lakhs of rupees) depend on the development option you choose and the results of the drilling as shown in the following table.

		q_1	q_2	q_3	q_4
	a_1	800	400	50	100
Decision	a_2	60	60	60	60
	a_3	300	100	0	0

Based on information about the site, you have estimated the prior probabilities of the four results to be

$$P(q_1) = 0.1\ P(q_2) = 0.15\ Pq_3 = (0.25)\ P(q_4) = 0.5$$

(a) Based on the prior probabilities, which development option should you choose?

(b) Rather than making a firm decision among the three options, you now have a fourth option of running a seismic test to get more information on the likelihood of success. From your experience in drilling in this area, you have estimated the conditional probabilities of the well the success given in the test result. These values are denoted by $P(x_i/q_j)$ and appear in the following table.

		Well success			
		q_1	q_2	q_3	q_4
	x_1	0.3	0.2	0.3	0.3
Test result	x_2	0.1	0.15	0.2	0.4
	x_3	0.6	0.5	0.3	0.1
	x_4	0.0	0.15	0.2	0.2

Using Bayes' law, compute the posterior probabilities $P(q_j/x_i)$. The cost of the seismic test is Rs. 2 lakhs. Perform a decision analysis to determine whether or not to do the test. Also determine which of the development options to pursue on the result of the test. What is the profit associated with the optimal decision?

3. A new oil field has been discovered with different prior probabilities of well success than those identified in the second exercise question. These values are listed below. Using the original conditional probabilities for the results of the seismic test. Find the optimal policy

   $P(q_1) = 0.20\ P(q_2) = 0.20\ P(q_3) = 0.30\ P(q_4) = 0.30.$

4. Use the original statement of the problem given in the second exercise question but let the conditional leasing arrangement change. The returns for conditional leasing are Rs. 60 lakhs for a 60-lakh barrel well, Rs. 30 lakhs for a 40-lakh barrel well, and no return for a 10-lakh barrel well or a dry well. Find the optimal policy.

5. The daily demand for a particular type of printed circuit board in an assembly shop can assume one of the following values: 100, 120 or 130 with probabilities 0.2, 0.3 and 0.5. The manager of the shop is thus limiting his alternatives to stocking one of the three levels indicated. If he prepares more boards that are needed in the same day, he must reprocess those remaining at a cost price of Rs. 550/board. Assuming that it costs Rs. 600 to prepare a board for assembly and that each board produces a revenue of Rs.1050. Find the optimal stocking level by using a decision tree model.

6. In the fifth exercise question, suppose that the owner wishes to consider his decision problem over a two-day period. His alternatives for the second day are determined as follows. If the demand in day one is equal to the amount stocked, he will continue to order the same quantity on the second day. Otherwise, if the demand exceeds the amount stocked, he will have the options to order higher levels of stock on the second day. Finally, if day one's demand is less than the amount stocked, he will have options to order any of the lower levels of stock for the second day. Express the problem as a decision tree and find the optimal solution using the cost data given in the fifth exercise question.

7. Megatronics has completed the design of a new graphic display unit for computer systems and is about to decide on whether it should produce one of the major components internally or subcontract it to another local firm. The advisability of which action to take, depends on how the market will respond to the new product. If the demand is high, it is worthwhile to make the extra investment for special facilities and the equipment needed to produce the component internally. For low demand, it is preferable to subcontract. The analyst assigned to study the problem has produced the following information on costs (in lakhs of rupees) and the probability estimates of the future demand for the next five-year period.

Future demand			
Action	Low	Average	High
Produce	14	12	9
Sub-contract	10	11	16
Probability	0.11	0.60	0.30

   (a) Prepare a decision tree that describes the structure of this problem.

   (b) Select the best action based on the initial probability estimates for the future demand.

   (c) Determine the expected cost with perfect information.

8. With reference to the seventh exercise question, the management of Megatronics is planning to hire Dr. Nagaraja Sharma, an economist to prepare an economic forecast for the computer industry. The reliability of his forecasts from previous assignments is provided by the following table of conditional probabilities.

Future demand			
Economic forest	Low	Average	High
Optimistic	0.1	0.1	0.5
Normal	0.3	0.7	0.4
Pessimistic	0.6	0.2	0.1

a) Select the best action for Megatronics, if Dr. Sharma submits a pessimistic forecast for the computer industry.

b) Prepare a decision tree diagram for the problem with the use of Dr. Sharma's forecasts.

c) What is the Bayes' strategy for this problem.

d) Determine the maximum fee that should be paid for the use of Dr. Sharma's services.

9. Mr. Krishna Gauder is an expert in decision-support systems and has been hired by a small software engineering firm to help plan their R & D strategy for the next 6 to 12 months. The company wishes to devote up to 3 person-years or roughly Rs. 20 lakh to the R & D projects. Show how Mr. Gauder can use the decision tree to structure his analysis. State all your assumptions.

Chapter 15

METAHEURISTICS

15.1 INTRODUCTION

We have discussed various kinds of techniques such as simplex method, transportation and assignment problems, game problems, etc. to obtain optimum solution in the previous chapters. These techniques are widely used to find the optimum solution for practical problems. It is observed from past experience that these techniques give compatible results to the problems having linear characteristics.

In practice, some problems are complicated and non-linear in nature. The OR (Operations Research) techniques discussed earlier will not give good feasible solution to non-linear problems. An Heuristic method is a technique used to find the feasible solution to non-linear problems.

Similar to other OR techniques, Heuristic method is also an iterative process. A well designed Heuristic procedure is used to find solution for very large problems. The solution obtained will be feasible but, need not be optimal. However, the solution obtained will be nearly optimum. Heuristic methods should be carefully designed to fit the specific problem. Hence, Heuristic methods are designed to fit a particular type of problem.

Metaheuristics is an OR technique developed to obtain structure and guidelines to design a Heuristic method to fit a specific practical problem.

15.2 NATURE OF METAHEURISTICS

Let us consider a non-linear programming problem to illustrate the nature of Metaheuristics.

$$\text{Maximize } 24X^5 - 1950X^4 + 56000X^3 - 690000X^2 + 3600000X$$

subject to

$$0 \leq X \leq 26$$

Figure 15.1 illustrates variation of objective function $f(x)$ with variable x. The given non-linear problem is a complicated problem. The curve has non-linear characteristics. The value of $f(x)$ increases and then reaches maximum at $x = 5$ and then decreases. Further with increase in x, value $f(x)$ increases once again and reaches maximum at $x = 20$ and decreases again. It is observed from the curve that the characteristics are non-linear and form a non-repetitive pattern.

It shows a special type of non-linear characteristics. A Heuristic method is used to find the better solution for difficult non-linear programming problems like this. Heuristic procedure searches for the better solution iteration by iteration. This procedure continues until no improvement can be obtained for the current solution. Heuristics procedure climbs up higher and higher to maximize the objective function $f(x)$.

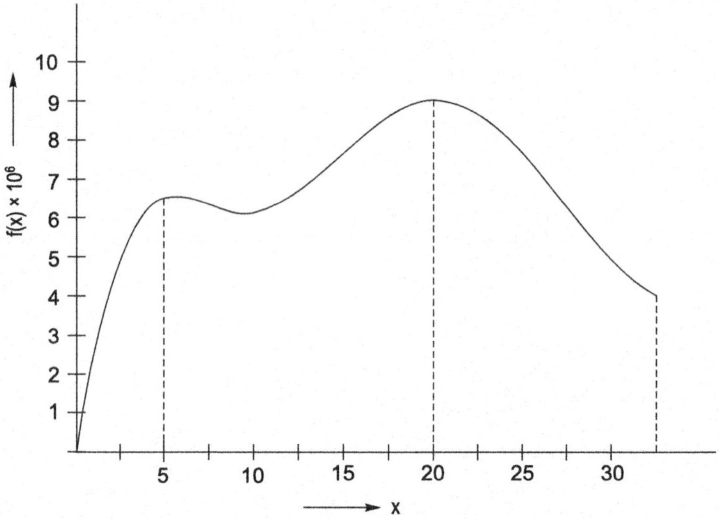

Figure 15.1 Objective function $f(x)$ vs variable x

Therefore, heuristic procedure is also called a hill climbing procedure. It continues to climb till it reaches the top of the hill. Further, if the value of $f(x)$ comes down, this maximum value is called local optimum. This iterative process continues and a number of local optimum values will be determined. Highest among the local optimum values is the global optimum. Value of the objective function in the present illustration $x = 5$ and $x = 20$ gives the maximum $f(x)$ values. They are called the local optimum conditions. Among the local optimum values, $x = 20$ is the highest. Hence, it is called the global optimum condition.

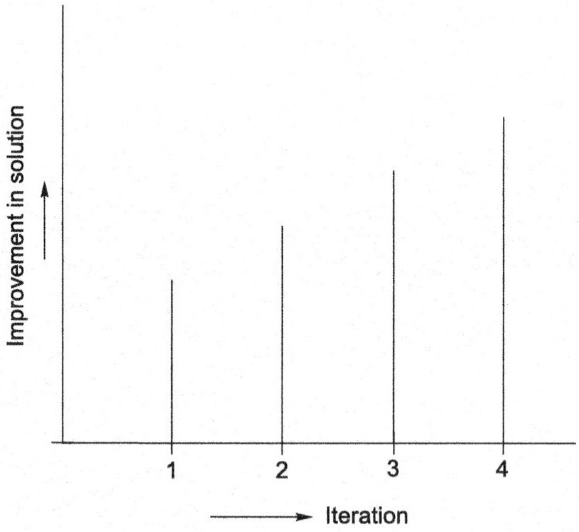

Figure 15.2 Improvement of solution with sequence of iteration

Figure 15.2 illustrates the local improvement in the value of $f(x)$ for increased value of x in successive iterations. Improvement in objective function may be tested for intermediate values such as $x = 1.5$,

$x = 2.5$, $x = 3.5$ etc. This method is called bisection method. Once the local function value reaches the top of the hill, it cannot climb further in the local limits.

A local improvement procedure will converge to one local optimum and then stop. When the local improvement procedure is designed to check the improvement in the neighbourhood region, the global optimum solution can be obtained. Local improvement procedure can be applied a number of times to cover the entire range. This approach gives good results for small problems with one variable, non-linear programming problem, but it may not give optimum solution for multi-variable, non-linear programming problems with complications.

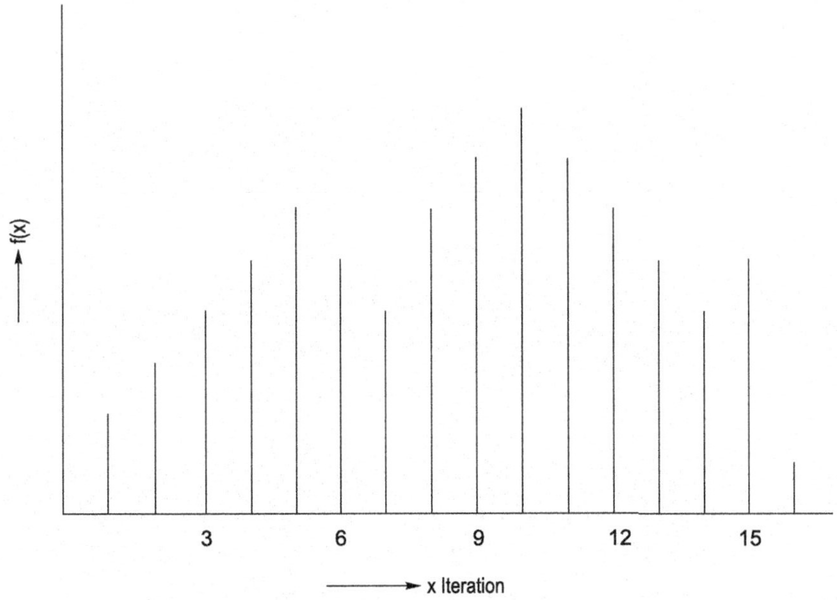

Figure 15.3 Improvement of $f(x)$ value with iterations

Figure 15.3 illustrates the global optimal solution obtained by moving from one local region to the other quickly. The disadvantage of this iterative process is that there is no guarantee of getting the optimum solution. A well-designed procedure will give feasible solution which is approximately optimal. The major area of application of metaheuristics is combinatorial optimization problems.

15.3 A TRAVELLING SALESMAN PROBLEM

Travelling salesman problems discussed in this chapter under, assignment problems are a category of combinatorial (i.e., combination of variables selected from a large group) optimization problems. As discussed in the Branch and Bound method of assignment problems, the objective of travelling salesman problems is to optimize the distance travelled by the salesman.

In travelling salesman problems, the salesman travels to a number of cities for sales. Starting from his home city, he wishes to find the route to visit the cities so as to minimize the total distance travelled.

***Example* 15.1**

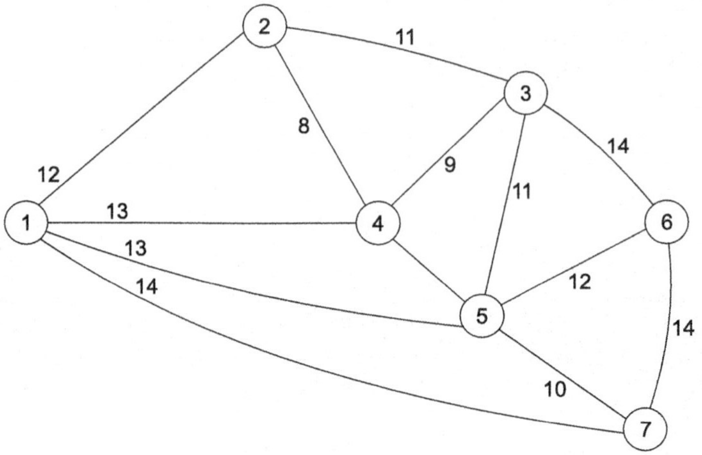

Figure 15.4 A travelling salesman problem—network representation

Figure 15.4 illustrates a travelling salesman problem. Here the salesman starts from his home city 6 and travels to all the 7 cities and comes back to his home city 6. Salesman is required to visit all the remaining cities once, before returning back to his home city. The distance between the two cities is given in Figure 15.4. The objective of the salesman is to determine the route for travel so that the total distance travelled is minimized.

The difficulty with the travelling salesman problem increases with increase in the number of cities to be travelled.

In a travelling salesman problem with n cities connected to one another, y the salesman can have $\frac{(n-1)!}{2}$ possible routes to travel from his home city and come back. For example, 10 city problems can have $\frac{(10-1)!}{2} = \frac{1\times 2\times 3\times 4\times 5\times 6\times 9}{2} = 1,81,440$ routes.

The Heuristic methods are commonly used to solve these problems. A sequence of solutions are generated using Heuristic procedure. Each new solution is obtained by making some adjustments.

For example, consider the route

$$6-7-5-1-4-2-3-6$$

In this route the distance travelled is 14 + 10 + 13 + 13 + 8 + 11 + 14 = 83

In the next step, if we consider to travel from 5 to 1 and reverse the journey, the new route will be

$$6-7-1-5-4-2-3-6$$

$$\text{The distance travelled} = 14 + 14 + 13 + 5 + 8 + 11 + 14$$
$$= 79$$

This small adjustment in the sequence of cities visited by the salesman is called a sub-tour reversal process.

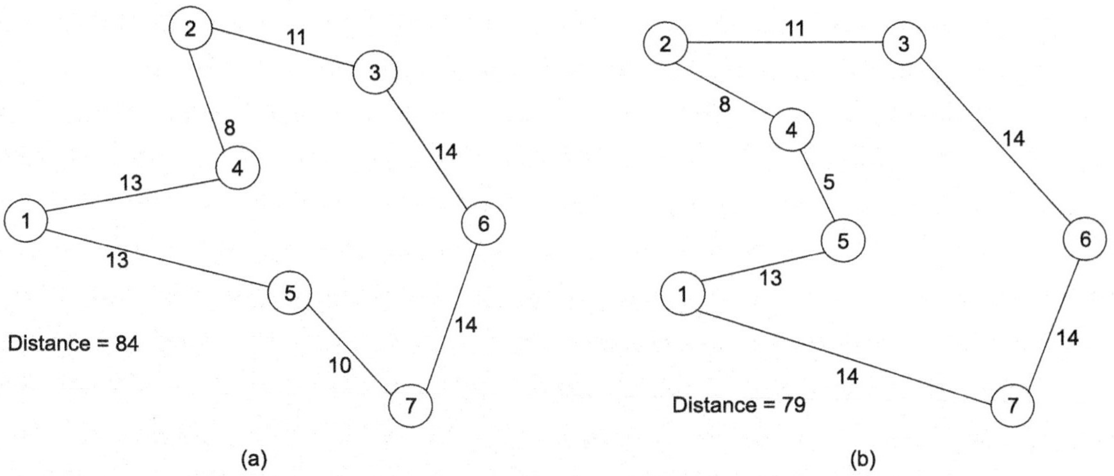

Figure 15.5 (a) & (b): Sub-tour reversal process

Figure 15.5 illustrates the sub-tour reversal process. In this process, Figure 15.5(a) shows the initial route 6 – 7 – 5 – 1 – 4 – 2 – 3 – 6 and Figure 15.5(b) shows the new solution 6 – 7 – 1 – 5 – 4 – 2 – 3 –6. Here, the link 7 – 5 is deleted and 1 to 7 is added. Similarly, 1 to 4 is deleted and 5 to 4 is added to reverse the movement from 5 to 1 as 1 to 5. Let us look for other possibilities to improve the solution.

Consider the route

6 – 7 – 5 – 1 – 4 – 2 – 3 – 6 Distance = 83

Iteration I

Reverse 7 – 5 to get

6 – 5 – 7 – 1 – 4 – 2 – 3 – 6

12 + 10 + 14 + 13 + 8 + 11 + 14 Distance = 82

Iteration II

Reverse 5 – 1 to get

6 – 7 – 1 – 5 – 4 – 2 – 3 – 6

14 + 14 + 13 + 5 + 8 + 11 + 14 Distance = 79

Iteration III

Reverse 1 – 4 to get

6 – 7 – 5 – 4 – 1 – 2 – 3 – 6

14 + 10 + 5 + 13 + 12 + 11 + 14 Distance = 80

Iteration IV

Reverse 4 – 2 to get

6 – 7 – 5 – 1 – 2 – 4 – 3 – 6

14 + 10 + 13 + 12 + 8 + 9 + 14 Distance = 80

The first iteration has improved the result and reduced the distance from 84 to 82 and the second iteration has reduced the distance from 82 to 79. Subsequent iterations have not improved the solution.

Consider the route

6 – 7 – 1 – 5 – 4 – 2 – 3 – 6 Distance = 79

Iteration V

Here, we can reverse the travel between three cities

Reverse 5 – 4 – 2 as 2 – 4 – 5

New route is

6 – 7 – 1 – 2 – 4 – 5 – 3 – 6

14 + 14 + 12 + 8 + 5 + 11 + 14 Distance = 78

Iteration VI

Reverse 4 – 5 – 3 as 3 – 4 – 5 (Refer Figure 15.4)

New route is

6 – 7 – 1 – 2 – 3 – 47 – 5 – 6

14 + 14 + 12 + 11 + 9 + 5 + 12 Distance = 77

Further reversal will not improve the solution. Hence sub-tour reversal process stops at this stage. The sub-tour reversal process is another type of local improvement process of Heuristics procedure.

15.4 TABU SEARCH

Tabu search is a popular Metaheuristic method used to search for optimum solution for a non-linear programming problem. Tabu search is a local search procedure similar to local improvement procedure discussed earlier in this chapter. The only difference is that it does not require the solution at each iteration and will be better than the preceding solution. This procedure accepts the solution at each iteration and continues to find local optimum solution. Once local optimum solution is found, this procedure will search for better solution in the neighbourhood.

The solution obtained for successive iterations will go up the hill, if upward trend is not available, select the solution that drops down the hill and follow a pattern and search for global optimum. The disadvantage of Tabu search procedure is that, there is a possibility of repetition of local optimum solution. To avoid the repeatability of the solution value, the Tabu search algorithm records all the solutions and forbids the repeatability of the solution. This procedure incorporates long-term memory to implement the concept.

15.4.1 Steps in Tabu Search Method

Step 1 Start with an initial feasible solution.

Step 2 Adopt a local search procedure to move into the neighbourhood of the present solution.

Step 3 Eliminate the solution of an iteration on the Tabu list, if it is not a better solution than the best solution on the list.

Step 4 Determine a move that determines the best solution.

Step 5 Use this solution as the next reference solution, whether it is a better solution than the present solution or not.

Step 6 Update the Tabu list and see that this best solution will not be repeated.

Step 7 If the Tabu list is full, eliminate the older solution and update the list towards the optimum solution.

Step 8 Adopt some stopping criterion.

(a) Fix the number of iterations.

(b) Fix the amount of CPU time.

(c) Fix the number of iterations without improvement in the best objective function value.

(d) Stop the search if there is no feasible move into the neighbourhood of the present solution.

Step 9 Accept the best solution determined on any iteration as the best solution.

15.5 SPANNING TREE

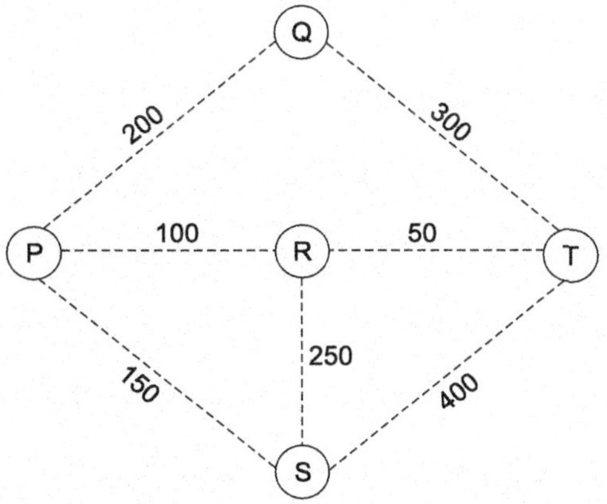

Figure 15.6 Spanning tree—a network representation

Figure 15.6 illustrates a network representation of a spanning tree. The spanning tree is a network with a number of nodes with dashed lines. The dashed lines between the nodes indicates the potential to insert a link into the network and the associated number represents the cost associated with the introduction of a particular link. Figure 15.6 has five nodes *P*, *Q*, *R*, *S* and *T* connected by dashed lines.

The main objective is to insert a link between every pair of node such that the total cost associated with it is minimum and no best solution is repeated.

Example 15.2

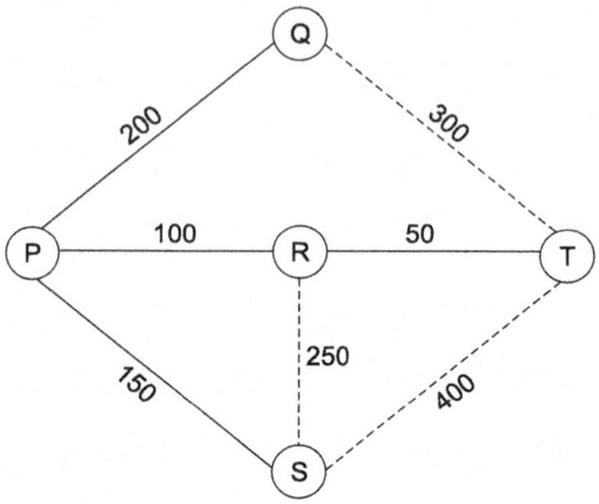

Figure 15.7 Minimum spanning tree

Figure 15.7 illustrates the minimum spanning tree. It is four links with total cost of Rs. 500.

Let

Constraint 1 Link *PS* can be inserted if link *ST* is inserted.

Constraint 2 Atmost, one of *PS*, *RS* and *PQ* can be inserted.

Let us charge a penalty for the violation of constraints

1. If constraint 1 is violated, charge Rs. 1000.
2. If constraint 2 is violated, charge Rs. 1000 for each of the additional links.

Determine the optimum solution using Tabu search method.

Solution

The given initial solution violates both the constraints given in the problem.

Penalty for violation of constraint	=	Rs. 1000
Penalty for addition of a link	=	Rs. 1000
Total penalty	=	Rs. 2000
Total cost	=	200 + 100 + 50 + 150 + 2000 (penalty)
	=	Rs. 2500

Iteration 1

To fulfill constraint 1, add link *ST* attached with cost Rs. 400.

To fulfill constraint 2, delete *PS* link attached with cost Rs. 150.

Now the spanning tree is as follows.

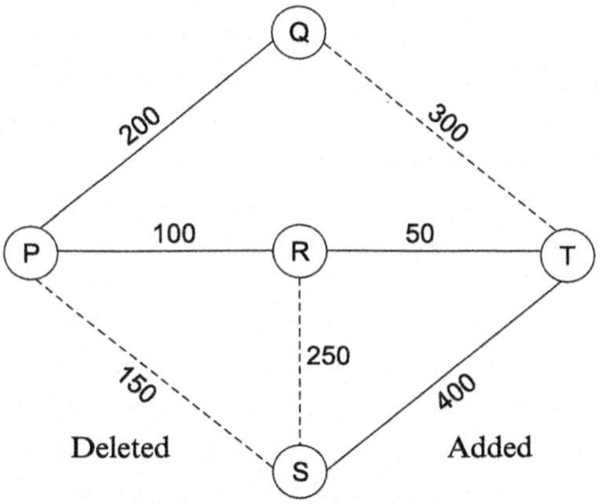

Figure 15.8 Local optimum

Now total cost = 200 + 100 + 50 + 400 = Rs. 750

Penalty is eliminated by fulfilling the constraints.

Iteration 2

Options for adding a link and deleting a link

Add	Delete	Cost	
PS	*ST*	—	Should not be repeated
PS	*RT*	850 + 1000	(Violates constraint 2)
PS	*PR*	800 + 1000	

Add	Delete	Cost	
QT	*RT*	1000	
QT	*PR*	950	
QT	*PQ*	850	Minimum
RS	*ST*	—	(should not be repeated)
RS	*RT*	950 + 1000	

From iteration 2, it is observed that, it is optimum to add link *QT* and delete link *PQ*.

The spanning tree is as follows

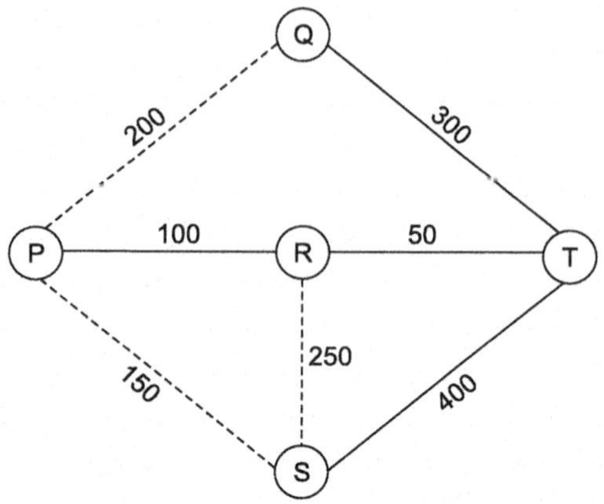

Figure 15.9 Modified spanning tree

Total cost = 300 + 100 + 50 + 400 = Rs. 850

Iteration 3

Options for adding a link and deleting a link

Add	Delete	Cost	
PQ	*QT*	–	Should not be repeated
PQ	*RT*	1000	
PQ	*PR*	950	
PS	*ST*	–	Should not be repeated
PS	*RT*	950	
PS	*PR*	900	
RS	*ST*	700	Minimum (repeated)
RS	*RT*	1050	

In iteration 3, deletion of *ST* and addition of *RS* is considered. Even though the cycle is repeated, the cost of the option is minimum. Now, the spanning tree is as follows

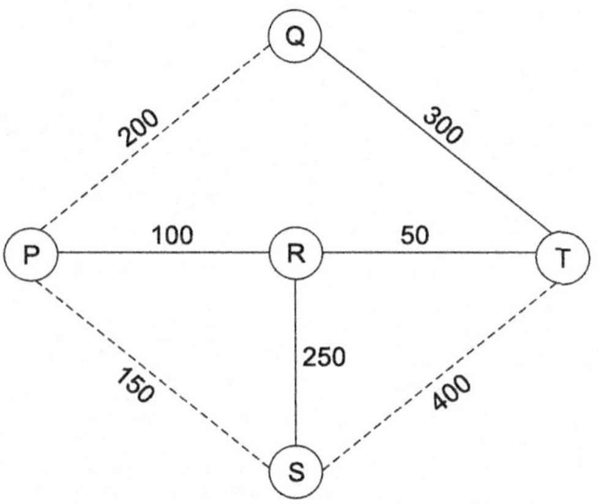

Figure 15.10 Optimum spanning tree

The total cost = 100 + 300 + 250 + 50 = Rs. 700

Iteration 4, 5 and 6 will continue, but the optimal or global solution will not improve.

15.6 SIMULATED ANNEALING

Simulated annealing is another Metaheuristic method used to determine the global optimum and feasible solution for a non-linear programming problem.

Annealing is a heat treatment process by which the material is heated to a certain temperature and then cooled with reference to time. Annealing is done to control the structure of lattice of atoms for improving the physical properties of the material. Simulated annealing is analogous to the physical annealing process used to improve the solution by iteration.

Unlike the Tabu search method, in this method, the solution is determined at successive iterations such that, this approach will accept some of the steps which have upward trend and reject some steps which have downward trend. In Tabu search method, most time is spent for climbing each hill. As soon as the top of a hill is reached, it will start climbing downward slowly searching for the tallest hill.

The simulated annealing technique will search for better solution with upward trend and continue the process till the top of the tallest hill is reached. Each iteration of the simulated annealing moves from local optimum to the neighbourhood of the solution. But unlike Tabu search, this process will accept the next neighbour, if it has upward trend, otherwise it will reject the next solution.

Let

Z_p be the objective function of present solution

Z_n be the objective function of next solution to be accepted

T be the tendency of acceptance

If $Z_n > n_p$ the next solution will be accepted.

If $Z_n < Z_p$, the solution will be accepted with the probability $P_{(a)}$.

Where

$$X = \frac{Zn - Zp}{T}$$

If the solution under consideration is better than the present solution it is accepted to be the next solution. If not, the probability of acceptance depends on the value of X.

Table 15.1 Normal distribution table

$X = \frac{Zn - Zp}{T}$	Probability of acceptance $P_{(a)}$
– 5	0.007
– 4	0.018
– 3	0.050
– 2	0.135
– 1	0.368
– 0.5	0.607
– 0.25	0.779
– 0.1	0.905
– 0.01	0.990

Table 15.1, shows the variation of probability of acceptance of the solution for an iteration with the variation in the value of X.

If the $P_{(a)}$ value for a particular iteration is less than the acceptable value, such iteration will be accepted for downward move.

Chapter 16

SIMULATION

16.1 INTRODUCTION

According to R.E Shennon, simulation is the process of designing a model of a real system and conducting experiments with this model for the purpose of either understanding the behaviour of the system or evaluating various strategies within the limits imposed by a set of criteria for the operation of the system.

The term simulation is used in different ways by different people. Simulation deals with the mechanism that controls the behaviour of a system. Simulation can be used to predict the future behaviour of a system and determine the course of action that influence the future behaviour.

Simulation is a powerful tool that provides evaluation procedure for alternative design, plans and policies, without any experimentation on a real system. The objective of simulation is to identify and quantify the risks associated with a particular design, plan or policy.

The simulation approaches are of two types, viz., deterministic and probabilistic. Deterministic simulation provides insight into the mechanism, but is not suited for making predictions to support decision-making. It cannot quantitatively address the risks and uncertainties present within the system.

Probabilistic simulation represents the uncertainties by specifying inputs as probabilistic distributions. If the inputs of a system are uncertain, the prediction of future performance of that system is also uncertain. The results of analysis of inputs based on probabilistic distribution are probabilistic in nature.

The result of a deterministic simulation of an uncertain system is a qualified statement whereas the results of a probabilistic simulation of a system is a quantified probability. The probabilistic approach of simulation method is useful for decision-making.

16.2 SIMULATION PROCESS

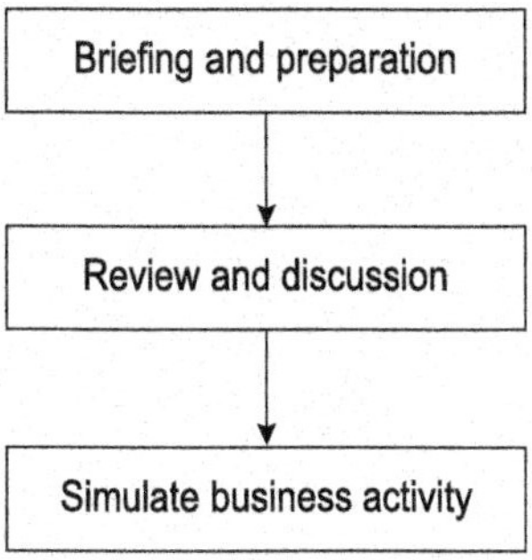

Figure 16.1 Simulation process

Figure 16.1 illustrates the simulation process. The basic simulation process consists of three stages.

In the first stage, all the inputs related to a system are collected and the problem is defined. The resources available with reference to all the parameters defined for a system are analysed.

In the second stage, depending on the nature of the system under study, the problem is solved using either deterministic or probabilistic approach of simulation method, and the results are obtained.

In the third stage, simulation results are reviewed and the consequences are discussed to take a decision.

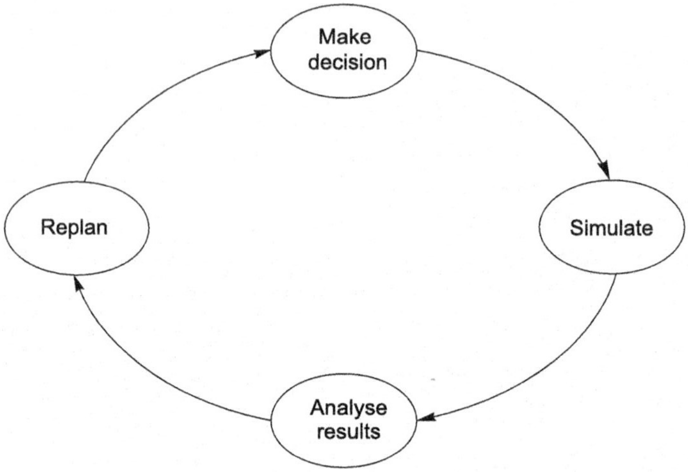

Figure 16.2 Decision cycle

From Figure 16.2, it is observed that simulation is a cyclic process. The decision taken in a certain period is reviewed, discussed and compared. For each period, depending on the changes in the requirements of the business, different input parameters are taken into consideration and subjected to the simulation process. The results are obtained using deterministic or probabilistic tools of the simulation method. The results are then analysed taking into consideration the pros and cons of the results, and the course of action is replanned. Based on the new plan, the decision-making is done. This is a cyclic process. As the future is always uncertain, this decision cycle is repeated at regular intervals of time.

16.3 TYPES OF SIMULATION

The simulation technique is divided into six categories.

1. Total enterprise simulation
2. Interactive simulation
3. Non-interactive simulation
4. Functional simulation
5. Concepts simulation
6. Planning simulation
7. Process simulation

16.3.1 Total Enterprise Simulation

Total enterprise simulation involves modelling of business functions such as marketing, finance operation, etc. Therefore, this type of simulation is called total enterprise simulation. Total enterprise simulation is divided into four sub-categories.

(a) Strategy simulation
(b) Appreciation/Acumen simulation
(c) Tactical simulation
(d) Totality simulation

(a) Strategy simulation

This type of simulation deals with strategic management of business. The simulation model developed is complex and requires more time for analysis. This simulation method is concerned with the management of the total enterprise. The decision involves marketing, product development and finance. Capacity and system improvement are focused in this type of simulation. Middle-level and senior managers use this simulation method to develop strategies for decision-making.

(b) Appreciation/Acumen simulation

These simulation methods are used to show how a business operates, and develops a basic financial knowledge. These types of simulations are of intermediate complexity. It takes a minimum period of time. They can be used every day for solving day-to-day problems. The problems handled are not challenging in nature. Problems related to accounts, sales, and demand are analysed using this method. Junior to medium-level managers use this method of simulation.

(c) Tactical simulation

Tactical simulation methods deal with the tactical management of a business. The strategic simulation focuses on the external environment, whereas tactical simulation concentrates on the internal management of the business. This type of simulation methods are complex. They are concerned with factory and financial measures of the efficient operation of a business, efficient use of resources, budgetary control, cash flow and meeting customer demand.

This type of simulation results are reviewed quarterly or once in a month. Tactical simulation approaches are deterministic in nature. These simulation methods are used by top-level managers.

(d) Totality simulation

Totality simulation replicates the operation of a real business. This simulation method combines the characteristics of both strategic and tactical simulations and is very complex. It involves a hundred or even more decisions to be made every time. Hence, it takes considerable time to yield results. These simulations are commonly used in academic institutions. The results are reviewed once in a semester.

16.3.2 Interactive Simulation

As the name implies, this type of simulation requires interaction between individuals or groups for collecting data and input information. Based on the interaction results, the analysis of the problem is carried out to find a course of action for the system. Interactive simulation methods are used in the marketing division and product design.

16.3.3 Non-interactive Simulation

In this simulation method, a group of workers work independently. Every individual or a group of individuals will be working on a specified work component on the shop floor. The individuals or groups do not require interaction as their decision areas, and the problems to be analysed are entirely independent and not related to one another.

16.3.4 Functional Simulation

Functional simulation focuses on a specific functional area such as sales, marketing, production, etc. Even though this type of simulation focuses on decisions related to functional areas, the financial aspects are also equally important. These simulations are complex. In this simulation method, multiple decisions are made. The level of importance of each decision is different.

Functional simulations are used by managers of all hierarchical levels in the organisational structure. Hence some decisions are of strategic nature, some of them are tactical and some are production-related.

The time required to obtain the results will vary and is of complex nature.

For example, production management simulation includes decision-making to setup the factory which requires strategic simulation. Next, decision-making is associated with running the factory. Related activities are scheduling of production, supply of material, etc. The simulation for making decision to setup the factory are done at the time of establishment of the factory. The scheduling decisions are made every month repeatedly.

Sales and marketing areas require tactical or strategic simulation which is interactive in nature to take decisions. Other functional areas do not require interaction.

16.3.5 Concepts Simulation

Concepts simulation focuses on business concepts. These decisions are simplex and involve three to four decisions to be taken each period. Here the results are obtained within a day. As these simulations are of short duration, they do not require interaction. New product launch simulation is an example of concepts simulation.

16.3.6 Planning Simulation

The preparation of a business plan such as development of an annual budget is an example of planning simulation. This type of simulation involves interaction. The planning simulation process consists of preparation, simulation and review as illustrated in Figure 16.2. There is no restriction to analyse and plan the simulation process. The number of plans are developed and analysed to take decision based on the careful examination of the results. Each group will develop a different plan to meet different objectives and produce a different plan. Planning simulations help managers to understand business implications. Even though a manager knows about the financial ratio, balance sheet and profit and loss, a knowledge of planning simulation helps managers to understand the managerial actions.

16.3.7 Process Simulation

Process simulation uses the mathematical techniques like sales performance analysis, inventory management, and forecasting, to determine the course of action. In this method, the quantitative and qualitative parameters are balanced to interpret the results. The different sets of data are taken into consideration to determine the outcome. The data sets considered for analysis illustrate different situations under study.

16.4 THE BASIC STEPS IN SIMULATION

Irrespective of the type of problem and its objective, the steps involved in the simulation process are the same. The basic steps involved in the simulation process are explained below.

Step 1 Problem Definition

In the first step of the simulation process, the goal of the study is defined with reference to the objectives and observations of the simulation process. The problem to be solved is determined.

Step 2 Project planning

The task to be carried out is divided into a number of work components. Each work component is assigned to a responsible person and a target is set for each individual to know the progress of the work. This work planning is essential for completion of the project in sufficient time with available resources.

Step 3 System Definition

In this step of simulation, the system is defined by identifying the components of the system model. The system defined is so complex and it requires an experienced and skilled manager to find the details and flexibility of the system defined. It is also required to analyse the measures of performance.

Step 4 Model formulation

In this step of simulation, basic requirements necessary to develop the simulation model are determined. The actual behaviour of the system is analysed and a flowchart of the operation of the system is developed. This flowchart is used to analyse the interaction of the variables involved in the simulation process.

Step 5 Input data collection and analysis

Once the model is formulated, the existing data are gathered and the new data are collected and used to determine the distribution pattern such as normal distribution curve. For example, the data related to the delivery of parts in a manufacturing plant.

Step 6 Model translation The model developed is translated into simulation programs.

Step 7 Verification and validation

In this step, the behaviour of the model is verified by debugging and animation, and compared with the expected results for its validation. The validation is done to ensure that there is no difference between the model and the real system. Statistical analysis helps to achieve the validation of the software results.

An expert may review the model to know the validity of the model.

Step 8 Experimentation and analysis

In this step, alternative models are developed and the simulation executed. The statistical results of all the alternative models are compared with the real system and the performance is analysed.

Step 9 Documentation and implementation

Documentation is the preparation of the written report and its presentation. This documentation gives the details of the results and discusses the implications of the simulation carried out. The best strategy is identified, recommended, and justified.

16.5 WHEN TO USE SIMULATION?

In early days, large and complex problems were analysed by specialized and trained programmers. Nowadays large number of software are available for simulation. Managers lacking proper training and experience are using these tools for solving problems inappropriately. Such inappropriate analysis will not give good results and the actual goal of simulation approach cannot be achieved.

The simulation approach cannot be used for solving all types of problems. The following factors should be evaluated before selecting the simulation process to find an amicable solution for a problem.

1. Type of problem
2. Availability of resources
3. Costs
4. Availability of data

16.5.1 Type of Problem

If any problem can be solved by an analytical method, use of simulation method is not required. Use of algorithms and mathematical equations for solving a problem is faster and not expensive. If a problem can be solved by conducting direct experimentation without disturbing the real system, experimentation is more appropriate than simulation study. Also, if the system is too complex, not understandable and cannot be defined, the simulation process will not give fruitful results.

16.5.2 Availability of Resources

A trained and experienced person is an important resource for conducting simulation analysis. Such a resource person is essential to develop a simulation model and to verify and validate the model. If the simulator is not experienced and well-trained, reliable results cannot be obtained.

Sufficient time should be allocated to carry out the simulation analysis. Enough time should be available for verification, validation and implementation of any changes required to get optimum results.

16.5.3 Costs

The computer and its accessories, and simulation software involves huge amount of money. If the savings with the application of simulation technique does not meet the amount required to purchase a computer and simulation software, it is not economical to opt for simulation.

16.5.4 Availability of Data

The data necessary for the analysis should be identified and located. If the data is not available, it should be collected. If the data does not exist, the simulation analysis will not give reliable results. The results cannot be compared with the performance of real system.

Figure 16.3 illustrates the steps and decision for conducting a simulation study.

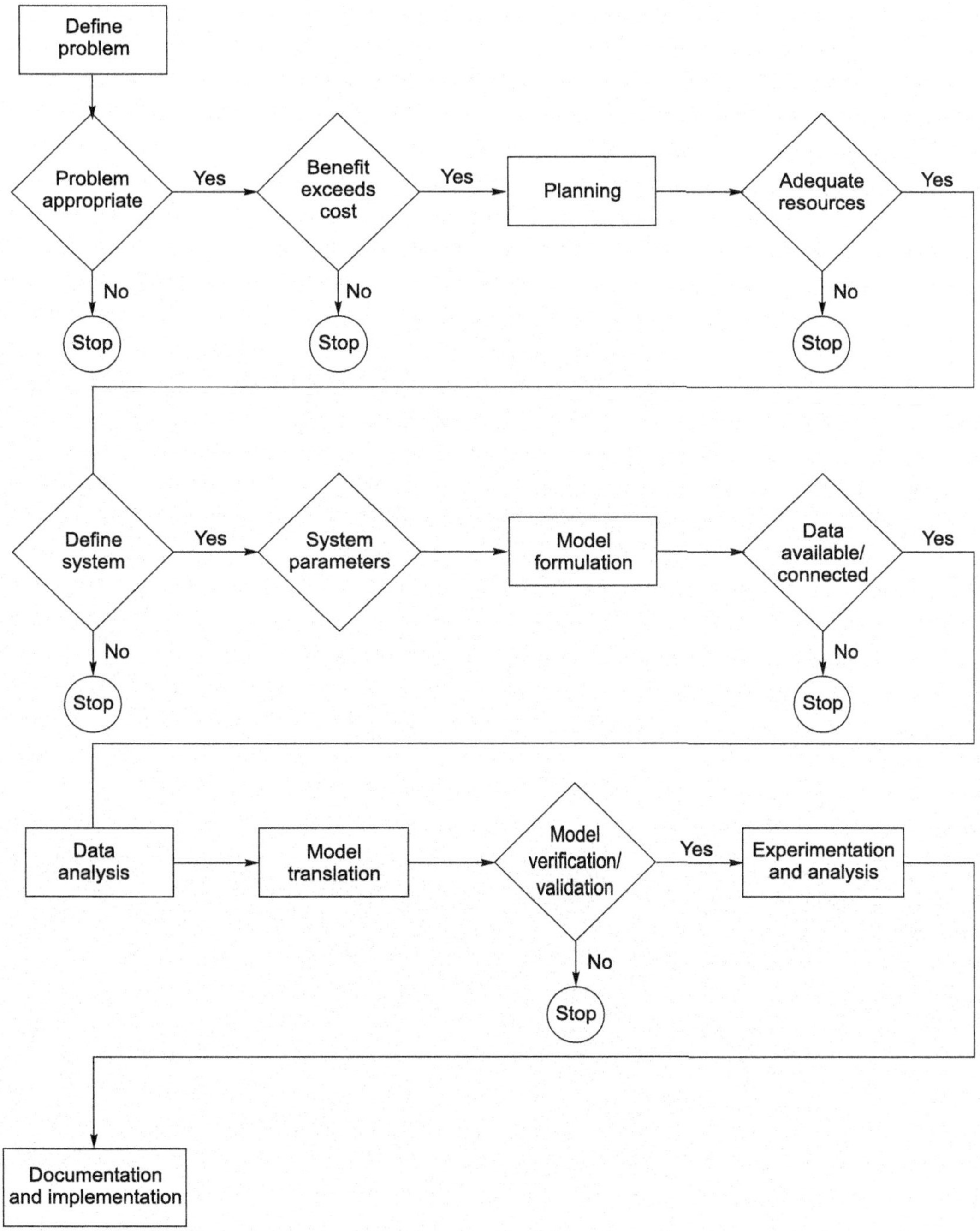

Figure 16.3 Flowchart of simulation analysis

16.6 MONTE CARLO SIMULATION

Monte Carlo simulation is a type of simulation technique used to find a solution for a problem whenever it is not possible to do experimentation and to create a mathematical equation. It is a simulation method based on sampling technique. In this method of simulation, probability distribution of the variables is determined. Then a set of values having characteristics similar to the real system are generated using a set of random numbers.

16.6.1 Steps in Monte Carlo Simulation

The following are the basic steps in the Monte Carlo simulation technique.

Step 1 Select the measure of effectiveness of the problem.

Step 2 Identify the variables influencing the measure of effectiveness.

Step 3 Determine the cumulative probability distribution of each variable identified above.

Step 4 Plot a curve with probability on the vertical axis and the value of variables on horizontal axis.

Step 5 Select a set of random numbers.

Step 6 Consider each random number selected above as a decimal value of the cumulative probability along the Y axis.

Step 7 With each decimal value, draw a horizontal line from the vertical axis to intersect the probability curve and then project a line down vertically on to horizontal axis.

Step 8 Record the value on horizontal axis obtained for all the decimal values.

Step 9 Using the value obtained above in the formula derived from measure of effectiveness, the solution is obtained. This value is the measure of effectiveness for that simulated value.

Step 10 Repeat steps 7, 8 and 9 until the satisfactory results are obtained.

Example 16.1

A general stores owner has bought new fountain pens. He sells them at Rs. 37.5 per unit. The cost of manufacturing each pen is Rs. 30 per unit. A market survey estimates the sales pattern as follows

Demand units/month	Probability
0 < 300	0.18
300 < 600	0.32
600 < 900	0.25
900 < 1200	0.15
1200 < 1500	0.06
1500 < 1800	0.04

A private school has contracted for 750 pens per month from the stores. The unsold pens are returned to the manufacturer. He will take them back at a cost less Rs. 4 per unit for handling charges. The market analyst wants to simulate the demand and profitability. The following random numbers may be used for simulation. 27, 15, 56, 17, 98, 71, 51, 32, 62, 83, 96 and 69.

You are required to

1. Allocate random numbers to the demand pattern forecast by the market analyst.
2. Simulate twelve months sales and calculate the monthly and annual profit or loss.
3. Calculate the loss on lost sales.

Solution

1. Allocation of random numbers

Demand	Probability	Cumulative probability	Allocated random number
0 < 300	0.18	0.18	00 – 17
300 < 600	0.32	0.50	18 – 49
600 < 900	.25	0.75	50 – 74
900 < 1200	0.15	0.90	75 – 89
1200 < 1500	0.06	0.96	90 – 95
1500 < 1800	0.04	1.00	96 – 99

2. Simulation: twelve months sales, monthly and annual profit/loss

Month	Random number	Demand	Sold	Return	Profit M sales	Loss on return	Net loss on lost sales	
1	27	450	450	300	3375	12,000	2175	
2	15	150	150	600	1125	2400	–1275	
3	56	750	750	–	5625	–	5625	
4	17	150	150	600	1125	2400	1275	
5	98	1650	750	–	5625	–	5625	900
6	71	750	750	–	5625	–	5625	
7	51	750	750	–	5625	–	2175	
8	32	450	450	300	3375	1200	5625	
9	62	750	750	–	5625	–	5625	300
10	83	1050	750	–	5625	–	5625	900
11	96	1650	750	–	5625	–	5625	
12	69	750	750	–	5625	–	5625	
					54,000	7200	46,800	2100

3. Loss on lost sale

2100 × 7.5 = Rs. 15,750

Example 16.2

Maruthi Suzuki motor company manufactures 40 cars per day, the sale of cars depending upon demand and which has the following distributions.

Sales of cars	37	38	39	40	41	42
Probability	0.10	0.15	0.20	0.35	0.15	0.05

The production cost and sales price of each car are Rs. 4 lakhs and Rs. 5 lakhs respectively. Any unsold car is to be disposed off at a loss of 2 lakh per car. There is a penalty of Rs. 1 lakh per car, if the demand is not met. Using the following random numbers, estimate the total profit/loss for the company for the next ten days random numbers : 9, 98, 64, 98, 94, 01, 78, 10, 15, and 19.

If the company decides to produce 39 cars per day, what will be its impact on profitability.

Solution

Allocate the random numbers from 00 to 99 in proportion to the probabilities associated with sales of cars as given below.

Sales of car	Probability	Cumulative probability	Range of random number
37	0.10	0.10	00 – 09
38	0.15	0.25	10 – 24
39	0.20	0.45	25 – 44
40	0.35	0.80	45 – 79
41	0.15	0.95	80 – 94
42	0.05	1.00	95 – 98

Based on the given random numbers, simulate the estimated sales and calculate the profit/loss on the basis of specified units of production.

Day	Random number	Estimated sale	Profit [Production 40 cars/day Rs. lakh]	Profit [Production 39 cars/day Rs. lakh]
1	9	37	$37 \times 1 - (3 \times 2) = 31$	$37 \times 1 - (2 \times 2) = 33$
2	98	42	$40 \times 1 - (2 \times 1) = 38$	$39 \times 1 - (3 \times 1) = 36$
3	64	40	$40 \times 1 = 40$	$39 \times 1 - (1 \times 1) = 38$
4	98	42	$40 \times 1 - (2 \times 1) = 38$	$39 \times 1 - (3 \times 1) = 36$
5	94	41	$40 \times 1 - (1 \times 1) = 39$	$39 \times 1 - (2 \times 1) = 37$
6	01	37	$37 \times 1 - (3 \times 2) = 31$	$37 \times 1 - (2 \times 2) = 34$
7	78	40	$40 \times 1 = 40$	$39 \times 1 - (1 \times 1) = 38$
8	10	38	$38 \times 1 - (2 \times 2) = 34$	$38 \times 1 - (1 \times 2) = 36$
9	15	38	$38 \times 1 - (2 \times 2) = 34$	$38 \times 1 - (1 \times 2) = 36$
10	19	38	$38 \times 1 - (2 \times 2) = 34$	$36 \times 1 - (1 \times 2) = 36$

If the company reduces production to 39 cars per day, there will be no additional profit or loss.

Example 16.3

A single-counter ticket booking centre in a cinema theatre has appointed one booking clerk. A customer on arrival immediately goes to the booking counter for being served, if the counter is free. If on the other hand, the counter is engaged, the customer has to wait. The customers are served on first come first served basis. The time of arrival and the time of service varies from one minute to six minutes. The distribution and service time is as follows

Arrival/service time (minute)	Arrival (Probability)	Service (Probability)
1	0.05	0.1
2	0.20	0.2
3	0.35	0.4
4	0.25	0.2
5	0.10	0.10
6	0.05	–

(a) Simulate the arrival and service of 10 passengers starting from 9 a.m. using the following random numbers in pairs respectively for arrival and service. Random numbers: 60, 09, 16, 12, 08, 18, 36, 65, 38, 25, 07, 11, 08, 79, 59, 61, 53, 77, 03, and 10.

(b) Determine the total duration of

i. Idle time of booking clerk and

ii. Waiting time of passengers.

Solution

Random allocation tables are as follows

Time minutes	Arrival (Probability)	Arrivals cumulative (Probability)	Random number allocated	Time minutes	Service (Probability)	Service cumulative (Probability)	Random number
1	0.05	0.05	00 – 04	1	0.10	0.10	00 – 09
2	0.20	0.25	05 – 24	2	0.20	0.30	10 – 29
3	0.35	0.60	25 – 59	3	0.40	0.70	30 – 69
4	0.25	0.85	60 – 84	4	0.20	0.90	70 – 89
5	0.10	0.95	85 – 94	5	0.10	1.00	90 – 99
6	0.05	1.00	95 – 99				

Simulation of 10 trials:

Ratio	Arrival minutes	Time	Start	Random number	Time minutes	Finish time	Waiting clerk	Time passenger
60	4	9.04	9.04	09	1	9.05	4	
16	2	9.06	9.06	12	2	9.08	1	
08	2	9.08	9.08	18	2	9.10		
36	3	9.11	9.11	65	3	9.14	1	
38	3	9.14	9.14	25	2	9.16		
07	2	9.16	9.16	11	2	9.18		
08	2	9.18	9.18	79	4	9.22		
59	3	9.21	9.22	61	3	9.25		1
53	3	9.24	9.25	77	4	9.59		1
03	1	9.25	9.29	10	2	9.31		4
	Total						6	6

In half an hour trial, the clerk was idle for 6 minutes and the customers had to wait for 6 minutes.

16.8 ADVANTAGES AND DISADVANTAGES OF SIMULATION

16.8.1 Advantages

1. Simulation testing is cheaper and faster than performing the multiple tests of the design each time.
2. High level of details can be obtained from the simulation process.
3. A simulation can give results that are not experimentally measurable with our present-day technology.
4. Simulation avoids danger and loss of life.
5. Input conditions can be varied and outcomes can be investigated.
6. Critical situations can be analysed without risk using the simulation process.
7. Simulation is cost-effective.
8. Simulation software is helpful to study the behaviour of a model over a long period of time for different input conditions.

16.8.2 Disadvantages

1. The major disadvantage is the simulation error.
2. It requires experimental results as base to generate accurate results for all other input conditions.
3. Simulation can be expensive to measure how one thing affects another to take the initial measurements, and to create the model itself.
4. To simulate a problem, knowledge of all the factors involved is essential.

EXERCISES

1. Generate 1000 daily "returns" $X_i = 1, 2, ...,1000$ from each of the two distributions, the Cauchy and the logistic. In each case, assume that $a = 0$, $b = 0.1$. Graph the total return over an n day period versus n. Is there a qualitative difference in the two graphs? Repeat with a graph of the average daily return.
2. Assume that you have available a uniform [0, 1] random number generator. Give a precise algorithm for generating observation from a distribution with probability density function $f(x) = e–x$, $0 \cdot x < \infty$. Record the time necessary to generate the sample mean of 10,00,000 random variables with this distribution. Compare with the sample mean. How large would the simulation need to be if we wanted to estimate the mean within 0.01 with a 95% confidence interval ?
3. Assume that you have available a uniform [0, 1] random number generator. Give a precise algorithm for generating observation from a distribution with probability density function $f(x) = x_3$, $0 < x < 0 < x < 1$ with this distribution. Determine the standard error of the sample mean. How large would the simulation need to be if we wanted to estimate the mean within 0.01 with a 95% confidence interval?
4. Assume that you have available a uniform [0, 1] random number generator. Give a precise algorithm for generating observations from a discrete distribution with probability function $x = 0, 1, 2, 3, 4, 5$, $P[X, x] = 0.1, 0.2, 0.25, 0.3, 0.1, 0.05$. Record the time necessary to generate the sample mean of 10,00,000 random variable with this distribution. Compare the sample mean and variance with their theoretical values. How large would the simulation need to be if we wanted to estimate the mean within 0.01 with a 95% confidence interval?

Chapter 17

PROJECT MANAGEMENT

17.1 INTRODUCTION

A project is a task with specific objectives that should be achieved in a certain period of time. A project has a performance specification to reach the target. Each project is limited by time, money, manpower and other resources.

Project management is the process of planning, scheduling and controlling projects. In the modern industrial scenario, product life cycle is reduced with increased globalization. Newer and newer products with variety of features are coming in the market to meet the ever-changing needs of the customers. Hence, it is very much essential to emphasise on efficient project management and allocate sufficient resources.

17.2 PHASES OF PROJECT MANAGEMENT

The following are the three phases of project management

(a) Planning

(b) Scheduling

(c) Controlling

17.2.1 Planning

Planning is the first phase of project management. The project planning process involves defining the project and predicting the outcome of the project. In this phase, goals and objectives of the project, quality specifications of the project, time required and budget estimated for the project are clearly defined. The required resources are estimated in this phase.

17.2.2 Scheduling

In this phase, the project is broken down into simple task components. This process is known as work breakdown structure. In scheduling, the component tasks are assigned to relevant personnel. The time, money and other resources required are allocated for each component task of the project. If any resource constraint exists, the optimum utilization of the resources for each component task is established.

17.2.3 Controlling

All the component tasks will be executed in the third phase. In this phase, there may be some deviation in the results achieved due to some unexpected events during the execution of the project. The feedback should be collected at regular intervals and careful control of the deviation in time required, quality to be met and financial requirements will help to achieve the goals and objectives set in the planning phase. The role of the project manager to achieve goals by optimizing the available resources is very important.

17.3 WORK BREAKDOWN STRUCTURE

The concept of work breakdown structure was used in Performance Evaluation and Review Technique (PERT) for defence applications in the United States in the year 1957. At that time, this concept was not called "work breakdown structure". In the year 1962, the PERT-related document was published by the department of defence. The term work breakdown structure was used for the first time in that document.

In 1968, the defence department of the United States issued the work breakdown structure for defence materials. In the modern industry applications, work breakdown structure concept is used to define discrete work elements of the projects. This helps to organize the resources to achieve the goal of the project. The work element may be a product, a part, data, a service or combination of any or all of the above.

The work breakdown structure helps the managers to estimate the cost, develop the project schedule and provide guidelines for controlling any deviation from reaching the target. The work breakdown structure is the process of discretization of the project into a number of small components in terms of size, duration, effort and responsibility. This is graphically represented by a tree structure.

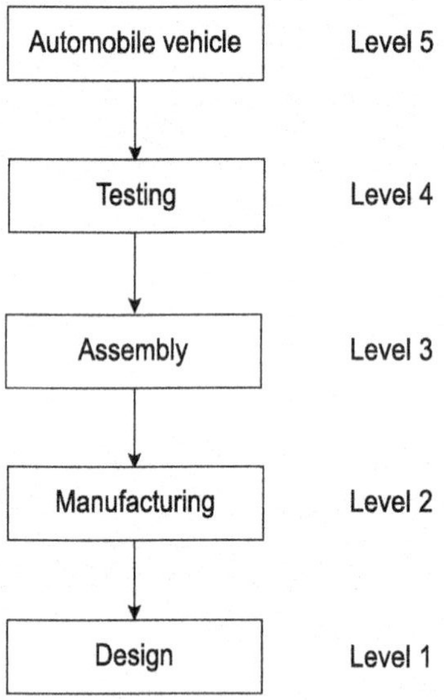

Figure 17.1 Work breakdown structure of an automobile

17.4 CONTROL CHARTS

Control charts are tools used to determine whether a manufacturing process is in statistical control. The control charts are also called process behaviour charts or Shewhart charts. The data collected from the process are used to determine the performance of the process. The results are plotted on a graph. The charts so plotted are analysed to determine whether the process is under control or not and what the source of variation is. The variation can be eliminated and the process can be brought under control using different techniques.

The control charts were developed for the first time by Walter A. Shewhart in the year 1920. He was working as a manager at Bell Electric company. He was also known as one of the gurus of total quality management. He taught Japanese managers in the 1920s about the concept of total quality management techniques. The Japanese managers who learnt the concepts of quality management from the gurus of TQM have achieved miraculous success in the industrial sector of Japan. These managers implemented the concept of TQM and used the tools like control charts and contributed for the revolution and economic growth of their country.

A control chart is a graphical representation of the collected information pertaining to measured quality characteristics or judged quality characteristics of samples selected from the lot of manufactured products. The control charts detect the variation in processing and determines and warns one variation from the specified tolerance limits.

The control charts are classified into two types.

1. Control charts for variables
2. Control charts for attributes

17.5 CONTROL CHARTS FOR VARIABLES

The control charts plotted using the data pertaining to measured quality characteristics are called control charts for variables. The following are the most commonly used control charts for variables

1. $\overline{X}$-chart with R-charts
2. $\overline{X}$-chart with σ-charts

17.5.1 Objectives of Control Charts for Variables

1. $\overline{X}$- and R-charts or $\overline{X}$- and σ-charts are used in combination in the process of quality control.
2. The control charts help us to determine whether the process is in control and if so with what dispersion.
3. The main objective of control charts is to determine variability that some product is made outside the tolerance limit.
4. A review of the situation will help to determine the tolerance limits tighter than the one that is required for the quality control of the product.
5. The control charts are used to provide information required to establish or change inspection procedure.
6. The control charts help to establish or change acceptance procedure or production procedure.
7. The control charts are effectively used to take decision on acceptance or rejection of manufactured or purchased products.
8. The control charts will help the quality control department for accepting the products and eliminates the costs associated with 100% inspection.

17.5.2 $\overline{X}$-Chart and R-chart

As we discussed earlier, the control charts for variables are economical tools compared to 100% inspection of all the manufactured products. In this process, the sample items are selected at random from the lot of products manufactured. These selected items are measured and the data are collected. For example if X_1, X_2, X_3, X_4, X_5 are measured values, the average or mean of these measured values are as follows

$$\overline{X} = \frac{X_1 + X_2 + X_3 + X_4 + X_5}{5}$$

On the same lines, many such samples are selected at random from the lot and $\overline{X}$ values are determined.

Then $$\overline{X} = \frac{X_1 + X_2 + X_3 + X_4 + X_5}{5}$$

i.e., Mean of the mean value is determined. R is the range of the value in each sub-group selected from the lot of manufactured items. i.e., $R = (X_{max} - X_{min})$ for all the sub-groups.

The mean of the range $\overline{R}$ is determined taking the average of all the R values

$$\overline{R} = \frac{\sum R}{N}$$

The $\overline{X}$- chart and R-chart are plotted using the data.

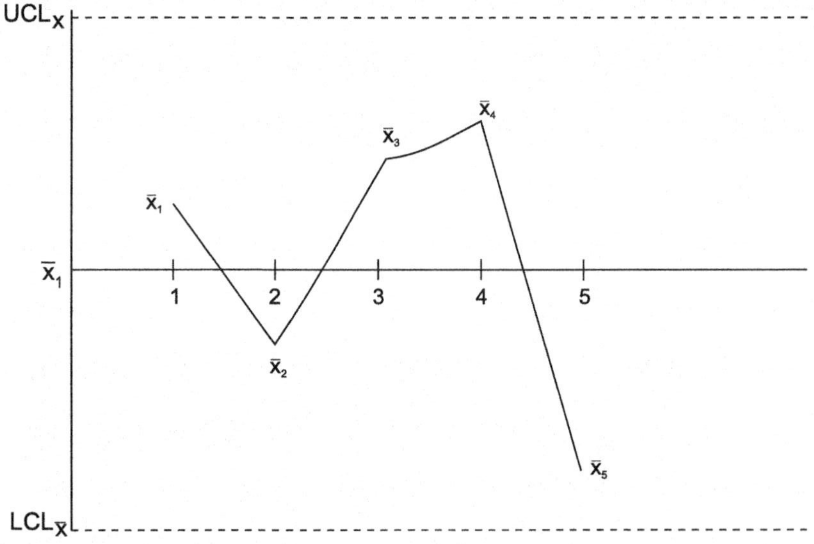

Figure 17.2 $\overline{X}$-chart

The upper control limit and lower control limit are determined using the formula

$$\text{UCL}_{\overline{X}} = \overline{\overline{X}} + A_2\overline{R}$$

$$\text{LCL}_{\overline{X}} = \overline{\overline{X}} - A_2\overline{R}$$

A_2 value is taken from Table-C for the sub-group of size n.

With reference to the UCL and LCL, the decision of acceptance or rejection will be taken.

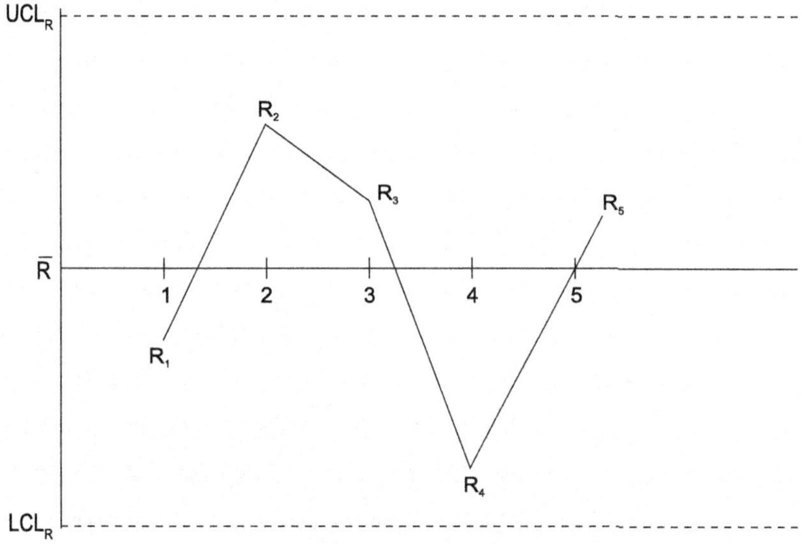

Figure 17.3 *R*-chart

Figures 17.2 and 17.3 illustrate the $\overline{X}$- chart and *R*-chart.

Range chart (*R*-chart) illustrated in Figure 17.3 helps to take decision of acceptance by analysing the variation of measured quality characteristics.

Example 17.1

In a capability study of a lathe used in turning cylindrical components to a diameter of 23.75 ± 0.1 mm, a sample of 6 consecutive pieces was taken each day for 8 days.

Serial No.	First Day	Second Day	Third Day	Fourth Day	Fifth Day	Sixth Day	Seventh Day	Eighth Day
1	23.77	23.80	23.77	23.79	23.75	23.78	23.76	23.76
2	23.80	23.78	23.78	23.76	23.78	23.76	23.78	23.79
3	23.78	23.76	23.77	23.79	23.78	23.73	23.75	23.77
4	23.73	23.70	23.77	23.74	23.77	23.76	23.76	23.72
5	23.76	23.81	23.80	23.82	23.76	23.74	23.81	23.78
6	23.75	23.77	23.74	23.76	23.79	23.78	23.80	23.78

Construct $\overline{X}$ and *R*-chart and find out the process capability for the machine.

Solution

Average diameter for the first day

$$\overline{X} = \frac{X_1 + X_2 + X_3 + X_4 + X_5 + X_6}{6}$$

$$\overline{X} = \frac{23.77 + 23.80 + 23.78 + 23.73 + 23.76 + 23.75}{6}$$

$$\overline{X} = 23.765$$

Similarly, the average for all the days are calculated and the results are listed as follows

$$\overline{X_1} = 23.765 \quad \overline{X_4} = 23.7767 \qquad \overline{X_7} = 23.7767$$

$$\overline{X_2} = 23.77 \quad \overline{X_5} = 23.7717 \qquad \overline{X_8} = 23.7667$$

$$\overline{X_3} = 23.7716 \quad \overline{X_6} = 23.7583$$

Range of diameter for the first day

$$R = X_{max} - X_{min}$$

$$R = 23.80 - 23.73$$

$$R = 0.07$$

Similarly, the range values for all the days are calculated and the results are listed as follows

$$R_1 = 0.07$$
$$R_2 = 0.11$$
$$R_3 = 0.06$$
$$R_4 = 0.08$$
$$R_5 = 0.04$$
$$R_6 = 0.05$$
$$R_7 = 0.06$$
$$R_8 = 0.07$$

For $\overline{X}$ chart, upper control limit

$$UCL_{\overline{X}} = \overline{\overline{X}} + A_2\overline{R}$$

$$= 23.7696 + 0.48(0.0675)$$

$$= 23.802$$

where $A_2 = 0.48$ from Table-C for $n = 6$.

Lower control limit

$$LCL_{\overline{X}} = \overline{\overline{X}} - A_2\overline{R}$$

$$= 23.7696 - [0.48(0.0675)]$$

$$LCL_{\overline{X}} = 23.7322$$

For R-chart, for $n = 6$ from table $D_4 = 2$, $D_3 = 0$,

$$UCL_R = D_4\overline{R} = 2 \times (0.0675) = 0.1350$$

$$LCL_R = D_3\overline{R}$$

$$= 0$$

Process capability $6\sigma' = 6.\frac{\overline{R}}{d_2}$ From Table-B

$$d_2 = 2.534 \text{ for } n = 6$$

$$6\sigma' = 6 \times \frac{0.0675}{2.534}$$

$$6\sigma' = 0.1598$$

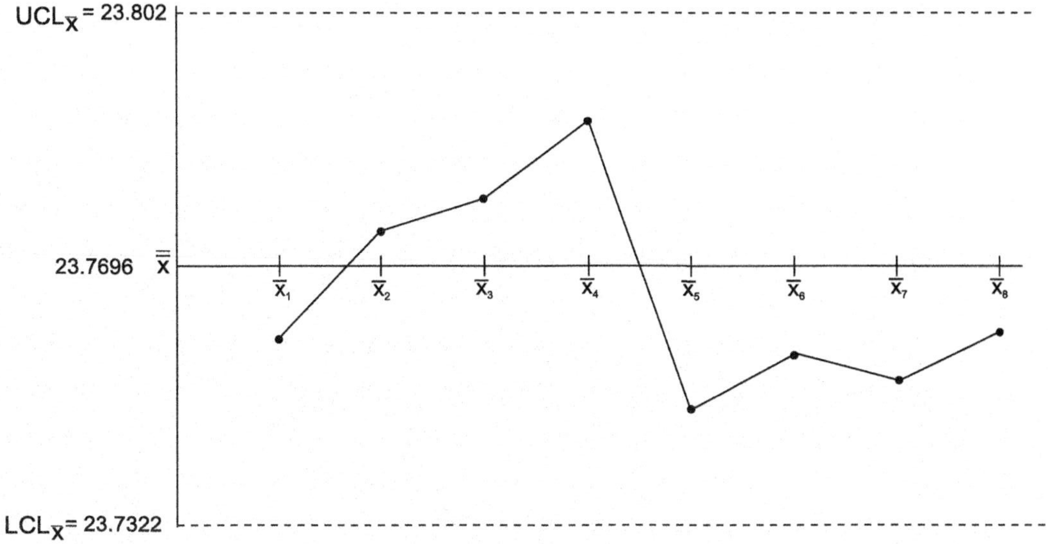

Figure 17.4 $\overline{X}$-chart

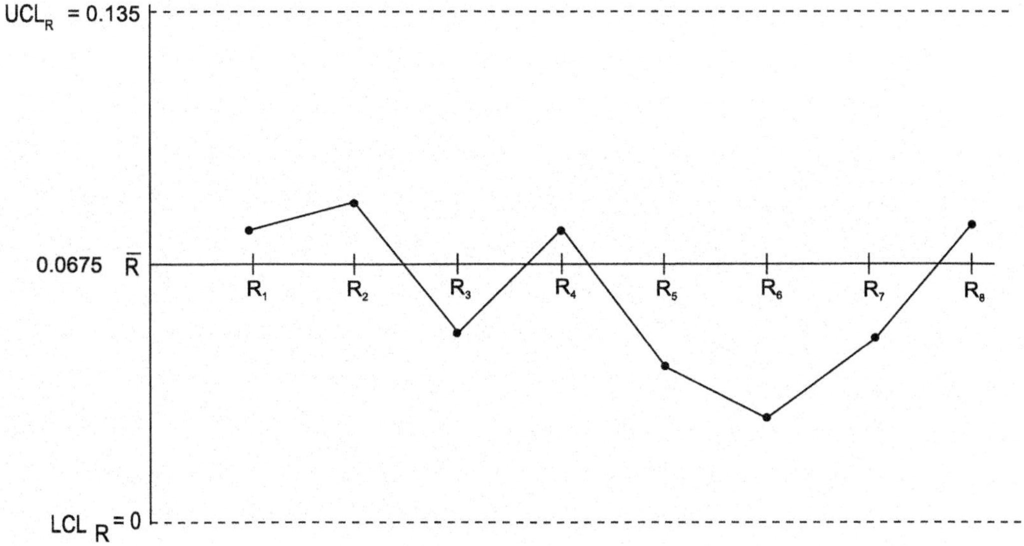

Figure 17.5 *R*-chart

Normal distribution curve below illustrates the deviation on both sides of the mean value.

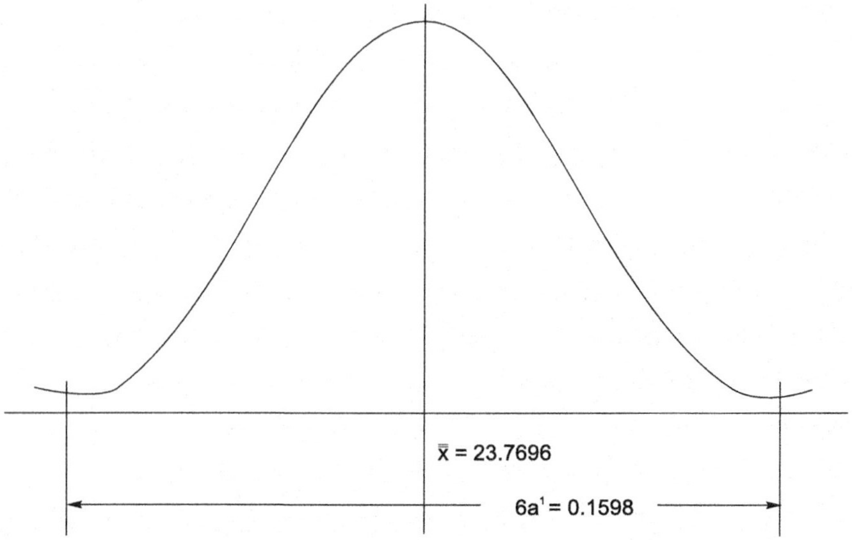

Figure 17.6 Normal distribution curve

$$(X_{max} - X_{min}) > 6\sigma'$$

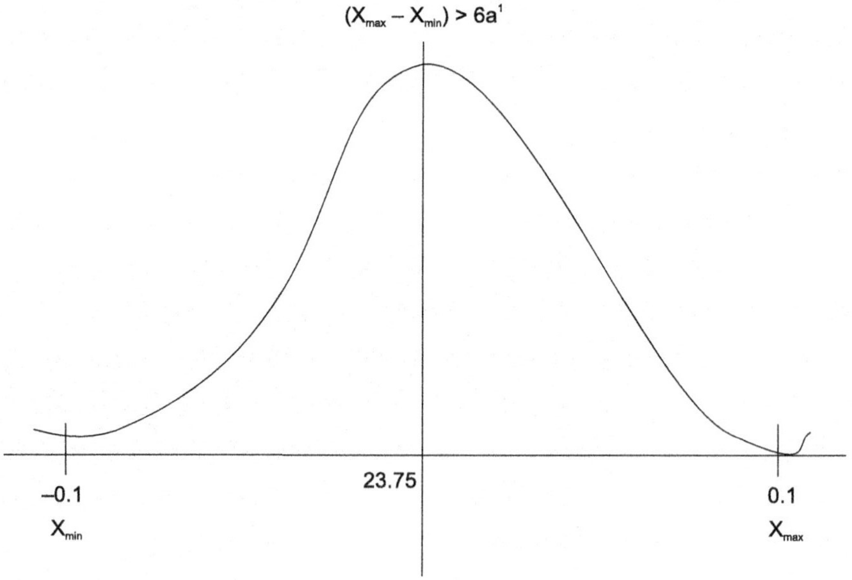

Figure 17.7 Given specification

Control limits taken are within the given specification and all the variables are within these limits; hence, the process is under control.

Example 17.2

The following table gives $\overline{X}$ and R values of 10 samples of 5 specimens. Draw $\overline{X}$- and R-chart. What proportion may be expected to the outside of the tolerance limit 0.420 ± 0.01?

It is possible to shift the mean without changing the average and range to what value should the mean be moved in order that the smallest proportion fall out of the limit.

Assume normal distribution.

Sub-group	$\overline{X}$	R
1	0.4150	0.007
2	0.4176	0.007
3	0.4156	0.004
4	0.4186	0.002
5	0.4160	0.011
6	0.4152	0.004
7	0.4720	0.005
8	0.4146	0.070
9	0.4172	0.008
10	0.4188	0.009

Solution

Mean of the mean values of 10 Sub-groups

$$\overline{\overline{X}} = \frac{\sum \overline{X}}{N} = \frac{4.1652}{10} = 0.41652$$

$$\overline{\overline{X}} = 0.4166$$

Mean of the range values of 10 Sub-groups

$$\overline{R} = \frac{\sum R}{N} = \frac{0.064}{10}$$

$$\overline{R} = 0.0064$$

Sample size $n = 5$

From table C for $n = 5$ $\quad A_2 = 0.58$

Upper control limit of $\overline{X}$

$$\text{UCL}_{\overline{X}} = \overline{\overline{X}} + A_2\overline{R}$$

$$= 0.4166 + 0.58(0.0064)$$

$$\text{UCL}_{\overline{X}} = 0.4203$$

Lower control limit of $\overline{X}$

$$\text{LCL}_{\overline{X}} = \overline{\overline{X}} - A_2\overline{R}$$

$$= 0.4166 - 0.58(0.0064)$$

$$\text{LCL}_{\overline{X}} = 0.4128$$

$$\text{UCL} = 0.4203$$

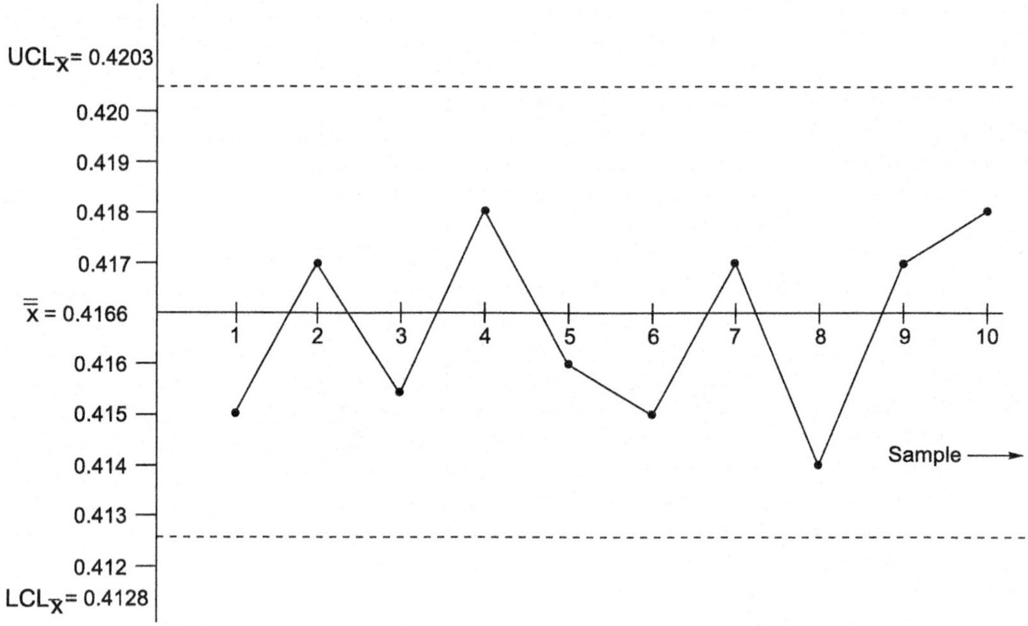

Figure 17.8 $\overline{X}$-chart

Upper control limit of Range 'R' from Table-C for $n = 5$

$$\text{UCL}_R = D_4\overline{R} = 2.11(0.0004) = 0.0135$$

$$D_4 = 2.11$$

$$D_3 = 0$$

Lower control limit of Range 'R'

$$\text{LCL}_R = D_3\overline{R} = 0$$

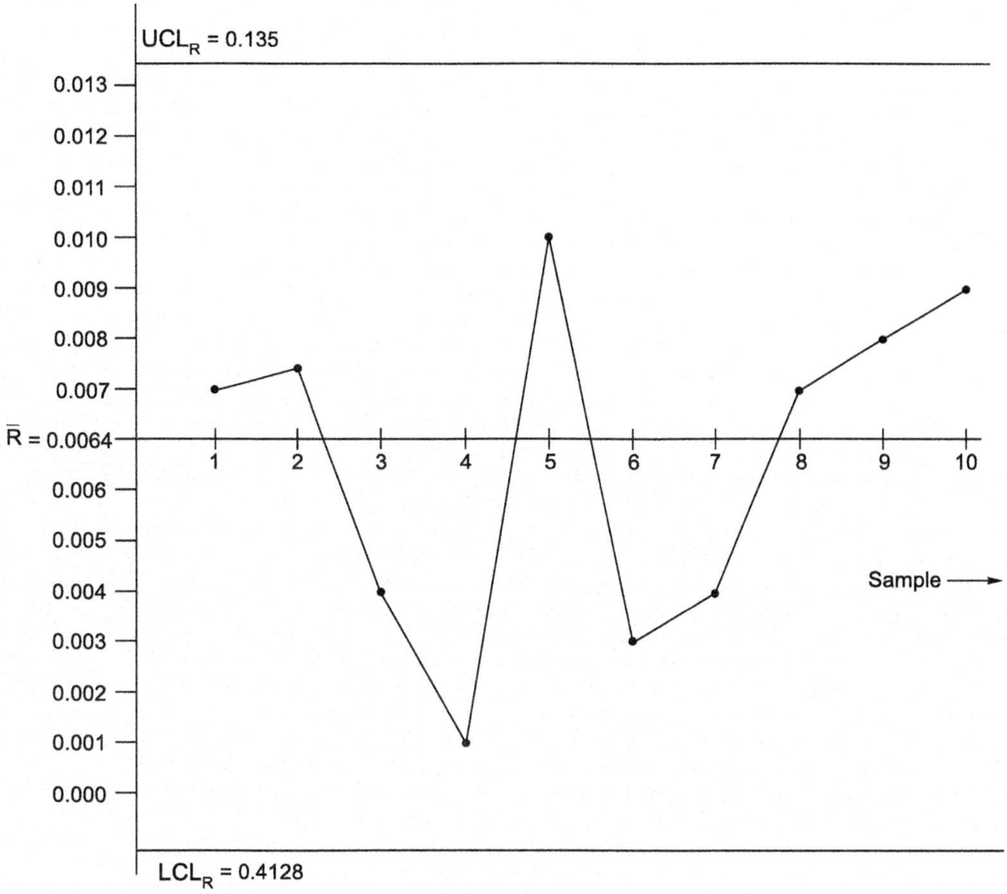

Figure 17.9 *R*-chart

Given specification 0.42 ± 0.01

$$\text{Upper Limit} = 0.43$$

$$\text{Lower limit} = 0.41$$

$$\sigma' = \frac{R'}{d_2}$$

$$= \frac{0.064}{2.326} \qquad \text{from Table-B For } n = 5,\ d_2 = 2.326$$

$$\sigma' = 0.00275$$

$$Z_1 \frac{X_i - \overline{\overline{X}}}{\sigma'} = \frac{0.43 - 0.4166}{0.00275} = 4.872 \text{ area under normal curve 100\%.}$$

$$Z_2 \frac{X_i - \overline{\overline{X}}}{\sigma'} = \frac{0.41 - 0.4166}{0.00275} = -2.18 \text{ area under normal curve 0.84\%.}$$

$$\overline{\overline{X}} = 0.4166$$

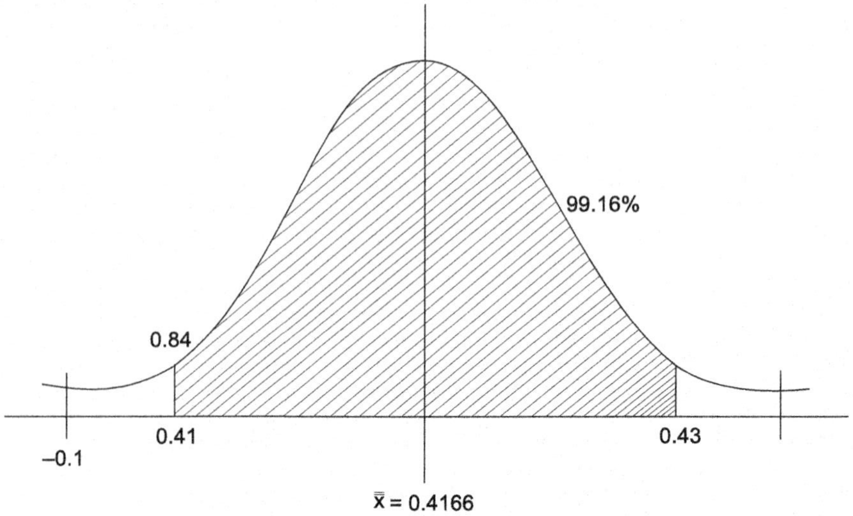

Figure 17.10 Area under normal curve

Now taking $\overline{\overline{X}}$ as 0.42 instead of 0.4166,

$$Z_1 \frac{0.43 - 0.42}{0.00275} = 3.36 \text{ 100\% area}$$

$$Z_2 \frac{0.41 - 0.42}{0.00275} = -3.36 \text{ Zero\% area}$$

Area beyond control is zero. Hence rejection is nil.

Example 17.3

Sub-group of 5 items each are taken from a manufacturing process at regular intervals. A certain quantity characteristics are measured and R values are computed for each Sub-group. After 25 Sub-groups, $\overline{X} = 357.50$ and $R = 8.80$. Compute the control chart limits. All points on both the charts fall within these limits. If the specification limits are 14.40 ± 0.40, what conclusion can you draw about the ability of the existing process to produce within these specifications?

Solution

Sub-group size = 5

No. of Sub-groups = 25

$$\Sigma\overline{X} = 357.50$$

$$\Sigma R = 8.80$$

Upper limit = 14.80

Lower limit = 14.0

To plot control charts for $n = 5$

From tables $A_2 = 0.58$

$$D_4 = 2.11$$
$$D_3 = 0$$
$$d_2 = 2.326$$

Mean of $\overline{X}$ values

$$\overline{\overline{X}} = \frac{357.50}{25}$$
$$\overline{\overline{X}} = 14.3$$
$$\overline{R} = \frac{88}{25} = 0.352$$

For $\overline{X}$ chart

Upper control limit

$$UCL_{\overline{X}} = \overline{\overline{X}} + A_2\overline{R}$$
$$= 14.3 + 0.58(0.352)$$
$$UCL_{\overline{X}} = 14.50$$

Lower limit of $\overline{X}$

$$LCL_{\overline{X}} = \overline{\overline{X}} - A_2\overline{X}$$
$$= 14.3 - 0.58(0.352)$$
$$LCL_{\overline{X}} = 14.10$$

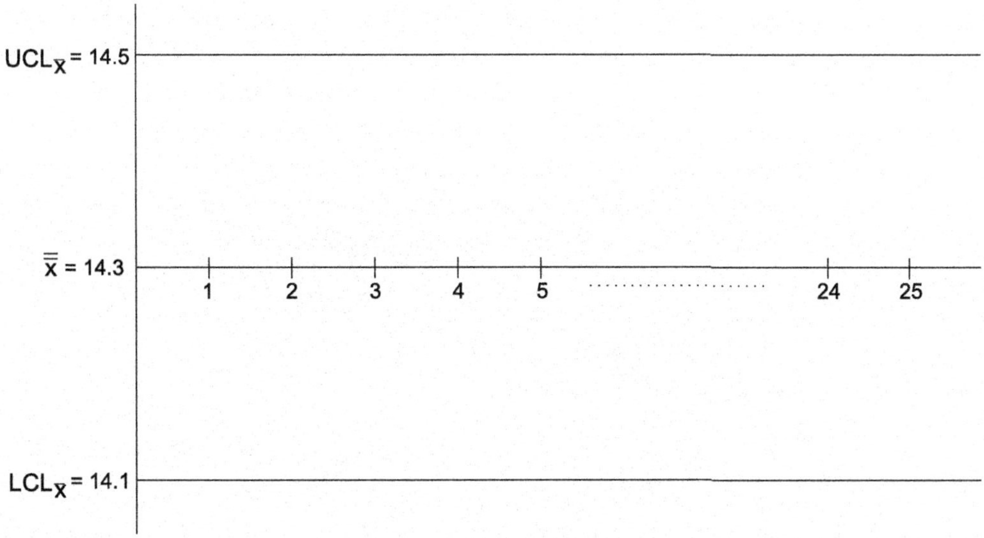

Figure 17.11 $\overline{X}$-Chart

For *R*-chart

$$\text{Upper control limit } \quad UCL_R = D_4 \overline{R}$$
$$= 2.11(0.352)$$
$$UCL_R = 0.74$$

$$\text{Lower control limit } LCL_R = D_3 \overline{R}$$
$$= 0(0.352)$$
$$LCL_R = 0$$
$$\sigma' = \frac{\overline{R}}{d_2} = \frac{0.352}{2.326}$$
$$\sigma' = 0.15$$

$$\text{Upper natural limit} = \overline{\overline{X}} + 3\,\sigma'$$
$$= 14.3 + 3(0.15)$$
$$= 14.75$$

$$\text{Lower natural limit} = \overline{\overline{X}} - 3\,\sigma'$$
$$= 14.3 - 3(0.15)$$
$$= 13.85$$

The process yields certain percentage of production which will have value, less than the lower limit $L = 14.00$.

Z value corresponding to upper limit

$$Z = \frac{X_1 - \overline{\overline{X}}}{\sigma'}$$
$$= \frac{14.8 - 14.3}{0.15}$$
$$Z = 3.3 \text{ Area under the normal curve}$$
$$= 0.99957 \text{ (From Table 'A')}$$
$$= 99.95\%$$
$$\text{The area beyond } 14.8 = 1 - 0.99957$$
$$= 4.3 \times 10^{-4}$$
$$= 0.043\% \text{ of production falls beyond upper limit}$$

Z value corresponding to lower limit

$$Z = \frac{X_1 - \overline{\overline{X}}}{\sigma'}$$
$$= \frac{14.0 - 14.3}{0.15}$$
$$Z = -2$$

Area under the normal curve from lower limit= 0.0228

= 2.28%

Figure 17.12 illustrates the normal curve.

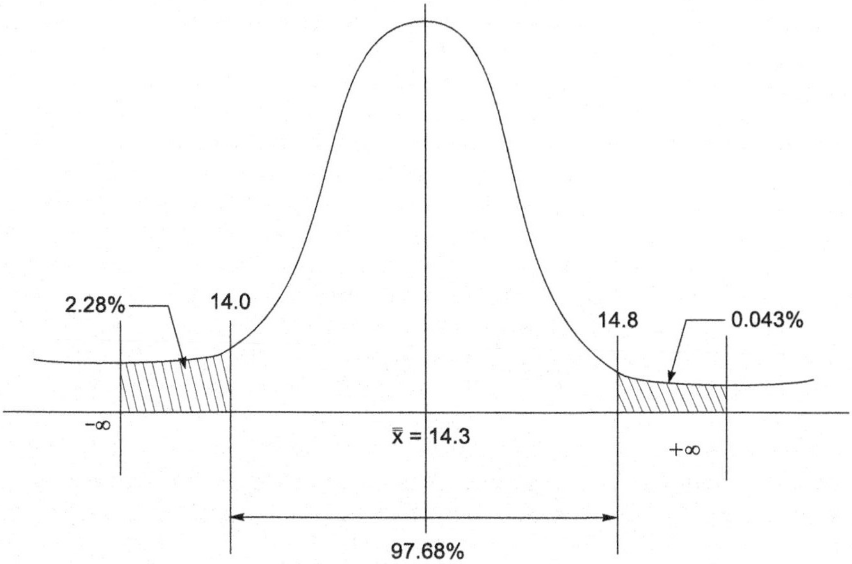

Figure 17.12 Normal distribution curve

17.5 LIMITATIONS OF VARIABLE CHARTS

1. $\overline{X}$ and *R*-charts are useful for quality characteristics that can be measured and expressed in numbers. However, some of the characteristics can be inspected as only attributes such as either conforming or non-conforming to the specifications. For example, blowholes, cracks, finish, under cuts, etc., will make an inspected component defective. The data of this type are analysed on the basis of number of products that conform to the specification.
2. $\overline{X}$ and *R*-charts can be used to analyse only one characteristic that can be measured at a time. Example for different dimensions of different parts of an assembly $\overline{X}$ and *R*-charts should be plotted one each. Therefore, it is advisable to prefer control charts for attributes such as defective and non-defective assembly.
3. The quality improvement such as reduction of spoilage, rework, etc., obtained from $\overline{X}$ and *R*-chart analysis leads to cost savings. This savings should compensate the cost of taking measurement, keeping charts and analysing them.

17.6 CONTROL CHARTS FOR ATTRIBUTES

The control charts plotted for conformity and conforming items inspected are called control charts for attributes. The attributes are blowholes, cracks, finish, etc.

Fraction defective 'P'

When a quality characteristic is measured only by attributes, a control chart based on fraction defective is used. Fraction defective is defined as the ratio of the number of defective articles present in any lot inspected to the total number of articles actually inspected. Fraction defective is always expressed as a decimal fraction.

17.7 *P*-CHART

The control chart constructed on the basis of fraction defective '*P*' is called *P*-chart.

Control limits on *P*-chart

$$\text{Upper control limit } UCL_P = \overline{P} + 3\sqrt{\frac{\overline{P}\left(1-\overline{P}\right)}{n}}$$

$$\text{Lower control limit } LCL_P = \overline{P} - 3\sqrt{\frac{\overline{P}\left(1-\overline{P}\right)}{n}}$$

Example 17.4

The following are the results of dial gauge inspection of components. Estimate the values of standard fraction defective. What value of $\overline{P}$ would you recommend for future construction of an appropriate chart?

Day	1	2	3	4	5	6	7	8	9	10	11	12
No. inspected	4100	4000	5800	5025	3370	4300	4450	7750	5630	5200	3600	5500
No. of defectives	48	107	176	37	35	50	45	95	55	55	36	59
Day	13	14	15	16	17	18	19	20	21	22	23	24
No. inspected	5676	5050	7850	6775	2730	8100	7200	7350	6550	6950	4875	4600
No. of defectives	71	60	92	81	56	167	85	105	80	76	66	60

Solution

Day	1	2	3	4	5	6	7	8	9	10	11	12
Fraction defective	0.117	0.026	0.03	0.0073	0.01	0.012	0.01	0.012	0.011	0.011	0.01	0.011
Day	13	14	15	16	17	18	19	20	21	22	23	24
Fraction defective	0.0134	0.012	0.0117	0.0113	0.013	0.021	0.012	0.014	0.0122	0.011	0.0135	0.013

Mean of fraction defective $\overline{P} = \frac{\sum P}{N} = \frac{0.3201}{24} = 0.13337$

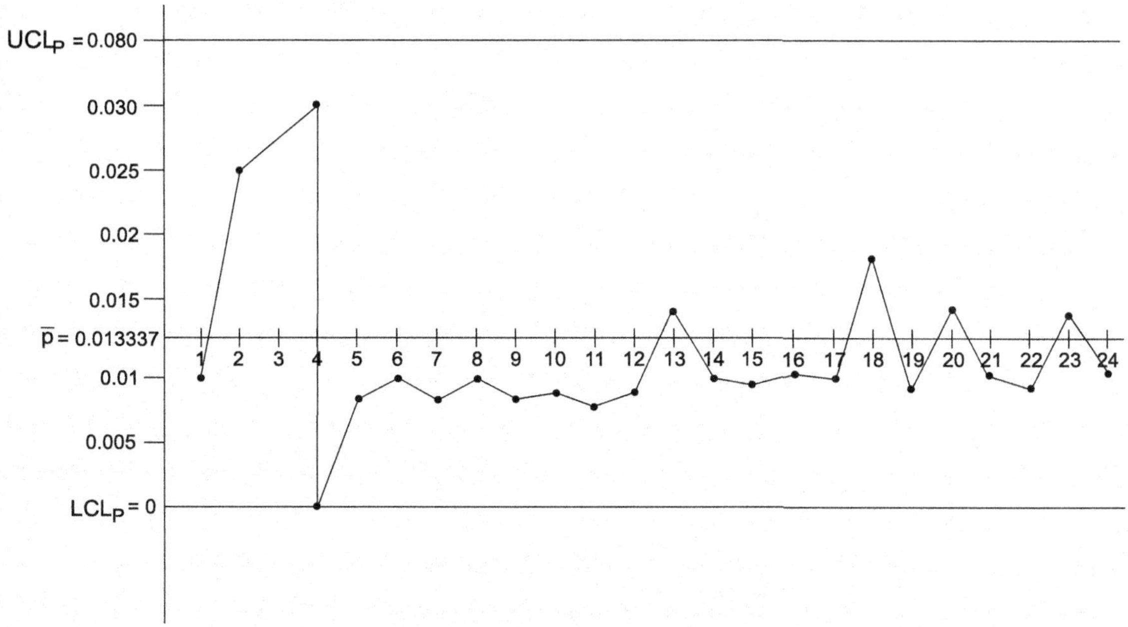

Figure 17.13 *P*-chart

Upper control limit

$$UCL_P = \bar{P} + 3\sqrt{\frac{\bar{P}(1-\bar{P})}{n}}$$

$$= 0.013337 + 3\sqrt{\frac{0.013337(1-0.013337)}{24}}$$

$$= 0.0835$$

Lower control limit

$$LCL_P = \bar{P} - 3\sqrt{\frac{\bar{P}(1-\bar{P})}{n}}$$

$$= 0.013337 - 3\sqrt{\frac{0.013337(1-0.013337)}{24}}$$

$$= 0.01337 - 0.070163$$

$$= 0$$

Example 17.5

In a factory producing spark plugs, the number of rejects found in the lots of 100 each after inspection is as follows. Construct an appropriate control chart.

Lot no.	No. rejected
1	5
2	10
3	12
4	8
5	6
6	5
7	6
8	3
9	3
10	5
11	4
12	7
13	8
14	2
15	3
16	4
17	5
18	8
19	6
20	10

Solution

Lot No.	1	2	3	4	5	6	7	8	9	10
Fraction rejected	0.05	0.1	0.12	0.08	0.06	0.05	0.06	0.03	0.03	0.05
Lot No.	11	12	13	14	15	16	17	18	19	20
Fraction rejected	0.04	0.07	0.08	0.02	0.03	0.04	0.05	0.08	0.06	0.1

$$n = 100$$

Mean of fraction rejected $\overline{P} = \frac{\sum P}{N} = \frac{120}{20 \times 100}$

$$= 0.06$$

Upper control limit $UCL_P = \overline{P} + 3\sqrt{\frac{\overline{P}(1-\overline{P})}{n}}$

$$= 0.06 + 3\sqrt{\frac{0.06(1-0.06)}{100}}$$

$$= 0.131$$

Lower control limit LCL_p $$= \bar{P} - 3\sqrt{\frac{\bar{P}(1-\bar{P})}{n}}$$

$$= 0.06 - 3\sqrt{\frac{0.06(1-0.06)}{100}}$$

$$= 0$$

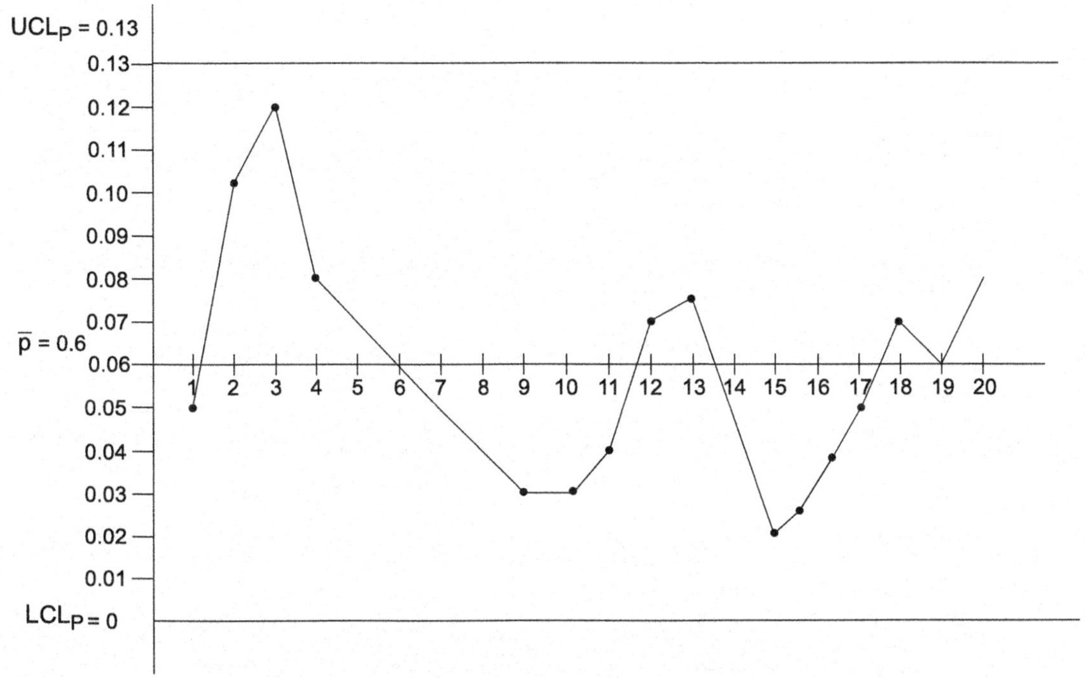

Figure 17.14 *P*-chart

17.7.1 Comparison of $\bar{X}$- and *R*-Charts with *P*-Chart

1. The *P*-charts are used for larger sample size whereas, $\bar{X}$- and *R*-chart are used for smaller sample size.
2. $\bar{X}$- and *R*-charts are sensitive to assignable causes of variation.
3. $\bar{X}$- and *R*-charts are best suited for critical dimensions. *P*-chart is best suited if the articles are either accepted or rejected.
4. Separate $\bar{X}$- and *R*-chart is required for each measured quality characteristics whereas the cost of computing charting may also be less since *P*-chart can be applied to any number of quality characteristics observed on one article.

5. The cost of collecting data for $\overline{X}$- and R-chart is more than that of P-chart. The time required for measuring dimension of one shaft is ten times the time required for inspecting the dimension of a shaft using Go and No-Go gauges.
6. P-chart is an attribute control chart, i.e., for quality characteristics that conform to the specifications. $\overline{X}$- and R-chart are used for quality characteristics that can be measured using measuring instrument.

17.7.2 Limitations of *P*-Chart

1. Control limits for each Sub-group show the limits on control chart.
2. If the average sub-group size is considered, one set of limits are used for the control chart.
3. If the actual sub-group size varies ± 25%, separate limits are calculated and separate charts are plotted for each Sub-group.
4. Analysis of control charts of different sub-groups is cumbersome.

17.8 *nP*-CHARTS

If the sample size varies, P-charts are used. If the sample size is constant, nP-charts are used. The nP-charts require less calculation.

Upper control limit of nP-chart

$$UCL_{nP} = n\overline{P} + 3\sqrt{n\overline{P}\left(1-\overline{P}\right)}$$

Lower limit of nP-chart

$$LCL_{nP} = n\overline{P} - 3\sqrt{n\overline{P}\left(1-\overline{P}\right)}$$

***Example* 17.6**

The following data is connected with the inspection of a mechanical part. Draw an nP-chart and calculate the control limits. Recommend a suitable value of P.

Lot size = 200

Lot no.	1	2	3	4	5	6	7	8	9	10
(nP) rejects	4	8	2	6	4	4	10	6	4	8

Solution

$$\Sigma nP = 56$$

$$n\overline{P} = \frac{56}{n} = \frac{56}{10} = 5.6$$

$$\therefore \overline{P} = \frac{5.6}{200} = 0.028$$

$$\text{Upper control limit } UCL_{nP} = n\overline{P} + 3\sqrt{n\overline{P}\left(1-\overline{P}\right)}$$

$$= 5.6 + 3\sqrt{5.6(1-0.028)}$$

$$UCL_{nP} = 12.59$$

$$\text{Lower control limit } LCL_{nP} = n\overline{P} - 3\sqrt{n\overline{P}(1-\overline{P})}$$

$$= 5.6 - 3\sqrt{5.6(1-0.028)}$$

$$LCL_{nP} = 0$$

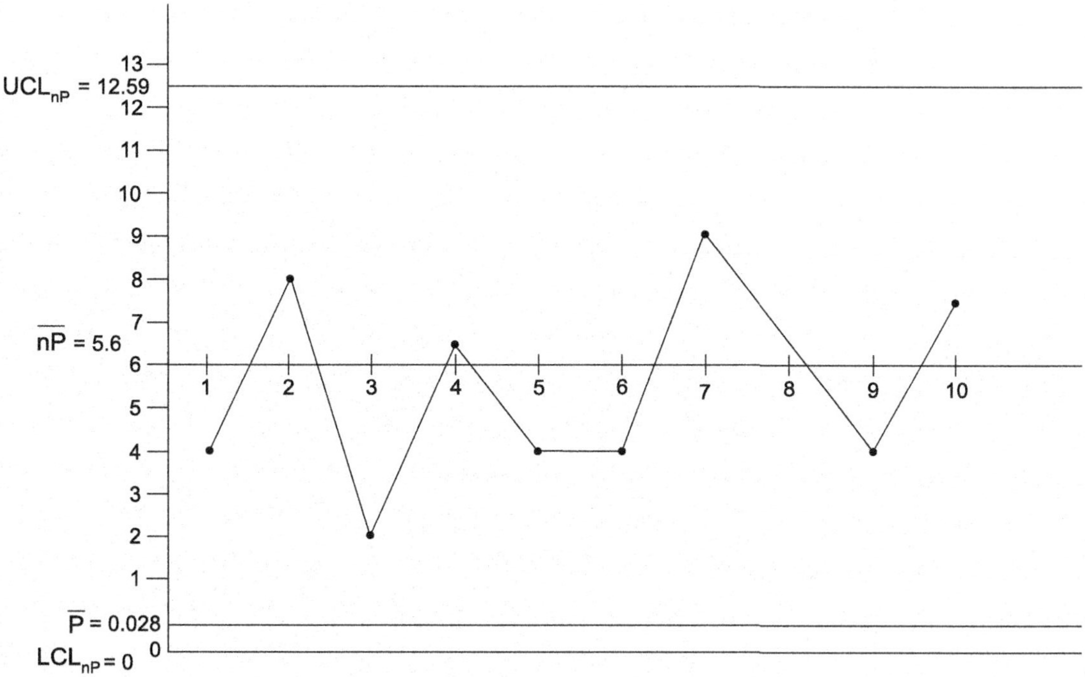

Figure 17.15 *nP*-chart

***Example* 17.7**

The following table gives the result of daily inspection of a product. The lot size is 200. Construct *nP*-chart.

Lot No.	1	2	3	4	5	6	7	8	9	10	11	12
Rejects	23	15	17	15	41	0	25	31	29	0	8	6

Solution

$$\sum n\overline{P} = 210$$

$$n\overline{P} = \frac{210}{12} = 17.5$$

$$\overline{P} = \frac{175}{n} = \frac{175}{200}$$

$$\overline{P} = 0.0875$$

$$\text{Upper control limit UCL}_{nP} = n\bar{P} + 3\sqrt{n\bar{P}\left(1-\bar{P}\right)}$$

$$= 17.5 + 3\sqrt{17.5(1-0.0875)}$$

$$\text{UCL}_{nP} = 29.488$$

$$\text{Lower control limit LCL}_{nP} = n\bar{P} + 3\sqrt{n\bar{P}\left(1-\bar{P}\right)}$$

$$= 17.5 - 3\sqrt{17.5(1-0.0875)}$$

$$\text{LCL}_{nP} = 0$$

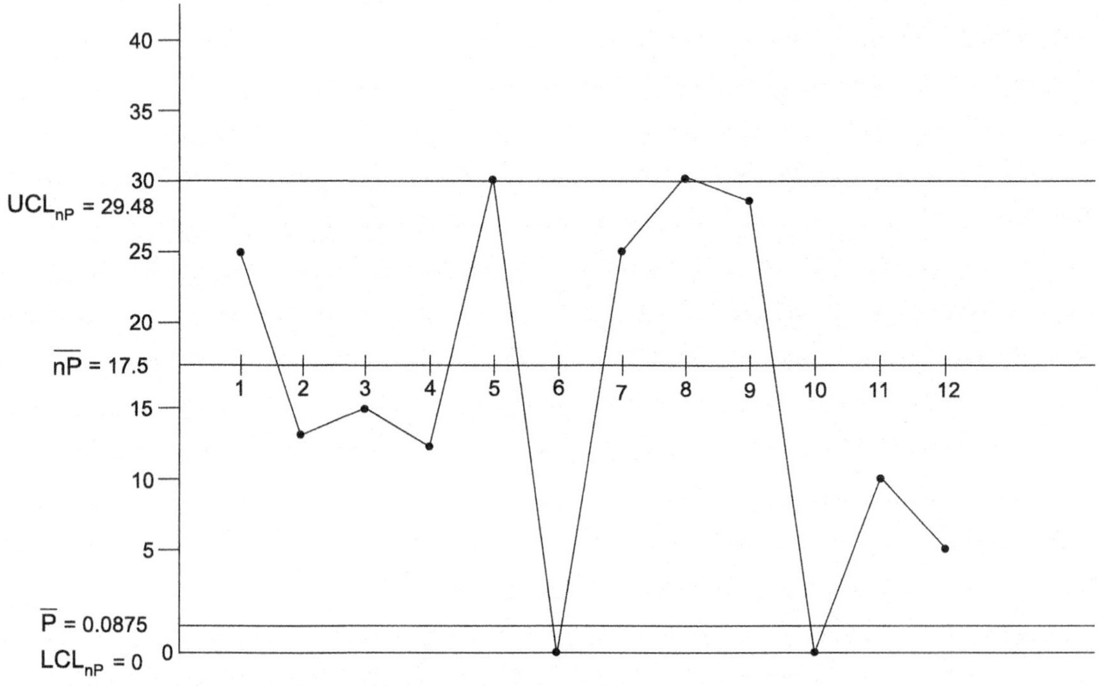

Figure 17.16 *nP*-chart

17.9 DIFFERENCE BETWEEN DEFECT AND DEFECTIVE

If an item fails to conform to the specifications of any of the quality characteristics, it is called defective. The quality characteristic that does not meet the specification is a defect. An item that contains at least one defect is defective. For example, hard spots and blowholes in a casting are the defects and the casting with these defects is defective.

17.9.1 Types of Defects

1. Defects such as blowholes, cracks, undercuts in a casting or a welded joint.
2. Imperfections in a cloth of unit area.

3. Small air holes in a glass bottle.
4. Breakdown at a weak spot in the insulation in an insulated wire under voltage.
5. Surface defects in a galvanized sheet.
6. Surface defect in a sheet of photographic film.

17.10 CONTROL CHARTS FOR DEFECTS

***C*-charts** The control charts plotted for the number of defects in the samples are called *C*-charts.

Upper control limit

$$\text{UCL}_C = \overline{C} + 3\sqrt{C}$$

Lower control limit

$$\text{LCL}_C = \overline{C} + 3\sqrt{C}$$

$$\text{Mean of the defects } \overline{C} = \frac{\text{Total number of defects}}{\text{Total number of samples}}$$

***Example* 17.8**

For the following data, find $\overline{C}$, LCL, UCL and plot *C*-chart.

Lot no.	No. of defects *C*	Lot no.	No. of defects *C*
1	33	13	28
2	27	14	33
3	17	15	26
4	22	16	43
5	19	17	28
6	28	18	30
7	31	19	20
8	28	20	17
9	24	21	23
10	12	22	27
11	47	23	12
12	18	24	15

Solution

$$\text{Mean of the defects } \overline{C} = \frac{\sum C}{n}$$

$$= \frac{612}{24}$$

$$\overline{C} = 25.52$$

$$\text{Upper control limit UCL}_C = \overline{C} + 3\sqrt{C}$$

$$= 25.52 + 3\sqrt{25.52}$$

$$\text{UCL}_C = 40.65$$

$$\text{Lower control limit } LCL_C = \bar{C} + 3\sqrt{C}$$

$$= 25.52 - 3\sqrt{25.52}$$

$$LCL_C = 10.35$$

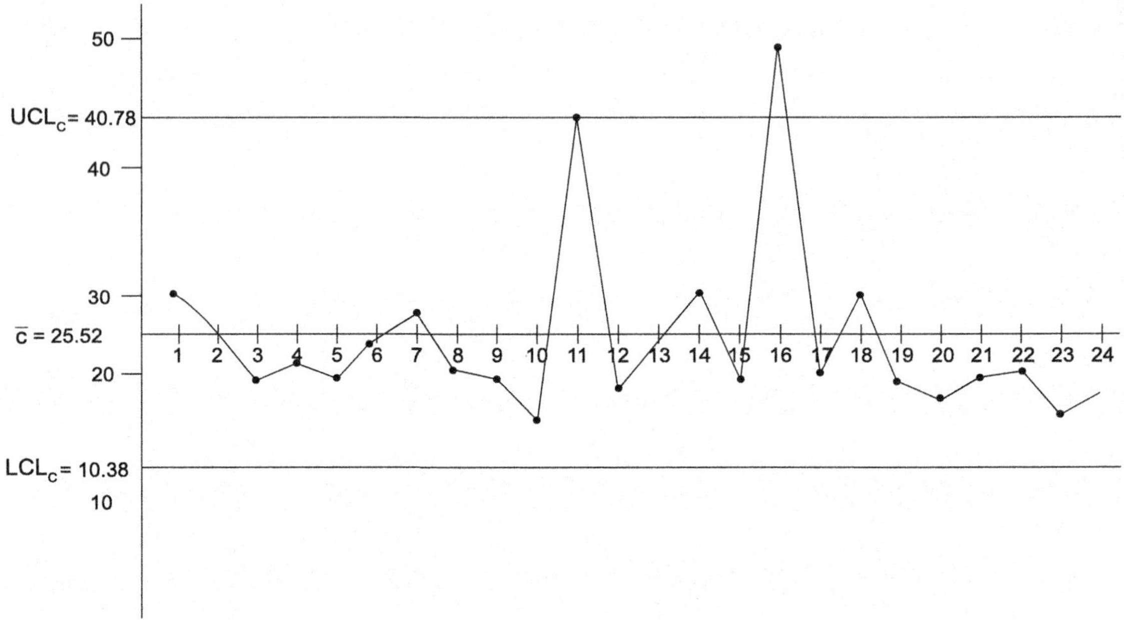

Figure 17.17 *C*-chart

In two lots, viz., lot numbers 11 and 16, the number of defects is more than UCL. Therefore, determine the new control limits excluding these two lots.

$$\text{New } \bar{c} = \frac{528}{22} = 24$$

$$\text{New } UCL_C = 24 + \sqrt{24}(3)$$

$$UCL_C = 38.69$$

$$\text{New } LCL_C = 24 - 3\sqrt{24}$$

$$LCL_C = 9.3$$

All the 22 lots fall within the new limits.

Example 17.9

The following table gives the number of errors of alignment, observed at the final inspection of a certain model of aeroplane. Prepare a *C*-chart and comment on the picture.

Acceptance No	No. of errors	Acceptance No.	No. of errors
201	7	214	9
202	6	215	8
203	6	216	15
204	7	217	6
205	4	218	4
206	7	219	13
207	8	220	7
208	12	221	8
209	9	220	15
210	9	222	6
211	8	223	6
212	5	224	10
213	5		

Solution

$$\text{Mean of the defects } \overline{C} = \frac{200}{25} = 8$$

$$\text{Upper control limit } UCL_C = \overline{C} + 3\sqrt{C}$$

$$= 8 + 3\sqrt{8}$$

$$= 16.48$$

$$\text{Lower control limit } LCL_C = \overline{C} - 3\sqrt{C}$$

$$= 8 - 3\sqrt{8}$$

$$= 16.48$$

$$\text{Lower control limit } LCL_C = \overline{C} - 3\sqrt{C}$$

$$= 8 - 3\sqrt{8}$$

$$LCL_C = 0$$

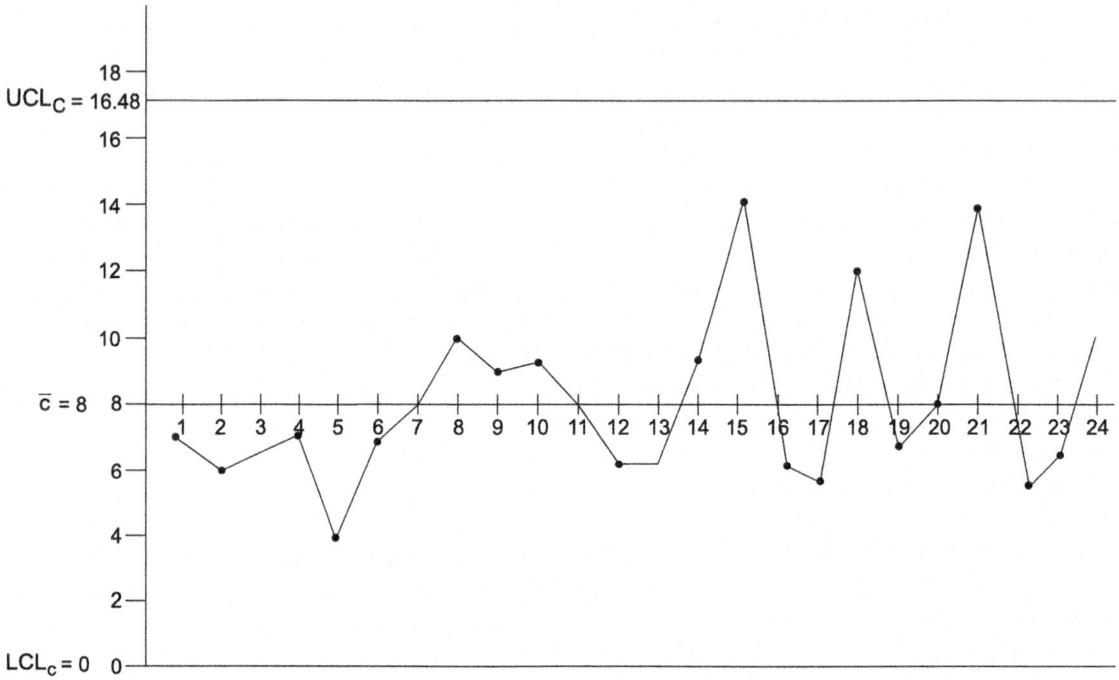

Figure 17.18 *C*-chart

17.11 CONTROL CHARTS FOR DEFECTS PER UNIT

For plotting *C*-charts, the number of defects in all the items should be the same in a sample. Whenever the number of defects vary from item to item, the *C*-chart cannot be used to find the solution and it will lead to confusion. The *U* charts are used to overcome this difficulty. *U* represents the number of defects per unit. *U* is obtained by dividing *C* by the number of units.

$$\therefore \quad U = \frac{C}{n}$$

$$\text{Mean of defects per unit } \overline{U} = \frac{\sum C}{n}$$

$$\text{Upper control limit of UCL}_U = \overline{U} + \sqrt{\frac{\overline{U}}{n}}$$

$$\text{Lower control limit of LCL}_U = \overline{U} - \sqrt{\frac{\overline{U}}{n}}$$

EXERCISES

1. The table below illustrates the thickness of paint on refrigerators for five refrigerators from each shift.

Sub-group shift no.	Thickness (in mm)				
1	2.7	2.3	2.6	2.4	2.7
2	2.6	2.4	2.6	2.3	2.8
3	2.3	2.3	2.4	2.5	2.4
4	2.8	2.3	2.4	2.6	2.7
5	2.6	2.5	2.6	2.1	2.8
6	2.2	2.3	2.7	2.2	2.6
7	2.2	2.6	2.4	2.0	2.3
8	2.8	2.6	2.6	2.7	2.5
9	2.4	2.8	2.4	2.2	2.3
10	2.6	2.3	2.0	2.5	2.4
11	3.1	3.0	3.5	2.8	3.0
12	2.4	2.8	2.2	2.9	2.5
13	2.1	3.2	2.5	2.6	2.8
14	2.2	2.8	2.1	2.2	2.4
15	2.4	3.0	2.5	2.5	2.0
16	3.1	2.6	2.6	2.8	2.1
17	2.9	2.4	2.9	1.3	1.8
18	1.9	1.6	2.6	3.3	3.3
19	2.3	2.6	2.7	2.8	3.2
20	1.8	2.8	2.3	2.0	2.9

Compute the values of 3 sigma control limits and draw the $\overline{X}$ and *R*-charts.

2. A company produces dials for a machine. These dials are supposed to have a constant diameter. To check on the production process, the first 4 dials are selected every half hour for 12 hours giving a total of 96 observations. It was found that $\overline{X} = 51.12$ mm and $R = 0.46$ mm. Find the upper and lower control limits and plot the $\overline{X}$ and *R*-chart.

3. In example 17.1, the out-of-control mean given for shift 11 was found to be due to an operator error. Recurrence has been prevented by giving the operator additional training. Delete this point from the data and recompute the centre line and the control limits. Are your conclusions any different?

4. Control charts $\overline{X}$ and *R* are maintained on certain dimensions of a manufactured part, measured in mm. The Sub-group size is 4. The values of *X* and *R* are computed for each sub-group. After 20 sub-groups, $\sum X = 412.83$ and $\sum R = 3.39$. Compute the value of the upper control limit, the control line and the lower control limit for the $\overline{X}$ and *R*-chart and estimate the value of σ' on the assumption that the process is in statistical control.

5. Control charts for $\overline{X}$ and *R*, σ are to be maintained on drawings from a bowl of chips, the distribution of which is approximately normal. The Sub-group size is 5. $\overline{X}$ is 60 and σ' is 8. Assume that 3 σ

control limits are to be based on $\overline{X}$ and σ'. Compute the value of the upper control limit, the contro line and the lower control limit for the $\overline{X}$, R and σ charts respectively.

6. Following are the inspection results of magnets for 19 observations.

Week no	No. of magnets inspected	No. of defective magnets	Week no.	No. of magnets inspected	No. of defective magnets
1	724	48	11	736	47
2	763	83	12	739	50
3	748	70	13	723	47
4	748	85	14	748	57
5	724	45	15	770	51
6	727	56	16	756	71
7	726	48	17	719	53
8	719	67	18	757	34
9	759	37	19	760	29
10	745	52			

Calculate the average fraction defective and 3 sigma control limits, construct the control chart and state whether the process is in statistical control.

7. A certain product is given 100% inspection as it is manufactured and the resultant daya are summarised by the hour. In the following table, 16 hours of data are recorded. Calculate the control limits using 3 sigma control limits and indicate the values that are out of control.

Hour	No. of inspected units	No. of defective units	Hour	No. of inspected units	No. of defective units
1	48	5	9	32	5
2	36	5	10	40	2
3	50	0	11	47	2
4	47	5	12	47	4
5	48	0	13	46	1
6	54	3	14	46	0
7	50	0	15	48	3
8	42	1	16	39	0

8. A manufacturer purchases small bolts in cartons that usually contain several thousand bolts. Each shipment consists of a number of cartons. As a part of the acceptance procedure for those bolts, 400 bolts are selected at random from each carton and are subjected to visual inspection for certain defects in a shipment of 10 cartons. The respective percentages of defectives in the samples from each carton are 0, 0, 0.5, 0.75, 0.20, 0.25, 0, 0.25 and 1.25. Does this shipment of bolts appear to exhibit statstical control with respect to the quality characteristics examined in the inspection?

9. In a manufacturing process, the number of defectives found in the inspection of 15 lots of 400 items each are as follows

Lot no	No. of defectives	Lot no	No. of defectives
1	2	9	18
2	5	10	8
3	0	11	6
4	14	12	0
5	3	13	3
6	0	14	0
7	1	15	6
8	0		

(a) Determine the trial control limits for *nP*-chart and state whether the process is in control.

(b) What will be the new value of mean fraction defective if some obvious points outside control limits are eliminated? What will be the corresponding upper and lower control limits and examine whether the process is still in control or not.

10. The following table gives the number of missing rivets noted at aircraft final inspection.

Airplane no.	No. of missing rivets	Airplane no.	No. of missing rivets	Airplane no.	No. of missing rivets
1	8	10	12	19	11
2	16	11	23	20	9
3	14	12	16	21	10
4	19	13	9	22	22
5	11	14	25	23	7
6	15	15	15	24	28
7	8	16	9	25	9
8	11	17	9		
9	21	18	14		

Find C. Compute trial control limits and plot control chart for C. What value of c' would you suggest for the subsequent period.

11. The following table refers to the average number of outlet leaks per radiator for 10 lots of 100 radiators each. Estimate *U* chart for the future production.

Lot no.	No. of leaks C	Leaks per radiator $d_n = U$	Lot no.	No. of leaks C	Leaks per radiator $d_n = U$
1	15	0.15	6	5	0.5
2	27	0.27	7	14	0.14
3	12	0.12	8	11	0.11
4	16	0.16	9	9	0.9
5	14	0.14	10	10	0.10

APPENDIX

Table A

Areas Under the Normal Curve

Proportion of total area under curve from $-\infty$ to $\dfrac{X_i - X'}{\sigma'}$

where σ' is the standard deviation of the universe.

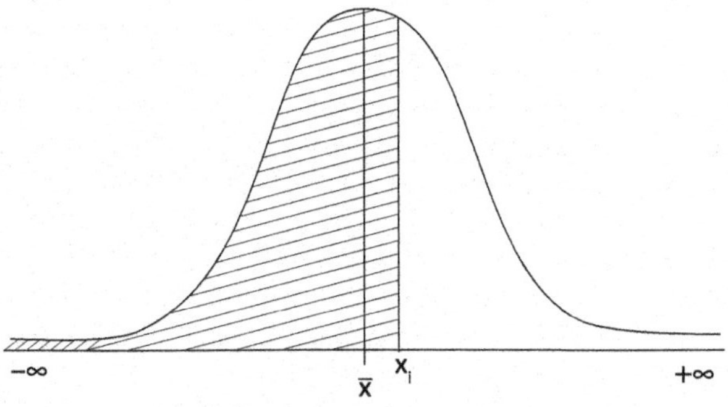

$\frac{X_i - X'}{\sigma'}$	0.09	0.08	0.07	0.06	0.05	0.04	0.03	0.02	0.01	0.00
−3.5	0.00017	0.00017	0.00018	0.00019	0.00019	0.00020	0.00021	0.00022	0.00022	0.00023
−3.4	0.00024	0.00025	0.00026	0.00027	0.00028	0.00029	0.00030	0.00031	0.00033	0.00034
−3.3	0.00035	0.00036	0.00038	0.00039	0.00040	0.00042	0.00043	0.00045	0.00047	0.00048
−3.2	0.00050	0.00052	0.00054	0.00056	0.00058	0.00060	0.00062	0.00064	0.00066	0.00069
−3.1	0.00071	0.00074	0.00076	0.00079	0.00082	0.00085	0.00087	0.00090	0.00094	0.00097
−3.0	0.00100	0.00104	0.00107	0.00111	0.00114	0.00118	0.00122	0.00126	0.00131	0.00135
−2.9	0.0014	0.0014	0.0015	0.0015	0.0016	0.0016	0.0017	0.0017	0.0018	0.0019
−2.8	0.0019	0.0020	0.0021	0.0021	0.0022	0.0023	0.0023	0.0024	0.0025	0.0026
−2.7	0.0026	0.0027	0.0028	0.0029	0.0030	0.0031	0.0032	0.0033	0.0034	0.0035
−2.6	0.0036	0.0037	0.0038	0.0039	0.0040	0.0041	0.0043	0.0044	0.0045	0.0047
−2.5	0.0048	0.0049	0.0051	0.0052	0.0054	0.0055	0.0057	0.0059	0.0060	0.0062
−2.4	0.0064	0.0066	0.0068	0.0069	0.0071	0.0073	0.0075	0.0078	0.0080	0.0082

Table A (Continued)

$\frac{X_i - X'}{\sigma'}$	0.09	0.08	0.07	0.06	0.05	0.04	0.03	0.02	0.01	0.00
−2.3	0.0084	0.0087	0.0089	0.0091	0.0094	0.0096	0.0099	0.0102	0.0104	0.0107
−2.2	0.0110	0.0113	0.0116	0.0119	0.0122	0.0125	0.0129	0.0132	0.0136	0.0139
−2.1	0.0143	0.0146	0.0150	0.0154	0.0158	0.0162	0.0166	0.0170	0.0174	0.0179
−2.0	0.0183	0.0188	0.0192	0.0197	0.0202	0.0207	0.0212	0.0217	0.0222	0.0228
−1.9	0.0233	0.0239	0.0244	0.0250	0.0256	0.0262	0.0268	0.0274	0.0281	0.0287
−1.8	0.0294	0.0301	0.0307	0.0314	0.0322	0.0329	0.0336	0.0344	0.0351	0.0359
−1.7	0.0367	0.0375	0.0384	0.0392	0.0401	0.0409	0.0418	0.0427	0.0436	0.0446
−1.6	0.0455	0.0465	0.0475	0.0485	0.0495	0.0505	0.0516	0.0526	0.0537	0.0548
−1.5	0.0650	0.0675	0.0682	0.0694	0.0606	0.0618	0.0630	0.0643	0.0655	0.0668
−1.4	0.0681	0.0684	0.0708	0.0721	0.0735	0.0740	0.0764	0.0778	0.0793	0.0808
−1.3	0.0823	0.0838	0.0853	0.0869	0.0885	0.0901	0.0918	0.0934	0.0951	0.0968
−1.2	0.0985	0.1003	0.1020	0.1038	0.1057	0.1075	0.1093	0.1112	0.1131	0.1151
−1.1	0.1170	0.1190	0.1210	0.1230	0.1251	0.1271	0.1292	0.1314	0.1335	0.1357
−1.0	0.1379	0.1401	0.1423	0.1446	0.1469	0.1492	0.1515	0.1539	0.1562	0.1587
−0.9	0.1611	0.1635	0.1660	0.1685	0.1711	0.1736	0.1762	0.1788	0.1814	0.1841
−0.8	0.1862	0.1894	0.1922	0.1949	0.1977	0.2005	0.2033	0.2061	0.2090	0.2119
−0.7	0.2148	0.2177	0.2207	0.2236	0.2266	0.2297	0.2327	0.2358	0.2389	0.2420
−0.6	0.2451	0.2483	0.2514	0.2546	0.2578	0.2611	0.2643	0.2676	0.2709	0.2743
−0.5	0.2776	0.2810	0.2843	0.2877	0.2912	0.2946	0.2981	0.3015	0.3050	0.3085
−0.4	0.3121	0.3156	0.3192	0.3228	0.3264	0.3300	0.3336	0.3372	0.3409	0.3446
−0.3	0.3483	0.3520	0.3557	0.3594	0.3632	0.3669	0.3707	0.3745	0.3783	0.3821
−0.2	0.3859	0.3897	0.3936	0.3974	0.4013	0.4052	0.4090	0.4129	0.4168	0.4207
−0.1	0.4247	0.4286	0.4325	0.4364	0.4404	0.4443	0.4483	0.4522	0.4562	0.4602
−0.0	0.4641	0.4681	0.4721	0.4761	0.4801	0.4840	0.4880	0.4920	0.4960	0.5000
+0.0	0.5000	0.5040	0.5080	0.5120	0.5160	0.5199	0.5239	0.5279	0.5319	0.5359
+0.1	0.5398	0.5438	0.5478	0.5517	0.5557	0.5596	0.5636	0.5675	0.5714	0.5753
+0.2	0.5793	0.5832	0.5871	0.5910	0.5948	0.5987	0.6026	0.6064	0.6103	0.6141
+0.3	0.6179	0.6217	0.6255	0.6293	0.6331	0.6368	0.6406	0.6443	0.6480	0.6517
+0.4	0.6554	0.6591	0.6628	0.6664	0.6700	0.6736	0.6772	0.6808	0.6844	0.6879
+0.5	0.6915	0.6950	0.6985	0.7019	0.7054	0.7088	0.7123	0.7157	0.7190	0.7224

$\frac{X_i - X'}{\sigma'}$	0.09	0.08	0.07	0.06	0.05	0.04	0.03	0.02	0.01	0.00
+0.6	0.7257	0.7291	0.7324	0.7357	0.7389	0.7422	0.7454	0.7486	0.7517	0.7549
+0.7	0.7580	0.7611	0.7642	0.7673	0.7704	0.7734	0.7764	0.7794	0.7823	0.7852
+0.8	0.7881	0.7910	0.7939	0.7967	0.7995	0.8023	0.8051	0.8079	0.8106	0.8133
+0.9	0.8159	0.8186	0.8212	0.8238	0.8264	0.8289	0.8315	0.8340	0.8365	0.8389
+1.0	0.8413	0.8438	0.8461	0.8485	0.8508	0.8531	0.8554	0.8577	0.8599	0.8621
+1.1	0.8643	0.8665	0.8686	0.8708	0.8729	0.8749	0.8770	0.8790	0.8810	0.8830
+1.2	0.8849	0.8869	0.8888	0.8907	0.8925	0.8944	0.8962	0.8980	0.8997	0.9015
+1.3	0.9032	0.9049	0.9066	0.9082	0.9099	0.9115	0.9131	0.9147	0.9162	0.9177
+1.4	0.9192	0.9207	0.9222	0.9236	0.9251	0.9265	0.9279	0.9292	0.9306	0.9319
+1.5	0.9332	0.0345	0.9357	0.9370	0.9382	0.9394	0.9406	0.9418	0.9429	0.9441
+1.6	0.9452	0.9463	0.9474	0.9484	0.9495	0.9505	0.9515	0.9525	0.9535	0.9545
+1.7	0.9554	0.9564	0.9573	0.9582	0.9591	0.9599	0.9608	0.9616	0.9625	0.9633
+1.8	0.9641	0.9649	0.9656	0.9664	0.9671	0.9678	0.9686	0.9693	0.9699	0.9706
+1.9	0.9713	0.9719	0.9726	0.9732	0.9738	0.9744	0.9750	0.9756	0.9761	0.9767
+2.0	0.9773	0.9778	0.9783	0.9788	0.9793	0.9798	0.9803	0.9808	0.9812	0.9817
+2.1	0.9821	0.9826	0.9830	0.9834	0.9838	0.9842	0.9846	0.9850	0.9854	0.9857
+2.2	0.9861	0.9864	0.9868	0.9871	0.9875	0.9878	0.9881	0.9884	0.9887	0.9890
+2.3	0.9893	0.9896	0.9898	0.9901	0.9904	0.9906	0.9909	0.9911	0.9913	0.9916
+2.4	0.9918	0.9920	0.9922	0.9925	0.9927	0.9929	0.9931	0.9932	0.9934	0.9936
+2.5	0.9938	0.9940	0.9941	0.9943	0.9945	0.9946	0.9948	0.9949	0.9951	0.9952
+2.6	0.9953	0.9955	0.9956	0.9957	0.9959	0.9960	0.9961	0.9962	0.9963	0.9964
+2.7	0.9965	0.9966	0.9967	0.9968	0.9969	0.9970	0.9971	0.9972	0.9973	0.9974
+2.8	0.9974	0.9975	0.9976	0.9977	0.9977	0.9978	0.9979	0.9979	0.9980	0.9981
+2.9	0.9981	0.9982	0.9983	0.9983	0.9984	0.9984	0.9985	0.9985	0.9986	0.9986
+3.0	0.99865	0.99869	0.99874	0.99878	0.99882	0.99886	0.99889	0.99893	0.99896	0.99900
+3.1	0.99903	0.99906	0.99910	0.99913	0.99915	0.99918	0.99921	0.99924	0.99926	0.99929
+3.2	0.99931	0.99934	0.99936	0.99938	0.99940	0.99942	0.99944	0.99946	0.99948	0.99950
+3.3	0.99952	0.99953	0.99955	0.99957	0.99958	0.99960	0.99961	0.99962	0.99964	0.99965
+3.4	0.99966	0.99967	0.99969	0.99970	0.99971	0.99972	0.99973	0.99974	0.99975	0.99976
+3.5	0.99977	0.99978	0.99978	0.99979	0.99980	0.99981	0.99981	0.99982	0.99983	0.99983

Table B

Factors for Estimating σ' from $\bar{R}$ or $\bar{\sigma}$

Number of observations in sub-group n	Factor for estimate from R $d_2 = \frac{R}{\sigma'}$	Factor for estimate from σ $c_2 = \frac{\bar{\sigma}}{\sigma'}$
2	1.128	0.5642
3	1.693	0.7236
4	2.059	0.7979
5	2.326	0.8407
6	2.534	0.8686
7	2.704	0.8882
8	2.847	0.9027
9	2.970	0.9139
10	3.078	0.9227
11	3.173	0.9300
12	3.258	0.9359
13	3.326	0.9410
14	3.407	0.9453
15	3.472	0.9490
16	3.532	0.9523
17	3.588	0.9551
18	3.640	0.9576
19	3.689	0.9599
20	3.735	0.9619
21	3.778	0.9638
22	3.819	0.9655
23	3.858	0.9670
24	0.895	0.9684
25	3.931	0.9696
30	4.086	0.9748
35	4.213	0.9784
40	4.322	0.9811
45	4.415	0.9832
50	4.498	0.9849
55	4.573	0.9863
60	4.639	0.9874
65	4.699	0.9884
70	4.755	0.9892
75	4.806	0.9900
80	4.854	0.9906
85	4.898	0.9912
90	4.939	0.9916
95	4.978	0.9921
100	0.015	0.9925

Estimate of $\sigma' = \frac{\bar{R}}{d_2}$ or $\frac{\bar{\sigma}}{c_2}$.

These factors assume sampling from a normal universe.

Table C

Factors for determining from $\bar{R}$ the 3-sigma control limits for $\bar{X}$ and R charts from $\bar{R}$

Number of observations in sub-group	Factor for $\bar{X}$ chart	Factor for R chart	
n	A_2	Lower control limit D_3	Upper control limit D_4
2	1.88	0	3.27
3	1.02	0	2.57
4	0.73	0	2.28
5	0.58	0	2.11
6	0.48	0	2.00
7	0.42	0.08	1.92
8	0.37	0.14	1.86
9	0.34	0.18	1.82
10	0.31	0.22	1.78
11	0.29	0.22	1.74
12	0.27	0.28	1.72
13	0.25	0.31	1.69
14	0.24	0.33	1.67
15	0.22	0.35	1.65
16	0.21	0.36	1.64
17	0.20	0.38	1.62
18	0.19	0.39	1.61
19	0.19	0.40	1.60
20	0.18	0.41	1.59

Upper control limit for $\bar{X} = UCL_{\bar{X}} = \bar{X} + A_2\bar{R}$

Lower control limit for $\bar{X} = LCL_{\bar{X}} = \bar{X} - A_2\bar{R}$

(If aimed-at or standard value $\bar{X}$ is used rather than $\bar{X}$ as the cerntral line on the control chart, $\bar{X}$ should be substituted for $\bar{X}$ in the preceding formulas.)

Upper control limit for $R = UCL_R = D_4\bar{R}$

Lower control limits for $R = LCL_R = D_3\bar{R}$

All factors in Table C are based on the normal distribution.

Table D

Factors for determining from $\bar{\sigma}$ the 3-Sigma control limits $\bar{X}$ for and σ charts

Number of observations in sub-group	Factor for chart $\bar{X}$	Factor for $\bar{\sigma}$ chart	
n	A_1	Lower control limit B_3	Upper control limit B_4
2	3.76	0	0.27
3	2.39	0	2.57
4	1.88	0	2.57
5	1.60	0	2.09
6	1.41	0.03	1.97
7	1.28	0.12	1.88
8	1.17	0.19	1.81
9	1.09	0.24	1.76
10	1.03	0.28	1.72
11	0.97	0.32	1.68
12	0.93	0.35	1.65
13	0.88	0.38	1.62
14	0.85	0.41	1.59
15	0.82	0.43	1.57
16	0.79	0.45	1.55
17	0.76	0.47	1.53
18	0.74	0.48	1.52
19	0.72	0.50	1.50
20	0.70	0.51	1.49
21	0.68	0.52	1.48
22	0.66	0.53	1.47
23	0.65	0.54	1.46
24	0.63	0.55	1.45
25	0.62	0.56	1.44
30	0.56	0.60	1.40
35	0.52	0.63	1.37
40	0.48	0.66	1.34
45	0.45	0.68	1.32
50	0.43	0.70	1.30
55	0.41	0.71	1.29
60	0.39	0.72	1.28
65	0.38	0.73	1.27

Table D (Continued)

Number of observations in sub-group n	Factor for chart $\overline{X}$ A_1	Factor for $\overline{\sigma}$ chart	
		Lower control limit B_3	Upper control limit B_4
70	0.36	0.74	1.26
75	0.35	0.76	1.25
80	0.34	0.75	1.24
85	0.33	0.77	1.23
90	0.32	0.77	1.23
95	0.31	0.78	1.22
100	0.30	0.79	1.21

Upper control limit for $\overline{X} = UCL_{\overline{X}} = \overline{X} + A_1\overline{\sigma}$

Lower control limit for $\overline{X} = LCL_{\overline{X}} = \overline{X} - A_1\overline{\sigma}$

(If aimed-at or standard value $\overline{X}$ is used rather than $\overline{X}$ as the central line on the control chart, $\overline{X}$ should be substituted for $\overline{X}$ in the preceding formulas.)

Upper control limit for $\sigma = UCL_{\sigma} = B_4\overline{\sigma}$

Lower control limit for $\sigma = LCL_{\sigma} = B_2\overline{\sigma}$

All factors in Table *D* are based on the normal distribution.

Table E

Factors for determining from the 3-Sigma control limits for $\overline{X}$, *R* and σ charts.

Number of observations in sub-group n	Factor for $\overline{X}$ A	Factor for *R* chart		Factor for a chart	
		Lower control limit D_1	Upper control limit D_2	Lower control limit B_1	Upper control limit B_2
2	2.12	0	3.69	0	1.84
3	1.73	0	4.36	0	1.86
4	1.50	0	4.70	0	1.81
5	1.94	0	4.92	0	1.76
6	1.22	0	5.08	0.03	1.71
7	1.13	0.20	5.20	0.10	1.67
8	1.06	0.39	5.31	0.17	1.64
9	1.00	0.55	5.39	0.22	1.61
10	0.95	0.69	5.47	0.26	1.58
11	0.90	0.81	5.53	0.30	1.56

Table E (Continued)

Number of observations in sub-group n	Factor for $\overline{X}$ A	Factor for R chart		Factor for a chart	
		Lower control limit D_1	Upper control limit D_2	Lower control limit B_1	Upper control limit B_2
12	0.87	0.92	5.59	0.33	1.54
13	0.83	1.03	5.65	0.36	1.52
14	0.80	1.12	5.69	0.38	1.51
15	0.77	1.21	5.74	0.41	1.49
16	0.75	1.28	5.78	0.43	1.48
17	0.73	1.36	5.82	0.44	1.47
18	0.71	1.43	5.85	0.46	1.45
19	0.69	1.49	5.89	0.48	1.44
20	0.67	1.55	5.92	0.49	1.43
25	0.60			0.55	1.39
30	0.55			0.69	1.36
35	0.51			0.62	1.33
40	0.47			0.65	1.31
45	0.42			0.67	1.30
50	0.39			0.68	1.28
60	0.36			3.71	1.26
70	0.34			0.74	1.24
80	0.32			0.75	1.23
90	0.30			0.77	1.22
100				0.78	1.20

$UCL_{\overline{X}} = \overline{X} + A_{\sigma'}$

$LCL_{\overline{X}} = \overline{X} - A_{\sigma'}$

(If actual average is to be used rather than standard or aimed at a average, $\overline{X}$ should be substituted for $\overline{X}$ in the preceding formulas.)

$UCL_R = D_{2\sigma'}$

Central line$_R = d_{2\sigma'}$

$LCL_R = D_{1\sigma'}$

$UCL_\sigma = B_{2\sigma'}$

Central line$_\sigma = c_{2\sigma'}$

$LCL_\sigma = B_{1\sigma'}$

Table F

3-Sigma control limit for control charts for percent defective

Standard percent defective (100p)	Upper control limit — Subgroup size																	
	100	150	200	300	400	500	600	800	1000	1500	2000	3000	4000	5000	10,000	20,000	50,000	1,00,000
0.1	1.05	0.87	0.77	0.65	0.57	0.52	0.49	0.44	0.40	0.34	0.31	0.27	0.25	0.23	0.90	0.17	0.14	0.13
0.2	1.54	1.29	1.15	0.97	0.87	0.80	0.75	0.67	0.62	0.55	0.50	0.44	0.41	0.39	0.33	0.29	0.26	0.24
0.4	2.29	1.95	1.74	1.49	1.35	1.25	1.17	1.07	1.00	0.89	0.82	0.75	0.70	0.67	0.59	0.53	0.48	0.46
0.6	2.92	2.49	2.24	1.94	1.76	1.64	1.55	1.42	1.33	1.20	1.12	1.02	0.97	0.93	0.83	0.76	0.70	0.67
0.8	3.47	2.98	2.69	2.34	2.14	2.00	1.89	1.74	1.65	1.49	1.40	1.29	1.22	1.80	1.07	0.99	0.92	0.88
1.0	3.98	3.44	3.11	2.72	2.49	2.33	2.22	2.06	1.94	1.77	1.67	1.54	1.47	1.42	1.30	1.21	1.13	1.09
1.2	4.47	3.87	3.51	3.09	2.83	2.66	2.53	2.35	2.23	2.04	1.93	1.80	1.72	1.66	1.53	1.43	1.35	1.30
1.4	4.92	4.28	3.89	3.43	3.16	2.99	2.84	2.65	2.51	2.31	2.19	2.04	1.96	1.90	1.75	1.65	1.57	1.51
1.6	5.36	4.67	4.26	3.77	3.48	3.28	3.14	2.93	2.79	2.57	2.44	2.29	2.20	2.13	1.98	1.87	1.77	1.72
1.8	5.79	5.06	4.62	4.10	3.79	3.58	3.43	3.21	3.06	2.83	2.69	2.53	2.43	2.36	2.20	2.08	1.98	1.93
2.0	6.20	5.43	4.97	4.42	4.10	3.88	3.71	3.48	3.33	3.08	2.94	2.77	2.66	2.59	2.42	2.30	2.19	2.13
2.5	7.18	6.32	5.81	5.20	4.84	4.59	4.41	4.16	3.98	3.71	3.55	3.36	3.24	3.16	2.97	2.83	2.71	2.65
3.0	8.12	7.18	6.62	5.95	5.56	5.29	5.09	4.81	4.62	4.32	4.14	3.93	3.81	3.72	3.51	3.36	3.23	3.16
3.5	9.01	8.00	7.40	6.68	6.26	5.97	5.75	5.45	5.24	4.92	4.73	4.51	4.37	4.28	4.05	3.89	3.75	3.67
4.0	9.88	8.80	8.16	7.39	6.94	6.63	6.40	6.08	5.86	5.52	5.31	5.07	4.93	4.83	4.59	4.42	4.26	4.19
5	11.54	10.34	9.62	8.77	8.27	7.92	7.67	7.31	7.07	6.69	6.46	6.19	6.03	5.92	5.65	5.46	5.29	5.21
6	13.12	11.82	11.04	10.11	9.56	9.19	8.91	8.52	8.25	7.84	7.59	7.30	7.13	7.01	6.71	6.50	6.32	6.32
7	14.65	13.25	12.41	11.42	10.83	10.42	10.12	9.71	9.42	8.98	8.71	8.40	8.21	8.08	7.77	7.54	7.34	7.24
8	16.14	14.65	13.75	12.70	12.07	11.64	11.32	10.88	10.57	10.10	9.82	9.49	9.29	9.15	8.81	8.58	8.36	8.26
9	17.59	16.01	15.07	13.96	13.29	12.84	12.50	12.04	11.71	11.22	10.92	10.57	10.36	10.21	9.86	9.61	9.38	9.27
10	19.00	17.35	16.36	15.20	14.50	14.02	13.67	13.18	12.85	12.32	12.01	11.64	11.42	11.27	10.90	10.64	10.40	10.28
12	21.75	19.96	18.89	17.63	16.87	16.36	15.98	15.45	15.08	14.52	14.18	13.78	13.54	13.38	12.97	12.69	12.44	12.31
14	24.41	22.50	21.36	20.01	19.20	18.66	18.25	17.68	17.29	16.69	16.33	15.90	15.65	15.47	15.04	14.74	14.47	14.33
16	27.00	24.98	23.78	22.35	21.50	20.92	20.49	19.89	19.48	18.84	18.46	18.01	17.74	17.56	17.10	16.78	16.49	16.35
18	29.53	27.41	26.15	24.65	23.76	23.15	22.71	22.07	21.64	20.97	20.58	20.10	19.82	19.63	19.15	18.81	18.52	18.36
20	32.00	29.80	28.49	26.93	26.00	25.37	24.90	24.24	23.79	23.10	22.68	22.19	21.90	21.70	21.20	20.85	20.54	20.38
25	37.99	35.61	34.19	32.50	31.50	30.81	30.30	29..59	29.11	28.35	27.90	27.37	27.05	26.84	26.30	25.92	25.58	25.41
30	43.75	41.22	39.72	37.94	36.87	36.15	35.61	34.86	34.35	33.55	33.07	32.51	32.17	31.94	31.37	30.97	30.61	30.43
35	49.31	46.68	45.12	43.26	42.15	41.40	40.84	40.06	39.52	38.70	38.20	37.61	37.26	37.02	36.43	36.01	35.64	35.45
40	54.70	52.00	50.39	48.49	47.35	46.57	46.00	45.20	44.65	43.79	43.29	42.68	42.32	42.08	41.47	41.04	40.66	40.06

Table F

3-Sigma control limit for control charts for percent defective (Continued)

Standard percent defective (100p)	Lower control limit																	
	Subgroup size																	
	100	150	200	300	400	500	600	800	1000	1500	2000	3000	4000	5000	10,000	20,000	50,000	1,00,000
0.1	0.00	0.00	0.00	0.00	0.00	0.00	0.00	0.00	0.00	0.00	0.00	0.00	0.00	0.00	0.01	0.03	0.06	0.07
0.2	0.00	0.00	0.00	0.00	0.00	0.00	0.00	0.00	0.00	0.00	0.00	0.00	0.00	0.01	0.07	0.11	0.14	0.16
0.4	0.00	0.00	0.00	0.00	0.00	0.00	0.00	0.00	0.00	0.00	0.00	0.05	0.10	0.13	0.21	0.27	0.32	0.34
0.6	0.00	0.00	0.00	0.00	0.00	0.00	0.00	0.00	0.00	0.00	0.08	0.18	0.23	0.27	0.37	0.44	0.50	0.53
0.8	0.00	0.00	0.00	0.00	0.00	0.00	0.00	0.00	0.00	0.11	0.20	0.31	0.38	0.42	0.53	0.61	0.68	0.72
1.0	0.00	0.00	0.00	0.00	0.00	0.00	0.00	0.00	0.06	0.23	0.33	0.46	0.53	0.58	0.70	0.79	0.87	0.91
1.2	0.00	0.00	0.00	0.00	0.00	0.00	0.00	0.05	0.17	0.36	0.47	0.60	0.68	0.74	0.87	0.97	1.05	1.10
1.4	0.00	0.00	0.00	0.00	0.00	0.00	0.00	0.15	0.29	0.49	0.61	0.76	0.84	0.90	1.05	1.15	1.24	1.29
1.6	0.00	0.00	0.00	0.00	0.00	0.00	0.06	0.27	0.41	0.63	0.76	0.91	1.00	1.07	1.22	1.33	1.43	1.48
1.8	0.00	0.00	0.00	0.00	0.00	0.02	0.17	0.39	0.54	0.77	0.91	1.07	1.17	1.24	1.40	1.52	1.62	1.67
2.0	0.00	0.00	0.00	0.00	0.00	0.12	0.29	0.52	0.67	0.92	1.06	1.23	1.34	1.41	1.58	1.70	1.81	1.87
2.5	0.00	0.00	0.00	0.00	0.16	0.41	0.59	0.84	1.02	1.29	1.45	1.64	1.76	1.84	2.03	2.17	2.29	2.35
3.0	0.00	0.00	0.00	0.05	0.44	0.71	0.91	1.19	1.38	1.68	1.86	2.07	2.19	2.28	2.49	2.64	2.77	2.84
3.5	0.00	0.00	0.00	0.32	0.74	1.03	1.25	1.55	1.76	2.08	2.27	2.49	2.63	2.72	2.95	3.11	3.25	3.33
4.0	0.00	0.00	0.00	0.61	1.06	1.37	1.60	1.92	2.14	2.48	2.69	2.93	3.07	3.17	3.41	3.58	3.74	3.81
5	0.00	0.00	0.38	1.23	1.73	2.08	2.33	2.69	2.93	3.31	3.54	3.81	3.97	4.08	4.35	4.54	4.71	4.79
6	0.00	0.18	0.96	1.89	2.44	2.81	3.09	3.48	3.75	4.16	4.41	4.70	4.87	4.99	5.29	5.50	5.68	5.77
7	0.00	0.75	1.59	2.58	3.17	3.58	3.88	4.29	4.58	5.02	5.29	5.60	5.79	5.92	6.23	6.46	6.66	6.76
8	0.00	1.35	2.25	3.30	3.93	4.36	4.68	5.12	5.43	5.90	6.18	6.51	6.71	6.85	7.19	7.42	7.64	7.74
9	0.41	1.99	2.93	4.04	4.71	5.16	5.50	5.96	6.29	6.78	7.08	7.43	7.64	7.79	8.14	8.39	8.62	8.73
10	1.00	2.65	3.64	4.80	5.50	5.98	6.33	6.82	7.15	7.68	7.99	8.36	8.58	8.73	9.10	9.36	9.60	9.72
12	2.25	4.04	5.11	6.37	7.13	7.64	8.02	8.55	8.92	9.48	9.82	10.22	10.46	10.62	11.03	11.31	11.56	11.69
14	3.59	5.50	6.64	7.99	8.80	9.34	9.75	10.32	10.71	11.31	11.67	12.10	12.35	12.53	12.96	13.26	13.53	13.67
16	5.00	7.02	8.22	9.65	10.50	11.08	11.51	12.11	12.52	13.16	13.54	13.99	14.26	14.44	14.90	15.22	15.51	15.65
18	6.47	8.59	9.85	11.35	12.24	12.85	13.29	13.93	14.36	15.03.	15.42	15.90	16.18	16.37	16.85	17.19	17.48	17.64
20	8.00	10.20	11.51	13.07	14.00	14.63	15.10	15.76	16.21	16.90	17.32	17.81	18.10	18.30	18.80	19.15	19.46	19.62
25	12.01	14.39	15.81	17.50	18.50	19.19	19.70	20.41	20.89	21.65	22.10	22.63	22.95	23.16	23.70	24.08	24.42	24.59
30	16.25	18.78	20.28	22.06	23.13	23.85	24.39	25.14	25.65	26.45	26.93	27.49	27.83	28.06	28.63	29.03	29.39	29.57
35	20.69	23.32	24.88	26.74	27.85	28.60	29.16	29.94	30.48	31.30	31.80	32.39	32.74	32.98	33.57	33.99	34.36	34.55
40	25.30	28.00	29.61	31.51	32.65	33.43	34.00	34.80	35.35	36.21	36.71	37.32	37.68	37.92	38.53	38.96	39.34	39.54

Table G

Summation of terms of Poisson's exponential binomial limit 1000 × probability of c or less occurrences of event that has average number of occurrences equal to c' or np'.

c / c' or np'	0	1	2	3	4	5	6	7	8	9	10	11	12
0.02	980	1000											
0.04	961	999	1000										
0.06	942	998	1000										
0.08	923	997	1000										
0.10	905	995	1000										
0.15	861	990	999	1000									
0.20	819	982	999	1000									
0.25	779	974	998	1000									
0.03	741	963	996	1000									
0.35	705	951	994	1000									
0.40	670	938	992	999	1000								
0.45	638	925	989	999	1000								
0.50	607	910	986	998	1000								
0.55	577	894	982	998	1000								
0.60	549	878	977	997	1000								
0.65	522	861	972	996	999	1000							
0.70	497	844	966	994	999	1000							
0.75	472	827	959	993	999	1000							
0.80	449	809	953	991	999	1000							
0.85	427	791	945	989	998	1000							
0.90	407	772	937	987	998	1000							
0.95	387	754	929	984	997	1000							
1.00	368	736	920	981	996	999	1000						
1.1	333	699	900	974	995	999	1000						
1.2	301	663	879	966	992	998	1000						
1.3	273	627	857	957	989	998	1000						
1.4	247	592	833	946	986	997	999	1000					
1.5	223	558	809	934	981	996	999	1000					

Table G (Continued)

c / c' or np'	0	1	2	3	4	5	6	7	8	9	10	11	12
1.6	202	525	783	921	976	994	999	1000					
1.7	183	493	757	907	970	992	998	1000					
1.8	165	463	731	891	964	990	997	999	1000				
1.9	150	434	704	875	956	987	997	999	1000				
2.0	135	406	677	857	947	983	995	999	1000				
2.2	111	335	623	819	928	975	993	998	1000				
2.4	091	308	570	779	904	964	988	997	999	1000			
2.6	074	267	518	736	877	951	983	995	999	1000			
2.8	061	231	469	692	848	935	976	992	998	999	1000		
3.0	050	199	423	647	815	916	966	988	996	999	1000		
3.2	041	171	380	603	781	895	955	983	994	998	1000		
3.4	033	147	340	558	744	871	942	977	992	997	999	1000	
3.6	027	126	303	515	706	844	927	669	988	996	999	1000	
3.8	022	107	209	473	668	816	909	960	984	994	998	999	1000
4.0	018	092	238	433	629	785	889	949	979	992	997	999	1000
4.2	015	078	210	395	590	753	867	936	972	989	996	999	1000
4.4	012	066	185	359	551	720	844	921	964	985	994	998	999
4.6	010	056	163	326	513	686	818	905	955	980	992	997	999
4.8	008	048	143	294	476	651	791	887	944	975	990	996	999
5.0	007	040	125	265	440	616	762	867	932	968	986	995	998
5.2	006	034	109	238	406	581	732	845	918	960	982	993	997
5.4	005	029	095	213	373	546	702	822	903	951	977	990	996
5.6	004	024	082	191	342	512	670	797	886	941	972	988	995
5.8	003	021	072	170	313	478	638	771	867	929	965	984	993
6.0	002	017	062	151	285	446	606	744	847	916	957	980	991

c / c' or np'	13	14	15	16
4.4	1000			
4.6	1000			
4.8	1000			
5.0	999	1000		
5.2	999	1000		
5.4	999	1000		
5.6	998	999	1000	
5.8	997	999	1000	
6.0	996	999	999	1000

Table G (Continued)

c / c' or np'	0	1	2	3	4	5	6	7	8	9	10	11	12
6.2	002	015	054	134	239	414	574	716	826	902	949	975	989
6.4	002	012	046	119	235	384	542	687	803	886	939	969	986
6.6	001	010	040	105	213	355	511	658	780	869	927	963	982
6.8	001	009	034	093	192	327	480	628	755	850	915	955	978
7.0	001	007	030	082	173	301	450	599	729	830	901	947	973
7.2	001	006	025	072	156	276	420	569	703	810	887	937	967
7.4	001	005	022	063	140	253	392	539	676	788	871	926	961
7.6	001	004	019	055	125	231	365	510	648	765	854	915	954
7.8	000	004	016	048	112	210	338	481	620	741	835	902	945
8.0	000	003	014	042	100	191	313	453	593	717	816	888	936
8.5	000	002	009	030	074	150	256	386	523	653	763	849	909
9.0	000	001	006	021	055	116	207	324	456	587	760	803	876
9.5	000	001	004	015	040	089	165	269	392	522	645	752	836
10.0	000	000	003	010	029	067	130	220	333	458	583	697	792
10.5	000	000	002	007	021	050	102	179	279	397	521	639	742
11.0	000	000	001	005	015	038	079	143	232	341	460	579	689
11.5	000	000	001	003	011	028	060	114	191	289	402	520	633
12.0	000	000	001	002	008	020	046	090	155	242	347	462	567
12.5	000	000	000	002	005	015	035	070	125	201	297	406	519
13.0	000	000	000	001	004	011	026	054	100	166	252	353	463
13.5	000	000	000	001	003	008	019	041	079	135	212	304	409
14.0	000	000	000	000	002	006	014	032	062	109	176	260	358
14.5	000	000	000	000	001	004	010	024	048	088	145	220	311
15.0	000	000	000	000	001	003	008	018	037	070	118	185	268

c / c' or np'	13	14	15	16	17	18	19	20	21	22	23	24	25
6.2	995	998	999	1000									
6.4	994	997	999	1000									
6.6	992	997	999	999	1000								
6.8	990	996	998	999	1000								
7.0	987	994	998	999	1000								
7.2	984	993	997	999	999	1000							
7.4	980	991	996	998	999	1000							
7.6	976	989	995	998	999	1000							
7.8	971	986	993	997	999	1000							

Table G (Continued)

c / c' or np'	13	14	15	16	17	18	19	20	21	22	23	24	25
8.0	966	983	992	996	998	999	1000						
8.5	949	973	986	993	997	999	999	1000					
9.0	926	959	978	989	995	998	999	1000					
9.5	898	940	967	982	991	996	998	999	1000				
10.0	864	917	951	973	986	993	997	998	999	1000			
10.5	825	888	932	964	978	988	994	997	999	999	1000		
11.0	781	854	907	944	968	982	991	995	998	999	1000		
11.5	733	815	878	924	954	974	986	992	996	998	999	1000	
12.0	682	772	844	899	937	963	979	988	994	997	999	999	1000
12.5	628	725	806	869	916	948	969	983	991	995	998	999	999
13.0	573	675	764	835	890	930	957	975	986	992	996	998	999
13.5	518	623	718	798	861	908	942	965	980	989	994	997	998
14.0	464	570	669	756	827	883	923	952	971	983	991	995	997
14.5	413	518	619	711	790	853	901	936	960	976	986	992	996
15.0	363	466	568	664	749	819	875	917	943	967	961	989	994
c / c' or np'	26	27	28	29									
12.5	1000												
13.0	1000												
13.5	999	1000											
14.0	999	999	1000										
14.5	998	999	999	1000									
15.0	997	998	999	1000									

BIBLIOGRAPHY

Besterfield, Dale H. *Total Quality Management*. Pearson Education India. 2011.

Buffa, Elwood S. and Sarin, Rakesh K. *Modern Productions / Operations Management*, 8th Edition. John Wiley. 2007.

Hillier, Frederick S. and Lieberman, Gerald J. *Introduction to Operations Research*, 5th Edition. McGraw-Hill. 2001.

Kerzner, Harold R. *Project Management: A Systems Approach to Planning, Scheduling and Controlling*, 11th Edition. John Wiley. 2013.

Lal H. *Organizational Excellence through Total Quality Management: A Practical Approach.* New Age International. 2008.

Lawrence, Beningston. *Project Management.* McGraw-Hill. 1970.

Leavenworth, Richard S. and Grant, Eugene L. *Statistical Quality Control.* Tata McGraw-Hill Education. 2000.

Levin, Richard I. and Rubin, David S. *Statistics for Management*, 7th Edition. Prentice Hall of India. 1998.

Moder, Joseph J. and Phillips, Cecil R. *Project Management with PERT and CPM.* Reinhold Publishing. 1967.

Montgomery, Douglas C. *Statistical Quality Control*, 4th Edition. John Wiley. 2000.

Wiest, Jerome D. and Levy, Ferdinand K. *A Management Guide to PERT and CPM.* Prentice Hall of India. 2002.

INDEX

N

O

P

Q

R

S

T

U

V

W

X

Z

CPSIA information can be obtained
at www.ICGtesting.com
Printed in the USA
BVOW04s2009081117
499763BV00063B/775/P

9 789380 381992